D0821516

cycling science

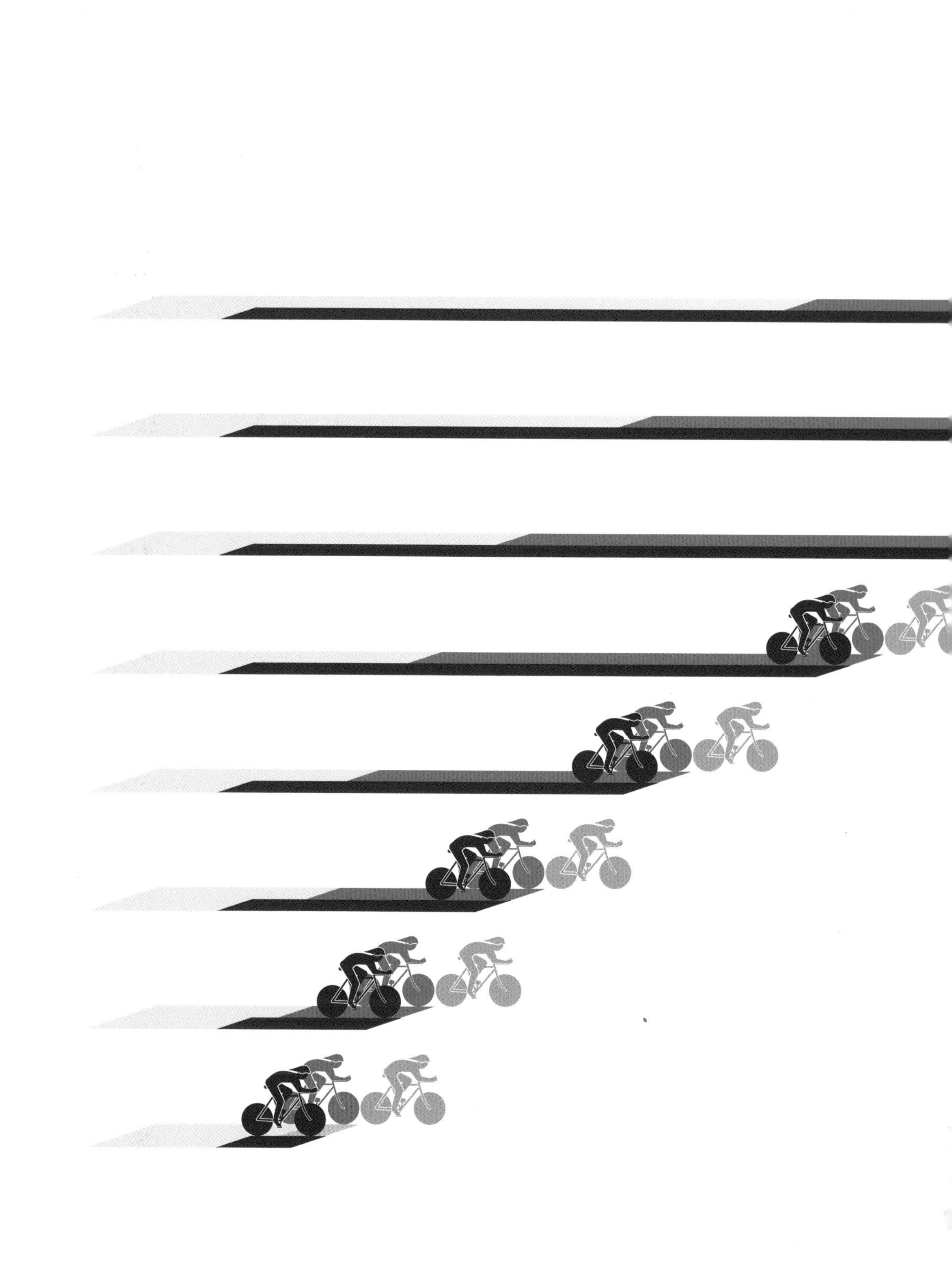

cycling science

MAX GLASKIN

How rider and machine work together

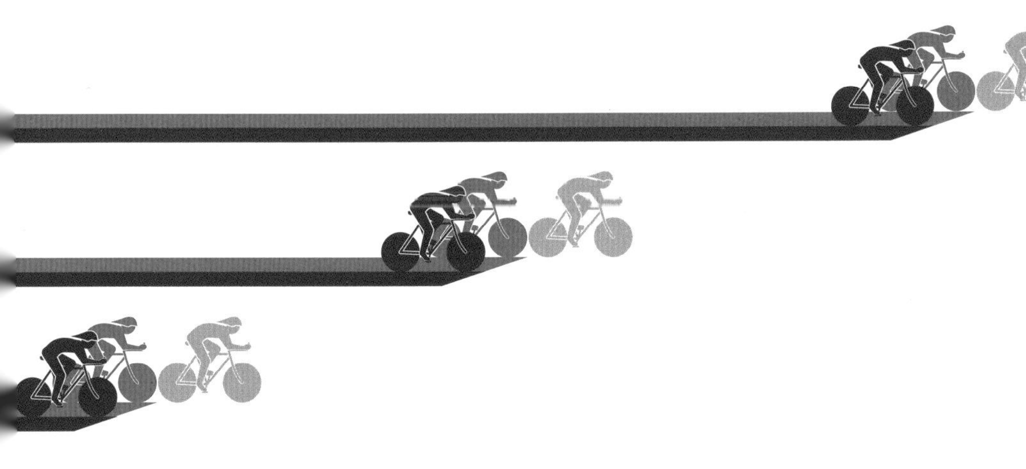

THE UNIVERSITY OF CHICAGO PRESS

Chicago and London

The University of Chicago Press, Chicago 60637
The University of Chicago Press, Ltd., London

Printed in China

20 19 18 17 16 15 14 13 12   1 2 3 4 5
ISBN-13: 978-0-226-92413-7 (cloth)
ISBN-10: 0-226-92413-0 (cloth)

Library of Congress Cataloging-in-Publication Data

Glaskin, Max.
Cycling science / Max Glaskin.
pages. cm.
ISBN-13: 978-0-226-92413-7 (cloth : alk. paper)
ISBN-10: 0-226-92413-0 (cloth : alk. paper)
ISBN-13: 978-0-226-92187-7 (e-book)
ISBN-10: 0-226-92187-5 (e-book) 1. Cycling. 2. Sports sciences.
I. Title.
GV1041.G53 2012
796.6—dc23
2012019386

The paper used in this publication meets the minimum requirements of the American National Standard for Information Sciences—Permanence of Paper for Printed Library Materials, ANSI Z39.48-1992.

Color origination by Ivy Press Reprographics.

This book was conceived, designed, and produced by
Ivy Press
210 High Street, Lewes
East Sussex BN7 2NS
United Kingdom
www.ivypress.co.uk

Creative Director Peter Bridgewater
Publisher Jason Hook
Editorial Director Tom Kitch
Art Director James Lawrence
Designer Lisa McCormick
Editor Jeremy Torr
Technical Editors John King & Sara Hulse
Assistant Editor Jamie Pumfrey
Illustrators Robert Brandt & Nick Rowland

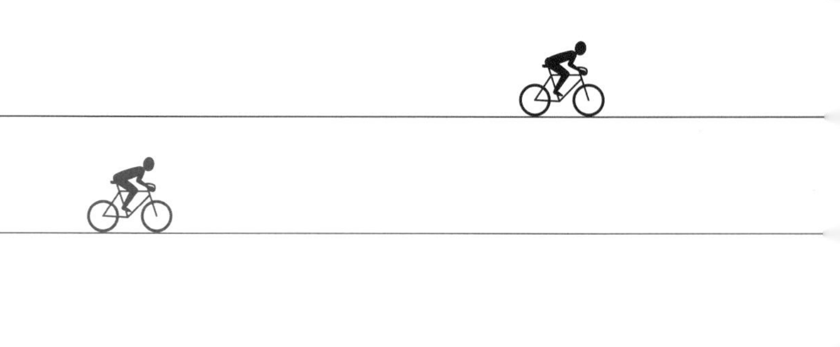

Contents

Introduction

Riding a bike is one of the most rewarding of human activities, whether it takes the form of those first tentative wobbles, a daily commute, occasional pleasure trips, or the adrenalin rush of elite competition. Understanding the science of cycling is equally gratifying because it enhances both the interest in, and the sheer enjoyment of, riding.

Cycling occupies a unique niche in the world. It satisfies concerns about the environment, sustainability, health, and fitness—while giving millions of people the freedom to travel independently. These benefits would be mere anecdotes if it were not for the fact that thousands of scientists have studied almost every aspect of this seemingly simple activity. This book brings together their significant findings. Some were made in the nineteenth century and others as recently as this year.

A wide range of sciences is embraced in the book because cycling involves a surprisingly diverse array of disciplines. The fundamental principles of physics lead on to the magic of engineering and technology, which has led to the development of phenomenally efficient bikes. There are the mysteries of balance, stability, and steering, as well as crucial explanations of aerodynamics. You, the rider, are also dissected and put under the microscope to reveal how you perform the ingenious and often underestimated trick of cycling.

This is not a manual or a training guide. It goes deeper than such books, but it will help everyone to get more from their cycling—whether for commuting, riding for pleasure, or racing to win—because it is accessible to all. Cycling is about traveling easily and using your energy efficiently. Without a doubt, your potential to do so can be boosted by your knowledge of the science.

Scientific discoveries are not always easy to comprehend, but this book presents them in a straightforward way. The *question-and-answer* pages contain unique info-graphics that convey scientific results very clearly. Every science question is mirrored by a question posed by a cyclist. The text and the info-graphics combine to answer them both.

▶ ***Science behind success*** *While cycling is a simple activity enjoyed by over a billion people worldwide, among the professionals it is an intensely scientific sport. Every technological improvement, from the efficiency of the bicycle and the aerodynamics of the rider's clothing and helmet, to the effectiveness of a team's training program, can potentially mean the difference between winning the Tour de France and making up the numbers in the peloton.*

sky
LCL
LCL
sky
adidas
IG
LCL

The *Equipment* pages show how and why crucial elements of the bicycle and cycling have evolved since 1817, when the first steerable two-wheeler was invented. It was a horse substitute introduced because feed prices had rocketed due to crop failure caused by volcanic ash clouds. So, events that are now properly understood through science were fundamental to the birth of the bicycle.

It is all very well explaining the what, why, how, and when, but, apart from riding a bike, little can bring a cyclist closer to their favorite activity than a world-class photograph. So, the photographic *Science in Action* feature pages reveal how the principle is applicable to the practice, while at the same time conveying the drama, excitement, and human endeavor that is cycling.

Assembling this mass of knowledge has been a challenge. The information within *Cycling Science* has been gathered from hundreds of scientific papers,

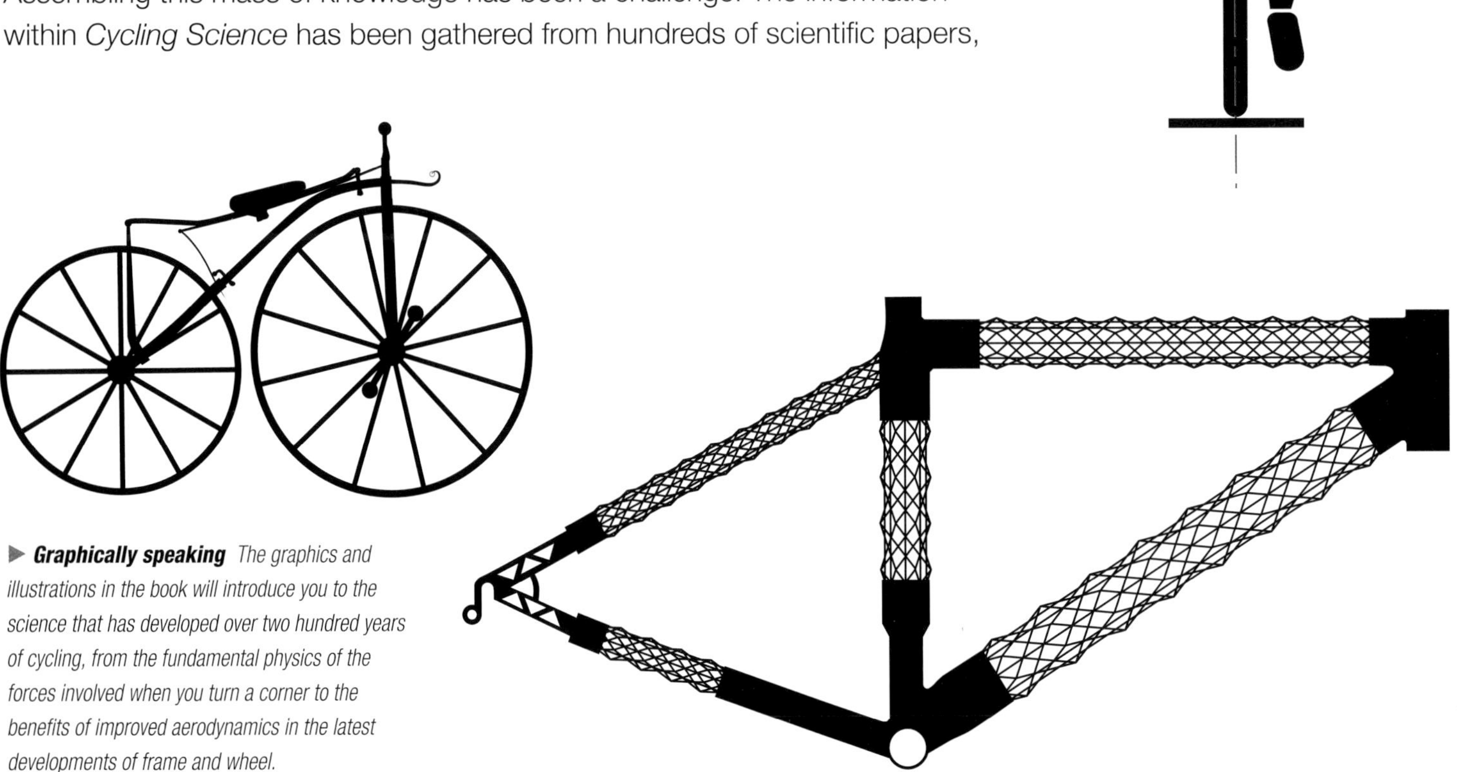

▶ ***Graphically speaking*** *The graphics and illustrations in the book will introduce you to the science that has developed over two hundred years of cycling, from the fundamental physics of the forces involved when you turn a corner to the benefits of improved aerodynamics in the latest developments of frame and wheel.*

conference proceedings, and official documents. Some were inspiring, others were real eye-openers, many were hidden away in academic journals. The most interesting and relevant findings and data have been harvested, translated into rider-friendly language, and improved with illuminating info-graphics so that they can be appreciated by nonscientists and scientists alike.

For cyclists who are eager to delve deeper, there is a comprehensive glossary that defines many of the terms and concepts. An extensive list of references points readers to the sources of the information, so that it is easy to engage more closely with the science.

It has been said that there are 1.2 billion cyclists in the world. Every single one began by wobbling tentatively, and now they cycle with ease. *Cycling Science* will make their ride even smoother.

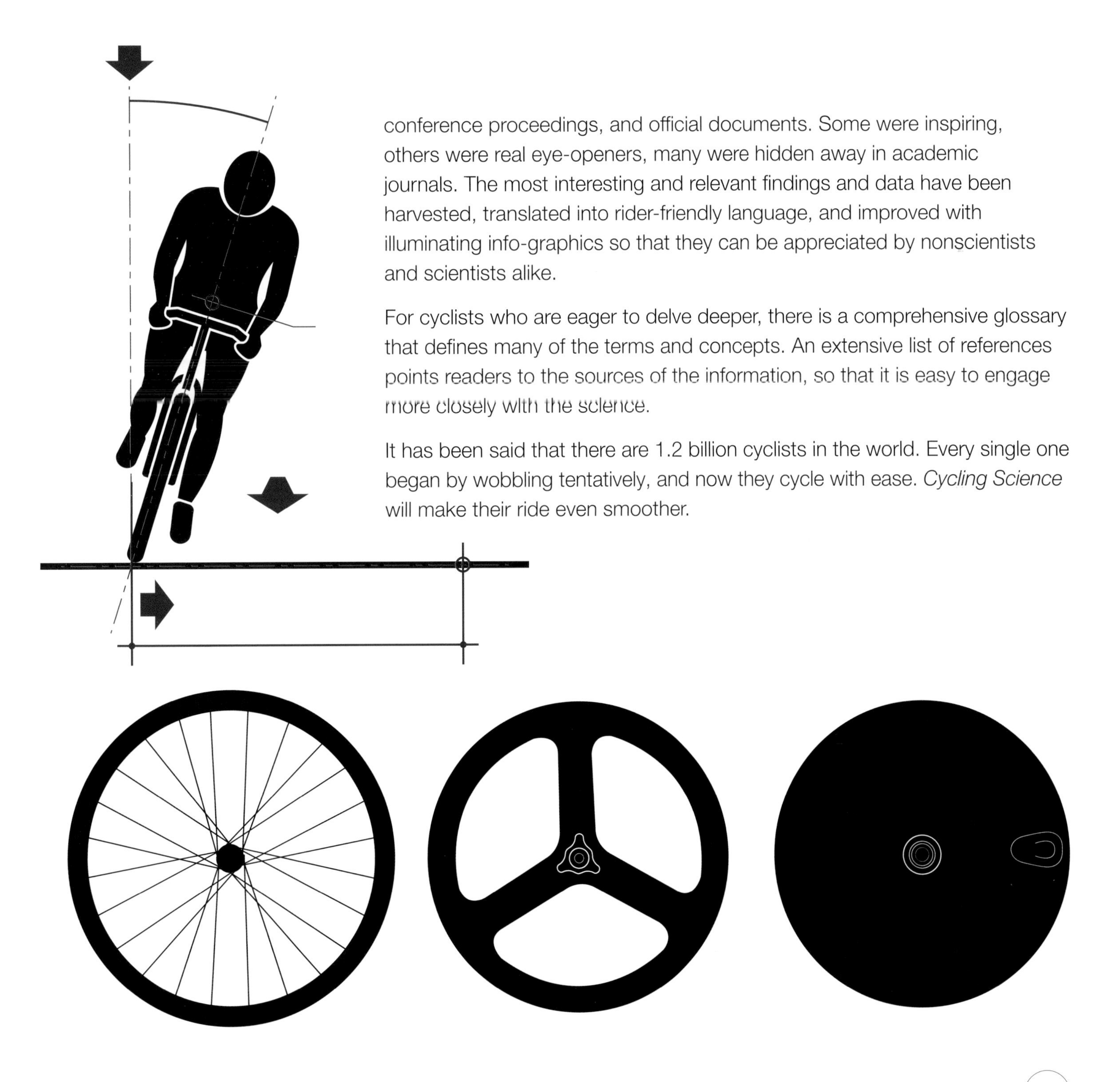

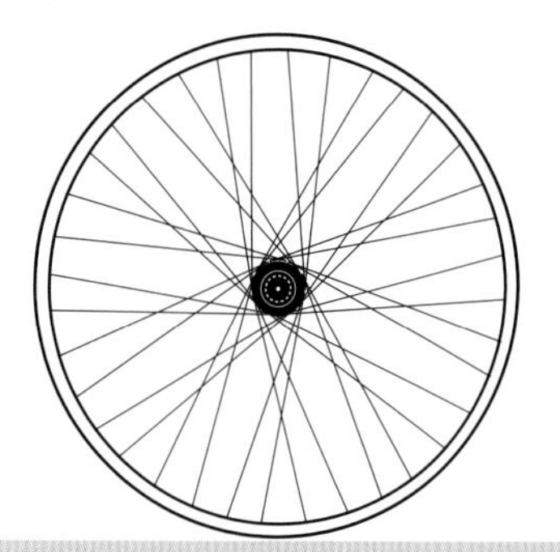

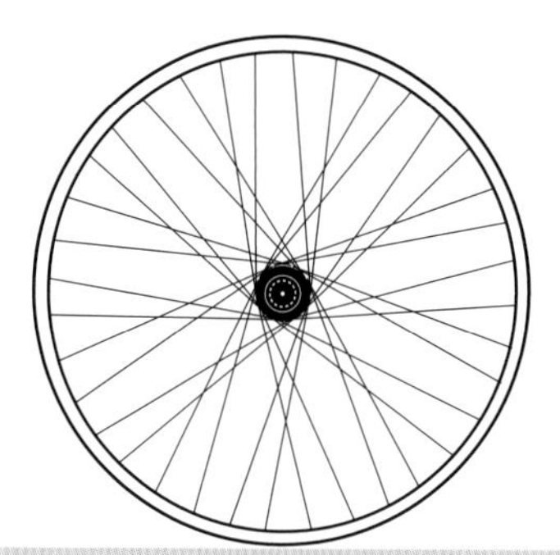

Cycling laws and regulations come and go but the laws of nature don't change, so cyclists and bikes must ride with them. Most of the fundamentals of science actually facilitate cycling and make it possible, although at least one does, in fact, make it harder. The key is to use science and technology to minimize the disadvantages and maximize the benefits so that cycling is easier. It must be reasonably easy because interest in cycling is booming to the point where bicycle sales far outstrip those of any other vehicle type. Cycling is not only a massively popular way to travel but has also become a political touchstone because of the impact that it can have on the environment, society, and the individual. Where once world leaders waved imperiously from the steps of jetliners, they are now eager to be seen on a bike. So, health, safety, climate, and other issues fundamental to human existence are brought into this chapter, which lays down a broad, smooth track for the journey ahead.

chapter one

fundamentals

What are the forces acting upon a bicycle?

Am I not alone on my bike?

There are four external forces that every cyclist must work with or against—gravity, air resistance, rolling resistance, and friction—and a fifth effect, referred to as inertia. None of them can be utterly vanquished (and it would not necessarily be desirable to eradicate them completely). However, it is wise to understand what you're up against so that you can minimize the negative consequences and harness the positive.

Gravity is the force that gives weight to matter. The Earth pulls everything to itself with a gravitational acceleration of about 32 ft/s^2 (9.8 m/s^2). In fact, gravity makes cycling possible by pressing the bike to the ground, but it makes riding uphill harder. Descending is made easier by the pull of gravity, but you never get back all the energy you put into climbing the same hill.

Air resistance generally works against the cyclist. The planet's gravity is strong enough to hold a blanket of air some 62 miles (100 km) thick to the Earth's surface. Cyclists couldn't breathe without it, but must push it aside continually to make progress. This same force can be helpful, too, if you've got a fair tailwind.

Rolling resistance results from the fact that, when a tire comes into contact with a road, both tire and road deform a little. The road and the tire do not spring back with the same energy that deformed them, with some energy always lost to heat. This has the effect of a resisting force.

Friction helps to move the bike forward by maintaining contact between tire and road. However, friction in the bearings of the bicycle's drivetrain—from the pedals through to the chain, gears, and hubs—can absorb up to 5 percent of the cyclist's energy.

Riders must also overcome inertia, which is not a force at all, but an innate property of matter—its resistance to any change in its state of motion. A bicycle's motion won't change if there are no forces acting on it.

Inertia

The principle of inertia is a way of saying that an object doesn't change its motion unless there is a force acting on it. The bigger the force, the greater the change in motion (in speed or direction). Steep hills, strong winds, muscular legs, and powerful brakes overcome inertia to the greatest degree. Mass determines how big the effect will be—under a particular force, a heavy bike will change its motion more slowly than a light model. Likewise, a rider who loses weight will accelerate more quickly than their former, fatter self.

inertia

Forces and inertia

Gravity

The Earth subjects bike and rider to a gravitational force that would make them accelerate downward at approximately 32 ft/s² (9.8 m/s²) if they weren't supported by the ground. Gravity makes cycling uphill harder, but without it you couldn't cycle at all—it keeps the bike on the ground and the rider on the saddle.

Air resistance

A cubic foot of dry air at 68°F (20°C) at sea level weighs about 0.076 lb (0.034 kg). When the cyclist and the atmosphere meet head on, some of a rider's energy is lost to pushing this air out of the way. If the difference in their speeds is more than about 9 mph (15 km/h) on a flat road, this becomes the biggest drain on the rider's energy.

Friction

The friction between the tire and the road surface is crucial for forward motion. Without it the wheel would spin on the spot, as if on ice. However, friction in the bearings of the bicycle's power train drains energy into wasted heat and noise.

Rolling resistance

Bike tires deform under the weight of bike and rider as the rubber comes into contact with the road surface. Because the tire doesn't spring back with quite the same energy as it is deformed, this shape changing absorbs a small amount of the energy which, in the main, has been put into the system by the cyclist pressing on the pedals. A hard tire on soft ground suffers from similar rolling resistance, although this time it's the ground that deforms, once again absorbing the rider's energy.

equipment: the bicycle

The modern bicycle is the result of two centuries of refinement, propelled by better understandings of science and technology. Bike builders and designers have repeatedly worked out novel ways to use established materials and to incorporate new materials to supplement or replace the old. Yet a cyclist from the late nineteenth century would have little difficulty in recognizing today's bicycle because the silhouette of frame, wheels, saddle, and handlebars has remained largely unaltered. They may be alarmed by the slender saddle or amused by the 27 gears, but they would certainly be reassured by the familiar chain and spokes.

There are hundreds of parts on a bicycle, the majority exposed to full view. On the most functional bikes they each serve a mechanical purpose. Without doubt, the most important part is the frame, often described as the heart of the bicycle by people whose grasp of anatomy should disqualify them from medical practice. It is, actually, more akin to the skeleton, holding everything together, supporting many of the components and also the rider. If the metaphor is to be continued, the front fork can reasonably be equated to a limb, articulated at the headset.

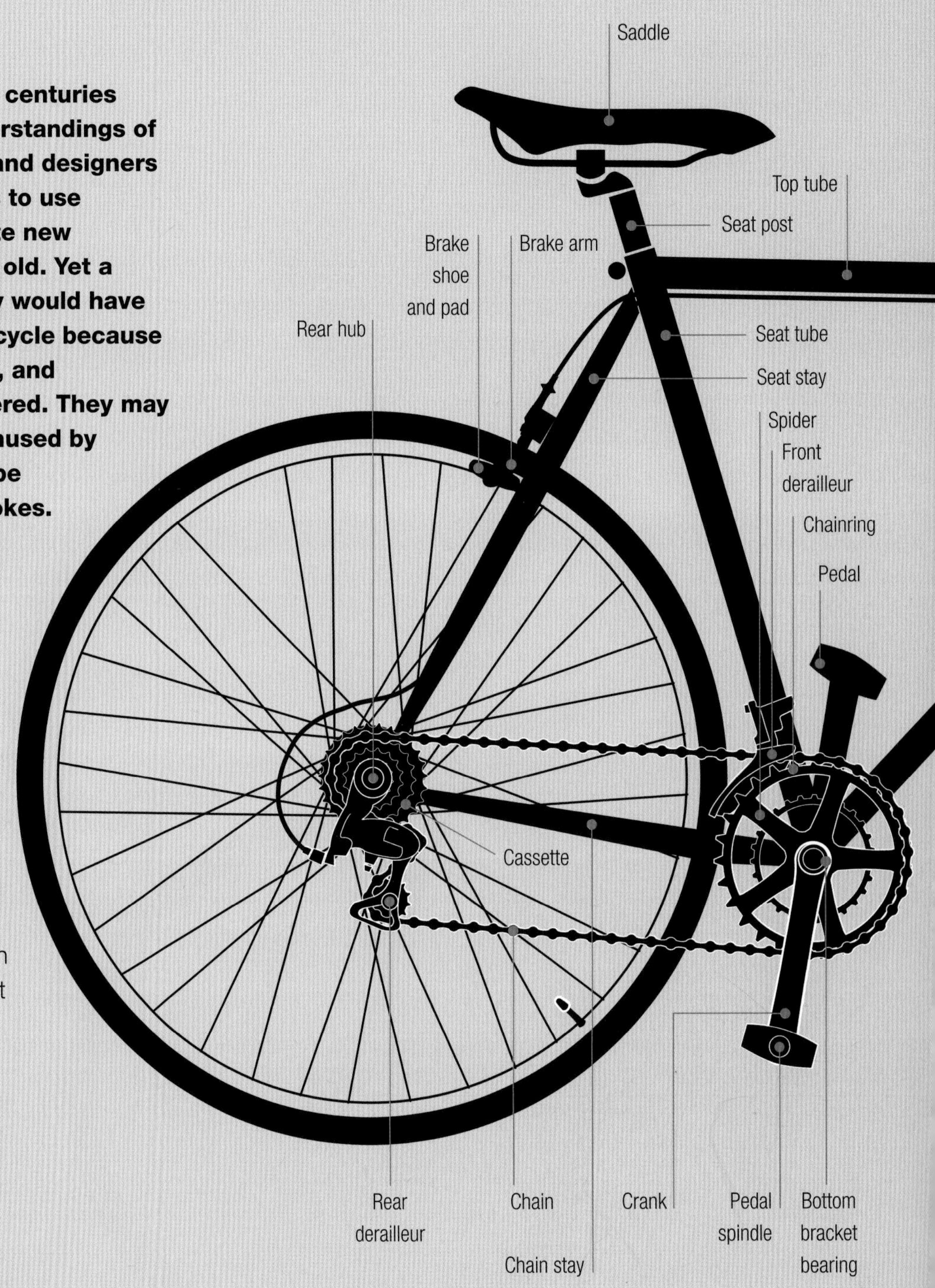

The major components

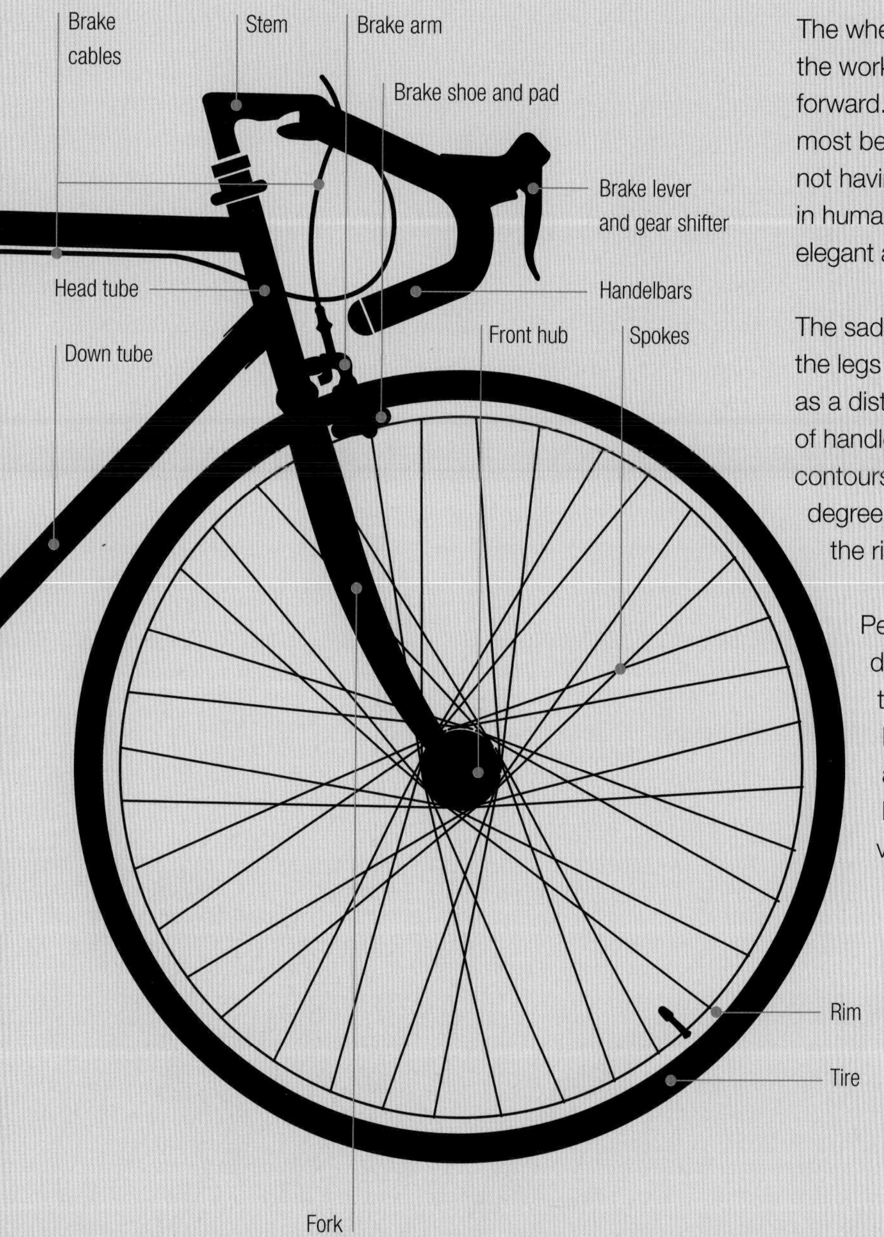

The wheels define it as a machine that is able to translate the work by the rider into motion along the ground, preferably forward. Even a Neanderthal would agree that the wheel is the most beautiful of human inventions, while kicking himself for not having thought of it first. The wheel has been fundamental in human development and its application in the bicycle is both elegant and minimalist, giving the bike its extraordinary efficiency.

The saddle's role of seating the rider comfortably while allowing the legs to move freely has determined its unique appearance as a distorted reflection of the body parts it supports. The lines of handlebars, however, are as varied as handshakes, their contours a compromise between the need for an appropriate degree of bike control and the flailing, versatile geometry of the rider's arms.

Pedals, cranks, chainsets, gears, and brakes are diverse in design and detail. They will continue to be refashioned as the market, marketeers, and technologies evolve. Only a handful of each year's fresh ideas will endure. There will always be the frame, wheels, saddle, and handlebars. For all the other hundreds of components and their various configurations, only the fittest will survive.

◀ ***Variations on a theme*** *Bicycles come in scores of varieties, with differences in design, components, and materials to optimize the balance between function and cost. Every one is an assembly of parts, each of which can be changed in line with the owner's desires, with replacements that may be mass-manufactured or handcrafted.*

How efficient is cycling?

Why is it easier to ride than to walk?

Walking is fine and running is faster but cyclists go farther, quicker, for longer. Cycling is the most efficient way for a human being to use their own energy for propulsion. The bicycle is an extraordinary machine capable (at best) of converting 98.6 percent of the cyclist's pedaling effort into spinning the wheels, while those who stride along with their feet on the ground waste a third of theirs.[1]

The average human walking speed is about 3 mph (5 km/h).[2] At this speed, humans can travel for hours. Nevertheless, the walking movement is wasteful and inefficient. At every single step, the knee of the grounded leg bends and flexes, lowering the entire body for a moment before levering it back up to normal height. While this is going on, the spine bends a little, the hips twist, arms move to and fro, and the swinging motion of the other leg stops with an energy-absorbing impact when the foot strikes the ground.[3] In fact, compared to an "ideal" walking machine, we are only about 65 percent efficient. In other words, in walking we lose about one-third of our energy to effects other than forward motion. No engineer would be proud of such an inefficient vehicle.

The bicycle, however, forces us to adopt a pose that maximizes the use of our calories. A walker moving at a particular speed could, if they rode a bike, travel three times faster without having to increase their effort at all, because the two-wheeled machine is phenomenally efficient.[4] There are two reasons for this remarkable improvement. The first is the way the bike is configured to make us use our bodies more efficiently. The second is that the machine acts as a lever, multiplying the distance our feet travel around the bottom bracket with each pedal stroke into the much greater distance the tires travel along the ground.

Energy efficiency

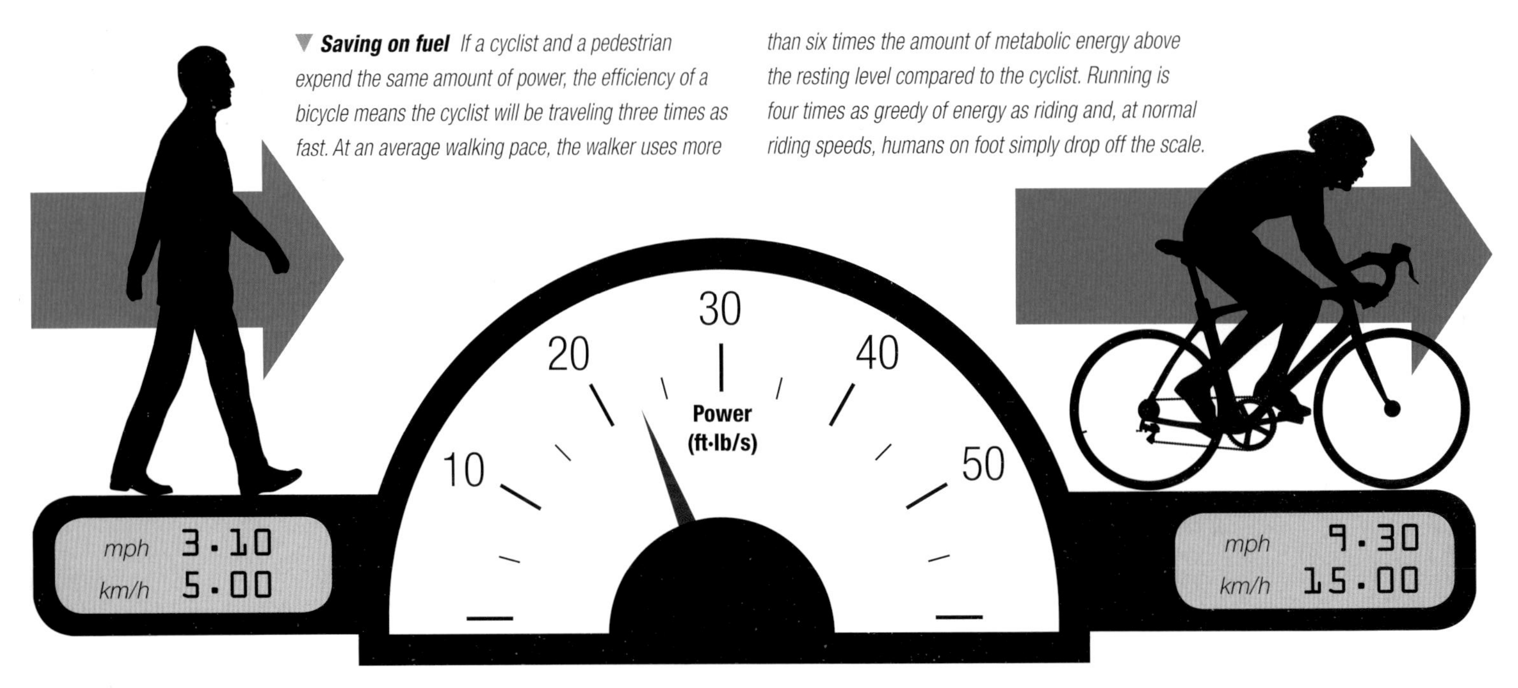

▼ ***Saving on fuel*** *If a cyclist and a pedestrian expend the same amount of power, the efficiency of a bicycle means the cyclist will be traveling three times as fast. At an average walking pace, the walker uses more than six times the amount of metabolic energy above the resting level compared to the cyclist. Running is four times as greedy of energy as riding and, at normal riding speeds, humans on foot simply drop off the scale.*

Speed records

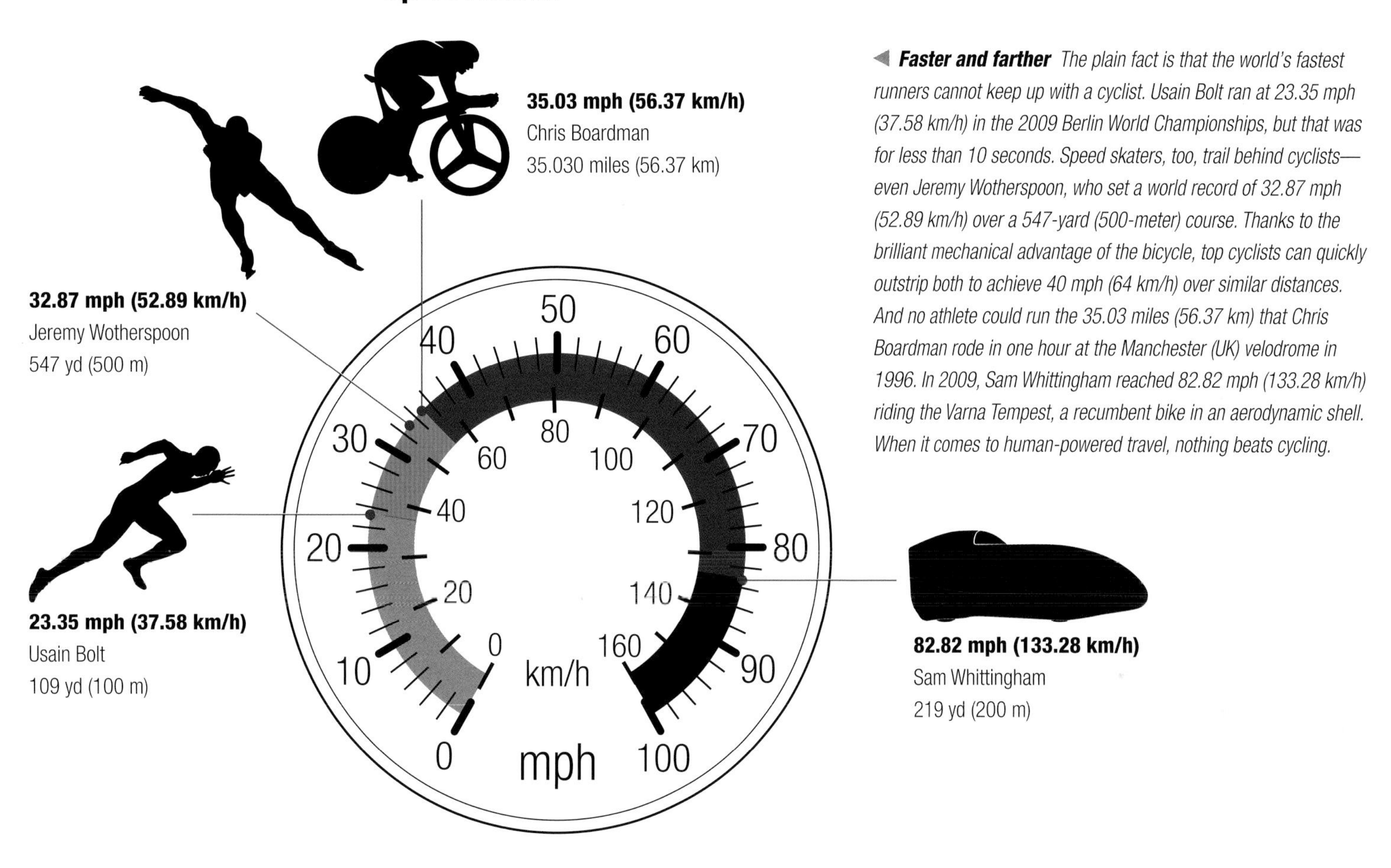

◀ ***Faster and farther*** *The plain fact is that the world's fastest runners cannot keep up with a cyclist. Usain Bolt ran at 23.35 mph (37.58 km/h) in the 2009 Berlin World Championships, but that was for less than 10 seconds. Speed skaters, too, trail behind cyclists—even Jeremy Wotherspoon, who set a world record of 32.87 mph (52.89 km/h) over a 547-yard (500-meter) course. Thanks to the brilliant mechanical advantage of the bicycle, top cyclists can quickly outstrip both to achieve 40 mph (64 km/h) over similar distances. And no athlete could run the 35.03 miles (56.37 km) that Chris Boardman rode in one hour at the Manchester (UK) velodrome in 1996. In 2009, Sam Whittingham reached 82.82 mph (133.28 km/h) riding the Varna Tempest, a recumbent bike in an aerodynamic shell. When it comes to human-powered travel, nothing beats cycling.*

Pedal power

▶ ***Wheel radius to crank length ratio*** *The pedal as a lever gives a mechanical advantage which, in its simplest form, can be calculated as the ratio of the wheel radius to the crank length. As an approximate example, a direct-drive High Wheeler with a wheel radius of 30 in (75 cm) and cranks 6 in (15 cm) long amplifies the distance moved by the ratio 30/6 = 5. That is, with one revolution of the pedals, the big wheel will travel 15 ft (471.25 cm) along the ground, five times greater than the 3 ft (94.25 cm) the Victorian rider's foot has traveled around the axle. The real magic of modern bikes is that they have gears, so the rider can choose their preferred ratio at any time.*

Distance the bicycle travels

Distance the foot/pedal travels

What is the most efficient design? → Which type of bike should I choose?

While most bicycles use the same basic design, there are many variations on the theme, each with a different level of efficiency. Most have pedals, a chain, brakes, a saddle, and handlebars; many have gears. Those are about the only common areas; like most adaptable creatures, bicycles have evolved into a multiplicity of designs, each excelling in a particular niche.

Some designs are conceived to be efficient for a single purpose. A track bike, with its stiff frame and no brakes, may not be very comfortable but at a velodrome it allows the rider to reach very high speeds. A mountain bike has suspension that smooths rough terrain, but its fat, heavy tires prevent it from breaking any speed records. A hybrid may be equally at home commuting in the city and roaming the trails. The key to choosing a bicycle is to identify what area of efficiency is your key criterion. That decision will dictate the choice of wheels and frame shapes.

The vast majority of bike frames rely on engineering's favorite shape—the triangle. It's the strongest two-dimensional geometric shape, which is why the main part of the frame is, or approximates to, a triangle. Admittedly, the top tube may not be horizontal and a short section of head tube may add a fourth side a few inches long, but designers try to get as close as possible to a triangle because it is the most stable and rigid structure for a given mass of material. In a triangle, the three fixed tubes share the transmission of forces between them when put under pressure or stretched. The triangle can be deformed only by changing the length of one of its sides or by breaking it.

Variations on this triangular principle can be seen at the core of each style of bike, playing a key part in the relative power requirement and comfort of each model. The cycling science of frame triangles is more fully explored on pages 48–49.

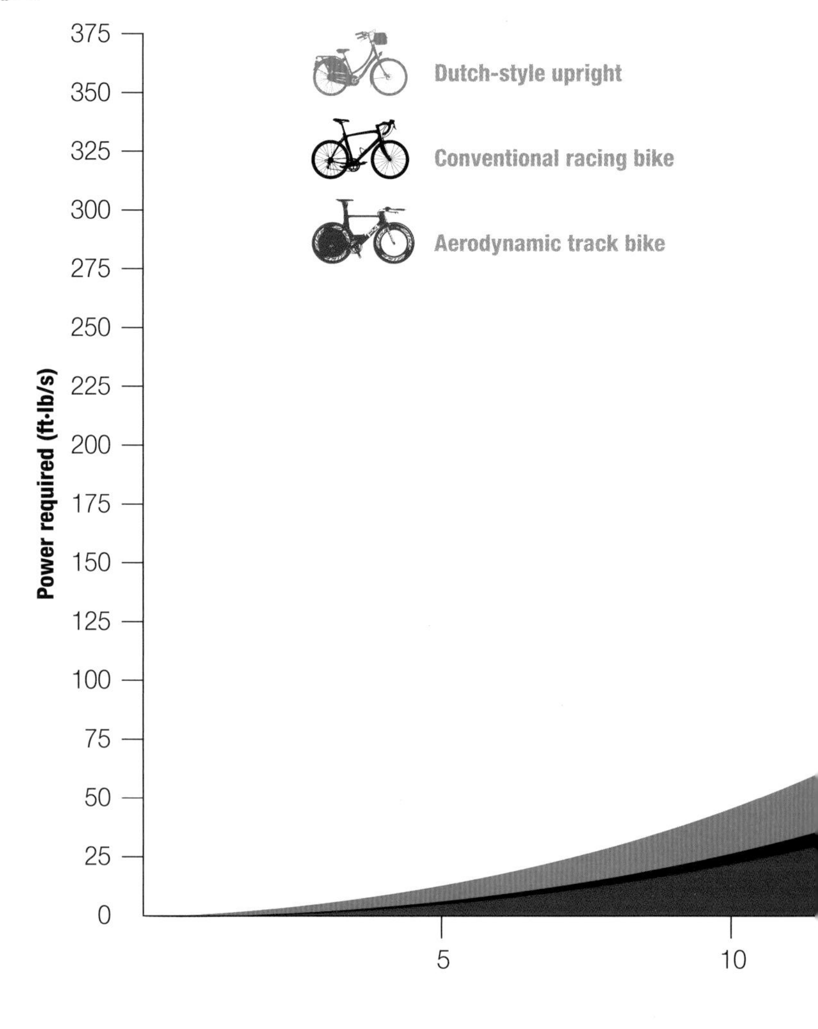

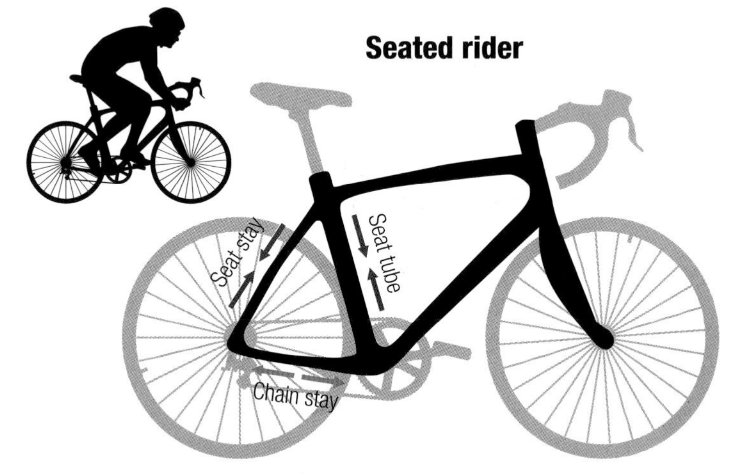

Frame and power

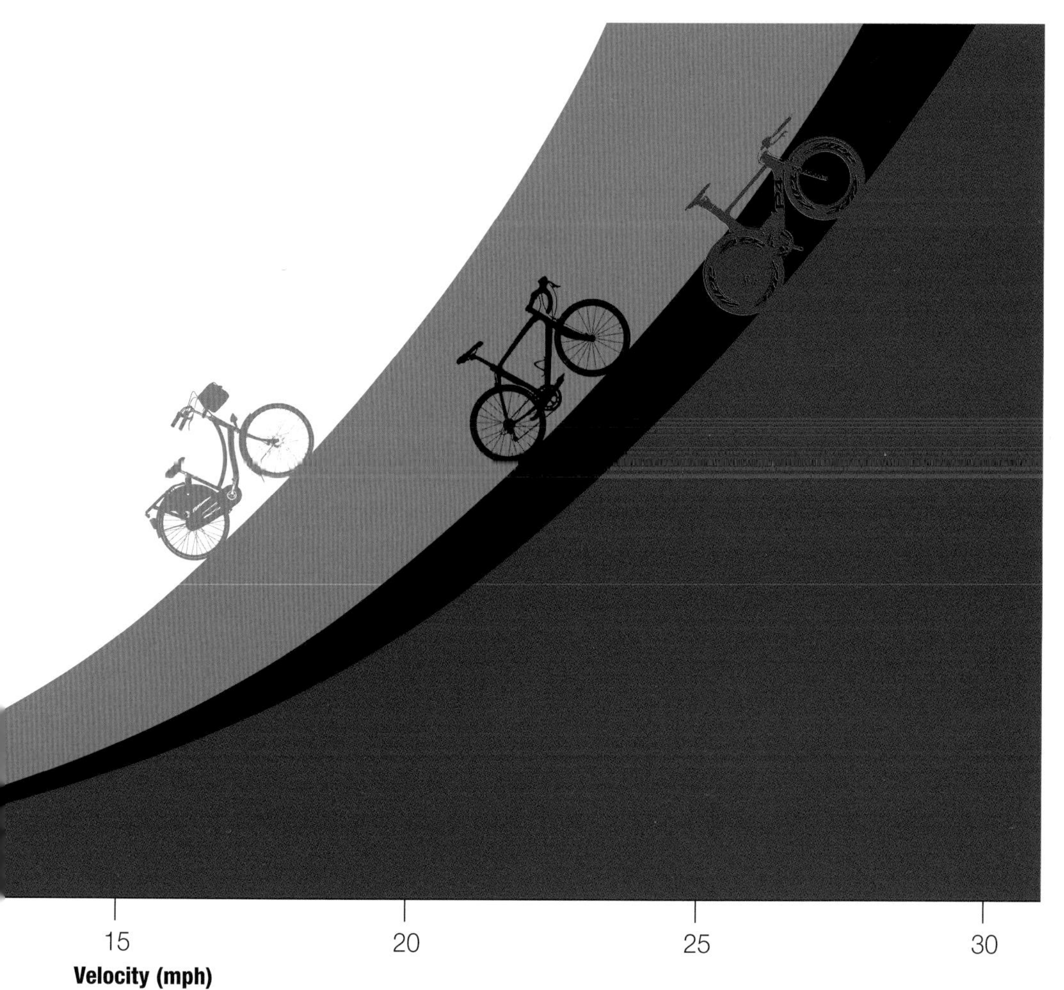

◀ ***Power requirements*** *This graph demonstrates how the power efficiency of different styles of bike becomes more significant at the top end of the speed curve. At speeds under 6 mph (10 km/h), the Dutch-style upright bicycle only needs a little more pedal power from its rider than does the sleek velo-raptor. As soon as the rider gets cranking, however, the efficiency of the upright slumps while the racing bikes start to deliver on what they were built for: speed. If you want to save energy at speed, then you have to sacrifice comfort and carrying capacity.*[5]

◀ ▶ ***The power of triangles*** *The triangle frame excels at the transmission of forces without distortion. When a rider sits on the saddle (left), the two adjacent sides (the seat tube and the seat stay) share that load, while the opposite side (the chain stay) is put under tension. When the rider stands up (right), the down tube and seat tube are stretched (and the bottom bracket supports their weight through the pedals), while the top tube is compressed and stops the bike from collapsing in the middle.*

science in action

conservation of energy

Energy cannot be created or destroyed—that is one of the fundamental concepts of physics. This is demonstrated every time you ride. The nutrients in the food and drink that you consume are metabolized to nourish your body and power your bike. Riders convert the chemical energy in food and drink into mechanical energy to propel them forward. Other energy conversions less useful to cyclists also take place—making them hot, moving the air around them, and creating sound from the chain and the tires on the road. If it was possible to measure all of the output energies and add them together, they would equal the energy stored, then consumed and expended, during that ride. The total energy put into the cyclist has been converted, but it is never lost.

Unfortunately, it is very hard to prove this empirically because cyclist and bike are not an isolated system as far as physics is concerned. There are other energy impacts that cannot be excluded—for example, gusts of wind, the roughness of the road, and what the rider ate for dinner the previous evening. Also, there are factors that cannot be readily quantified, such as fingernail growth and eyelid blinks. Yet the law of the conservation of energy remains true—energy may be converted from one form to another, but in total it cannot be created or destroyed.

Manuel Gárate (pictured right) was first to cross the finish line on Mont Ventoux during the 20th stage of the 2009 Tour de France. The famous mountain summit is 6,273 ft (1,912 m) high so, given that he and his Rabobank team bike had started the day in Montélimar at just 328 ft (100 m) above sea level, and estimating that they weighed 159 lb (72 kg) in total, he'd have successfully converted some 300 calories of physiological energy into gravitational potential energy. If he had freewheeled back down, much of this would have been converted into kinetic energy.

▶ ***Pit stops*** *Riders become depleted as they convert their own resources into kinetic energy to propel their bikes. They have expended energy metabolized from the nutrients in food and drink and this must be replenished if they are to continue. Water contains no calories, but without it, metabolization, muscles, and the entire human body would cease to function.*

LCI
Azpiru
Excavaciones
Rabobank
GIANT
Direction Générale

How does bike geometry relate to gender? → Why are men's and women's bikes different?

You've probably noticed that men and women are different. So are their bikes. Physiological differences mean that many models are configured differently for the two sexes. The main design changes are in frame dimensions and in the detailing of the main contact points—the handlebars and saddle.

Historically, women's bikes did not feature the horizontal top tube, leaving an "open" frame. This allowed women to step on without having to lift a leg and expose an ankle. It also better accommodated long skirts. The fact that many women's bikes are still designed like this is due less to science than to convention. This open-frame design does not offer the structural integrity of a closed triangle, so it is usually strengthened using extra tubing—with the penalty of increased weight. A hybrid bike without a top tube can weigh 7 oz (200 g) more than a closed-triangle frame.

Today, however, the main issues with women's bikes are the dimensions and geometry. Women are, on average, 15 percent lighter and 8 percent shorter than men. Because one of the main functions of the frame is to place the rider in a stable, comfortable position between the wheels while pedaling at their optimum efficiency, it is clear that women's frames should not just be smaller but should also have different proportions. The first consideration is to bring the saddle and the handlebars closer together to reduce the degree of back and arm stretch required of a woman rider. Narrower bars are used to complement the female rider's narrower shoulders, often with a shallower drop and shorter reach. The female rider's smaller hands are best matched to smaller gear shifters and brake levers that rest closer to the bar. Another contact point, the saddle, should also differ from that designed for a man. It needs to be wider to support the sit bones of the pelvis and have a shorter nose to prevent uncomfortable pressures on soft tissue.

Saddle science

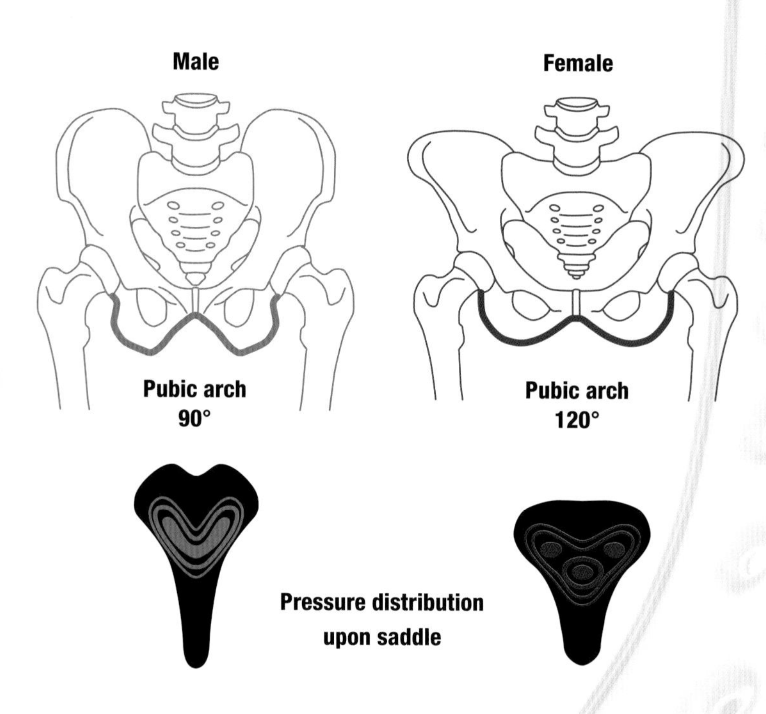

▲ ***Fundamental difference*** *The gluteus maximus muscle covers the sit bones until one sits down—then it leaves part of the pelvis, called the ischial tuberosity, unpadded and bearing the rider's weight. It's the saddle's job to support this comfortably. Men don't give birth, so these swellings of the pelvis are narrower than they are for women and the saddles are designed accordingly.*

Anatomy of the rider

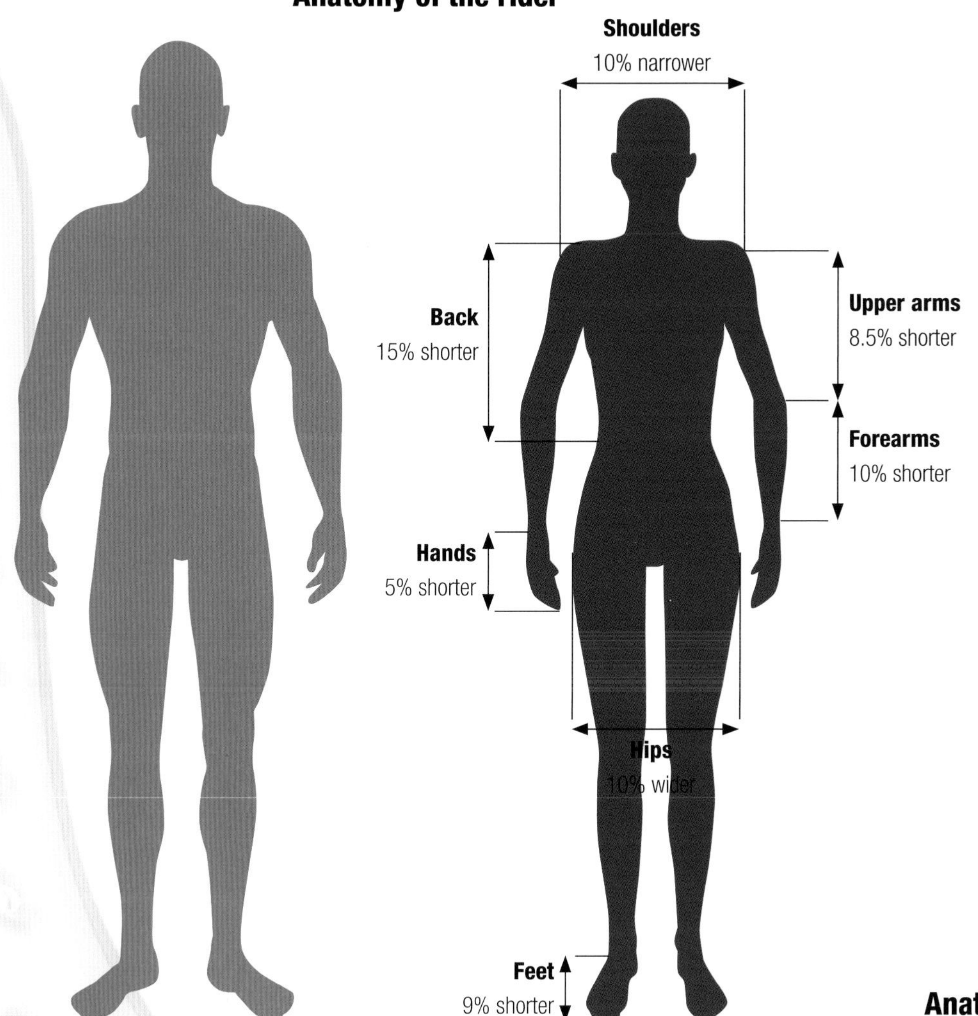

◀ ***Body of evidence*** *Research has shown that the average weight of men 20 years old or over is about 195 lb (88.3 kg), while for women it is about 165 lb (74.7 kg). Mean height for men is 5 ft 9 in (176 cm), and for women 5 ft 4 in (162 cm).[6] While these general figures might suggest that women simply need smaller, less sturdy bikes than men, more important results are revealed when the anatomy is measured segment by segment. Women's skeletons develop sooner than men's and are fully mature at about 18 years old. Men's skeletons continue growing, maturing about three years later than women's—so their frames tend to be bigger and their limbs longer and thicker. Cycling is best when the bike is tailored to closely match the rider's physique, so gender-specific models are not a ploy of marketing departments—they make riding more comfortable and efficient.*

Anatomy of the bike

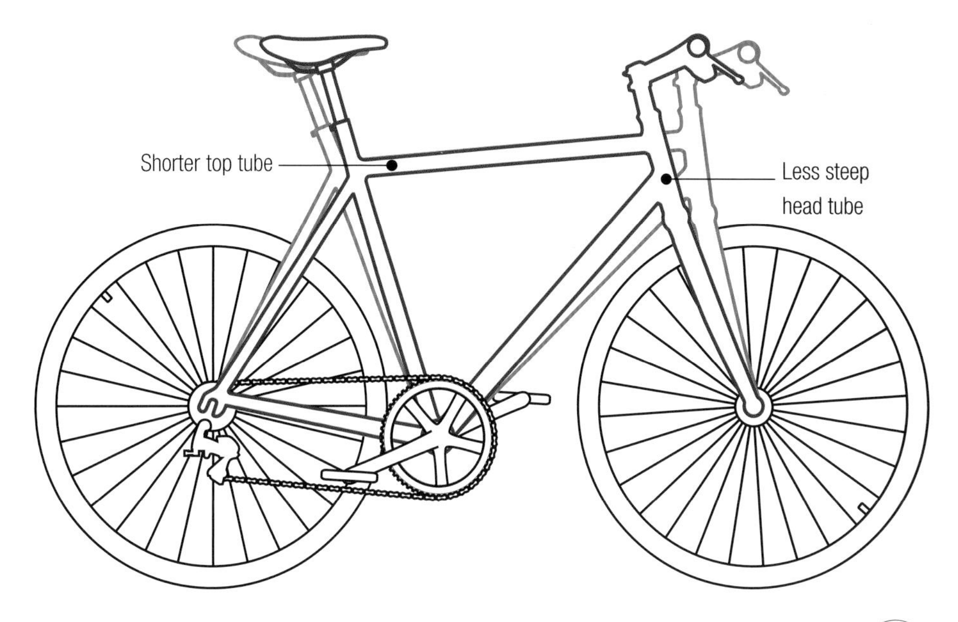

▶ ***Comparative bike frames*** *The primary difference between the male and female bike frame is the distance between saddle and handlebars. When designing a frame for a female rider, simply making the top tube shorter can have unwanted side effects: for example, it would normally make the steering more sensitive than before. The problem can be avoided by making the angle of the head tube—through which the steerer tube runs, connecting the fork to the handlebars via the stem—less steep. Assuming the geometry of the fork has not been changed, this effectively pushes the front axle farther forward, reinstating the distance between the two wheels.*

What is the environmental impact of cycling?

Will cycling save the planet?

The science endorses the gut feeling that cycling has a low impact on the environment, compared to other modes of transport. For every $1,000 spent by manufacturers to make bicycles, 1 short ton of carbon is generated.[7] Admittedly, this is 45 percent worse than the auto industry's figure, but $1,000 will buy many more new bikes than cars. So, per vehicle, making bicycles is many times cleaner than making cars.

When new bikes hit the road, they are friendlier to the Earth, too. For a given journey, the energy consumed by a driver is at least 42 times more than by a cyclist, a bus passenger uses 34 times as much, and a train passenger 27 times as much. The cyclist requires less space than all but the train passenger and pedestrian. For journeys of up to 6 miles (10 km), the bike is definitely best for both rider and the planet. In terms of lifespan, life-cycle analysis at MIT shows that the bicycle consumes the lowest energy per passenger-mile across its entire life, compared to other forms of transport.[8] If annual cycling distance is increased, revised life-cycle analysis calculations will boost the green credentials of the bicycle even further. And there is one more element of impact that vehicles have on the environment—noise pollution. Is there any vehicle quieter than a bicycle?

It's all very well drawing comparisons with other modes of travel but there can be no doubt that riders do use some of the Earth's resources to be able to cycle—in the form of food. Food production and transportation generate carbon dioxide and other greenhouse gases, the total weight of which is used as a measure of the impact on our planet. A rider cycling 1 mile (1.6 km) and metabolizing roughly 50 calories entirely from bananas would be responsible for the creation of 2¼ oz (65 g) of greenhouse gases. If instead they metabolized these calories from breakfast cereal with milk, the toll on our atmosphere would rise to 3⅛ oz (90 g) of greenhouse gases. If those figures are hard to stomach, consider the impact of refueling with a cheeseburger—that's 9⅛ oz (260 g) of greenhouse gases.[9]

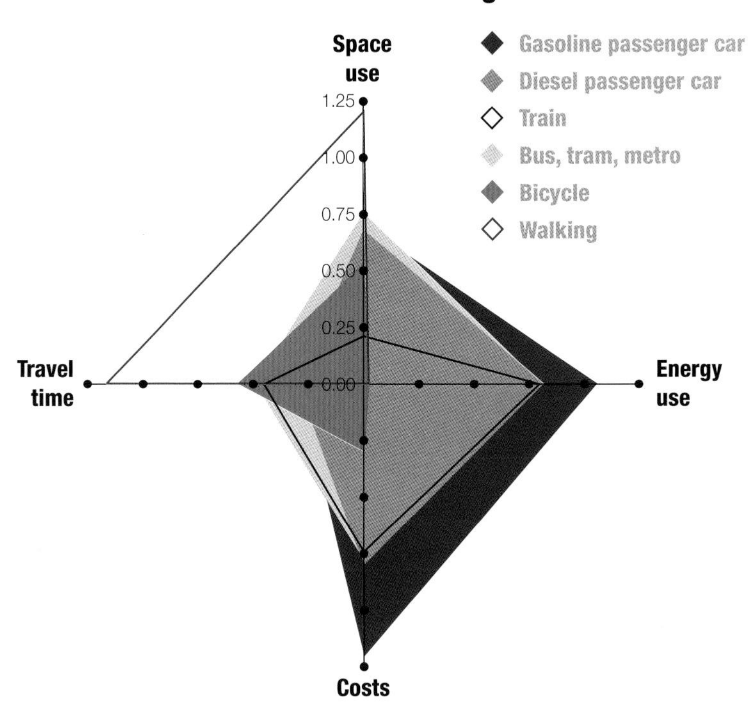

▲ ***Cheaper, sustainable*** *Researchers in the Netherlands studied four elements of urban journeys of 1.6 to 3.1 miles (2.5 to 5 km). The greater the distance from the center of the graph, the worse the score. The space-use figure was calculated by dividing the area occupied by the infrastructure by the annual distance traveled. The costs were just to the person traveling, not the cost of the infrastructure. Cycling is only marginally slower than the powered modes in these urban journeys, yet significantly cheaper and more energy-thrifty.*[10]

▶ ***Lifetime greenhouse gas emissions*** *For every mile it is pedaled in the course of its life, a bicycle uses the equivalent of 76 calories (319 kJ) of the Earth's resources. The next most frugal is a peak-hour bus, which uses more than three times as much. This also translates into higher greenhouse gas emissions. Bicycling creates about 73 lb of greenhouse gases for every mile of its life. On average, a train is roughly five times more polluting across its lifetime, a bus almost 10 times worse, a car 11 times, an SUV 13 times, and a pickup truck 18 times dirtier.*[11]

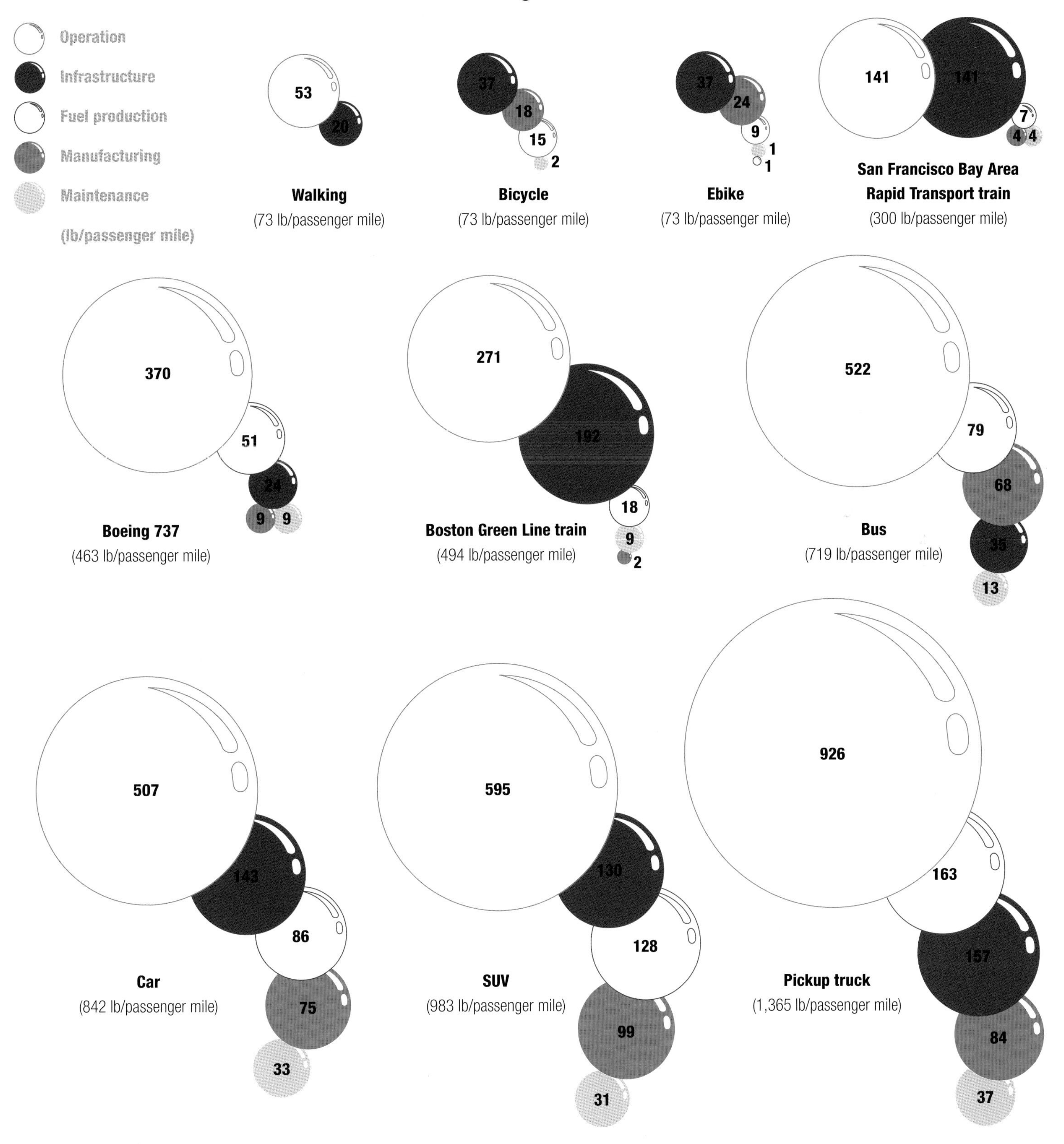
Greenhouse gas emissions
Operation
Infrastructure
Fuel production
Manufacturing
Maintenance
(lb/passenger mile)
53
20
Walking
(73 lb/passenger mile)
37
18
15
2
Bicycle
(73 lb/passenger mile)
37
24
9
1
1
Ebike
(73 lb/passenger mile)
141
141
7
4
4
San Francisco Bay Area
Rapid Transport train
(300 lb/passenger mile)
370
51
24
9
9
Boeing 737
(463 lb/passenger mile)
271
192
18
9
2
Boston Green Line train
(494 lb/passenger mile)
522
79
68
35
13
Bus
(719 lb/passenger mile)
507
143
86
75
33
Car
(842 lb/passenger mile)
595
130
128
99
31
SUV
(983 lb/passenger mile)
926
163
157
84
37
Pickup truck
(1,365 lb/passenger mile)

How does cycling benefit the cyclist?

Can cycling help me live longer?

Over an average lifetime, the expected increase in life expectancy for a modal shift from commuting by car to bicycle in urban areas was estimated to be from 3 to 14 months, or an average of 8½ months, per person, due to increased physical activity.[18]

Cycling does not guarantee immortality but it can increase your life expectancy by two years or more. Scientists worked this out by monitoring almost 17,000 Harvard alumni for up to 16 years. By their eightieth birthdays, those who had exercised adequately had extended their expected lifespan by an average of two years.[12] An extensive study in Copenhagen, Denmark, lasting 14 years and involving 30,000 adults, showed that those who did not cycle to work had a mortality rate 39 percent higher than those who did.[13] Blue-collar workers who cycled were found, in another study, to be as fit as their non-pedaling colleagues who were 10 years younger.[14]

While all exercise burns calories and so helps individuals to keep their weight down, there are few such activities that can be done while traveling. Bicycle commuting burns an average of 542 calories per hour, and a study of nearly 2,400 adults found that those who biked to work were fitter, leaner, and less likely to be obese. They had better triglyceride, blood pressure, and insulin levels than those who didn't actively commute to work.[15]

The sensible cyclist will want to compare the relative risks of traveling daily by car with those of riding a bike. This has been analyzed by a team in the Netherlands, as shown on the right. They demonstrated that the health benefits of cycling relative to car driving substantially outweigh the risks created by air pollution and traffic pollution.[16] With those risks understood, it's worth emphasizing that cycling to work has benefits beyond personal fitness. A survey by the UK's Chartered Management Institute found that cyclists are more likely to arrive at work on time, are more productive, and are less prone to stress than colleagues arriving by car or public transport.[17] For many, it's not practical to commute by bike, but for those who are lucky enough, the science appears to show that life is not only longer but also far more enjoyable.

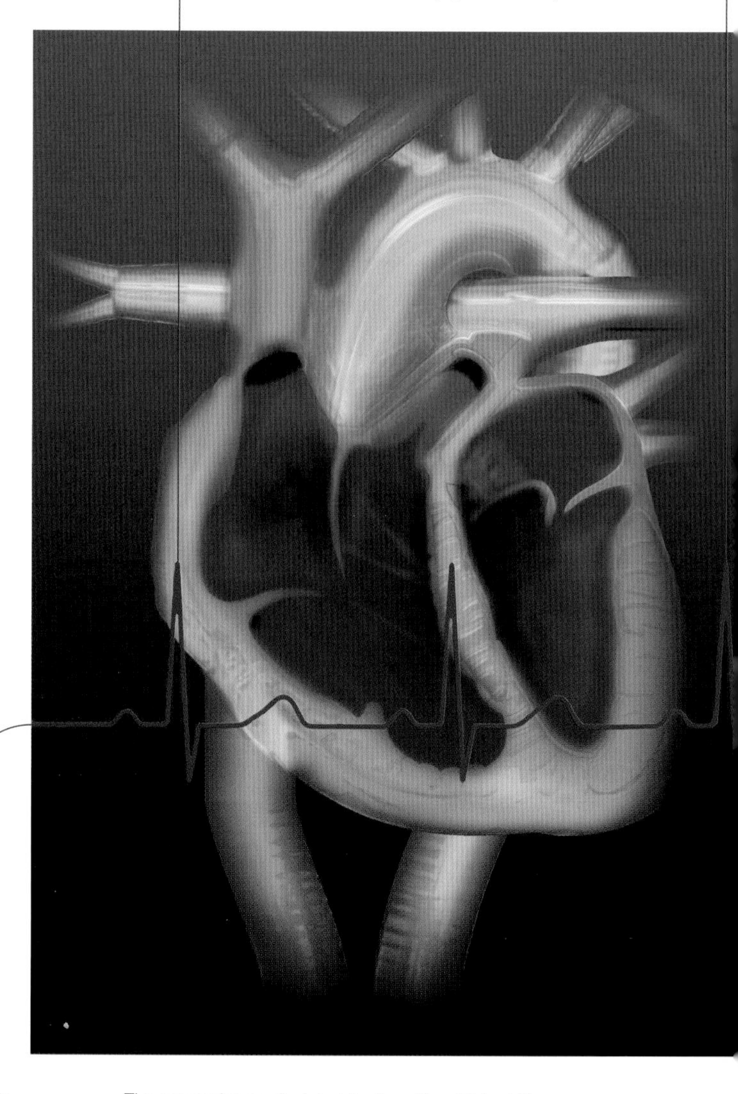

The researchers calculated the benefits of this shift from driving to cycling were nine times greater than the potential risks. Furthermore, they argued that there were benefits to the wider population due to lower pollution and fewer traffic accidents caused by cars.[19]

The bike and the heart

The reduction in life expectancy due to increased air pollution was expected to be from 0.8 to 40 days per person, while the increase in traffic accidents would result in another five to nine days lost on average, or around two weeks combined.[20]

▲ ***Cycling and a healthy heart*** *A 2010 Dutch study estimated the comparative health benefits and risks if 500,000 out of a total population of over 16 million switched to traveling short distances by bicycle instead of car on a daily basis. In the Netherlands, approximately half of all car journeys are under 4½ miles (7.5 km).*[21]

Oxygen uptake

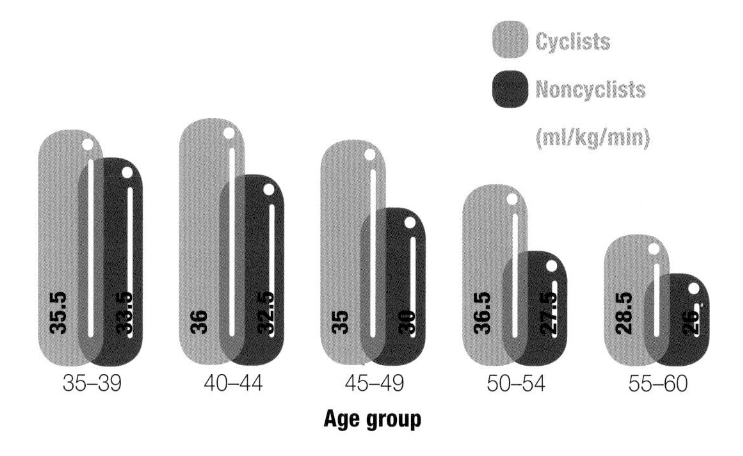

▲ ***Oxygen uptake, cyclists versus noncyclists*** *A survey of factory workers compared the fitness of cyclists with noncyclists. They measured their maximum oxygen uptake—VO_2 max (milliliters of oxygen per kilogram of body weight per minute). The comparative scores indicate that cyclists are fitter than noncyclists, and suggest that the older you get, the better riding a bike is for you.*[22]

▼ ***The comparative effects of cycling and driving to work*** *In a study, 58 percent of cyclists said their journeys were never disrupted by traffic, compared to only 4 percent of drivers. Some 9 percent of cyclists were stressed by their journeys, compared to nearly 40 percent of drivers. Almost one-quarter of motorists felt their productivity was affected by the stress of their commute, compared to none of the cyclists.*[23]

Commuting advantages

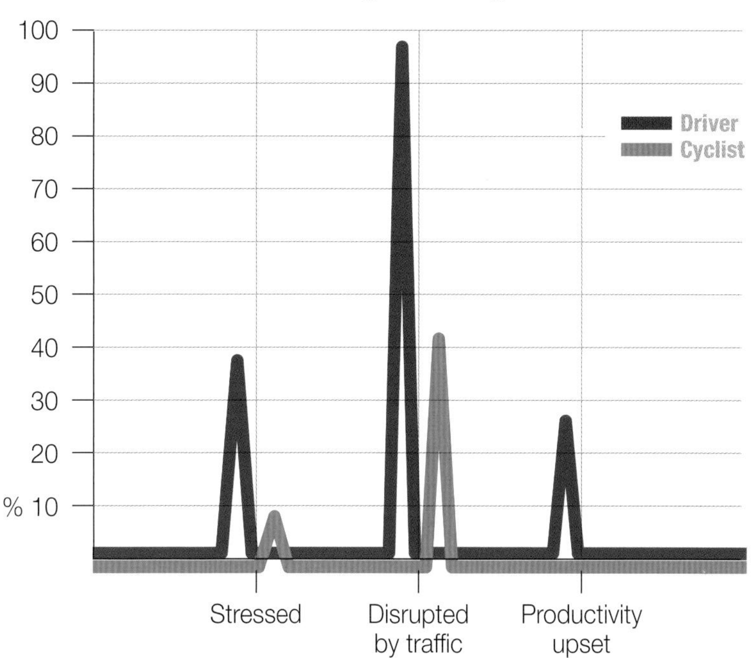

How risky is traveling by bike? Is there safety in numbers?

Motorists are no safer than cyclists. Figures from the UK show that there are about 0.4 fatalities for every million person-hours of cycling—a dramatically low figure considering it would take one cyclist more than 2,000 years riding nonstop to build up that many hours. This is about the same as an average European driver's risk of death, taken over their lifetime. The annual risk for drivers and cyclists is also very similar, and cycling is statistically at least 10 times safer than riding a motorcycle.

The more people cycle, the safer the roads seem to become, not just for cyclists but for all road users. In Portland, Oregon, all deaths from traffic accidents declined from 46 to 28 each year between 1997 and 2007, while the number of cycling commuters quadrupled to 6 percent of journeys. Similarly, cycle use in the Netherlands increased by 45 percent in the two decades to 1997, while cyclists' deaths fell by almost 40 percent. In Berlin, between 1990 and 2007, the share of bicycle trips doubled to 10 percent, while serious injuries to cyclists fell by 38 percent.[24]

Is there a direct correlation between these statistics? The phenomenon of "safety in numbers" is not so hard to understand. A growth in the number of cyclists makes them more visible and drivers change their own behavior accordingly. Cities are more likely to provide safer road designs and facilities for cyclists when there are more of them about. And when some drivers switch to cycling, it means that there are fewer cars on the road and, hence, a reduced chance of anyone colliding with a high-speed chunk of metal. And drivers become more aware of what being a cyclist is all about, so they drive more considerately as a result. There is no doubt that cycling has the power to improve road safety for everybody. Unlike the drivers of motorized vehicles, cyclists—because they are associated with a much lower level of kinetic energy—almost never injure or kill other road users or pedestrians.

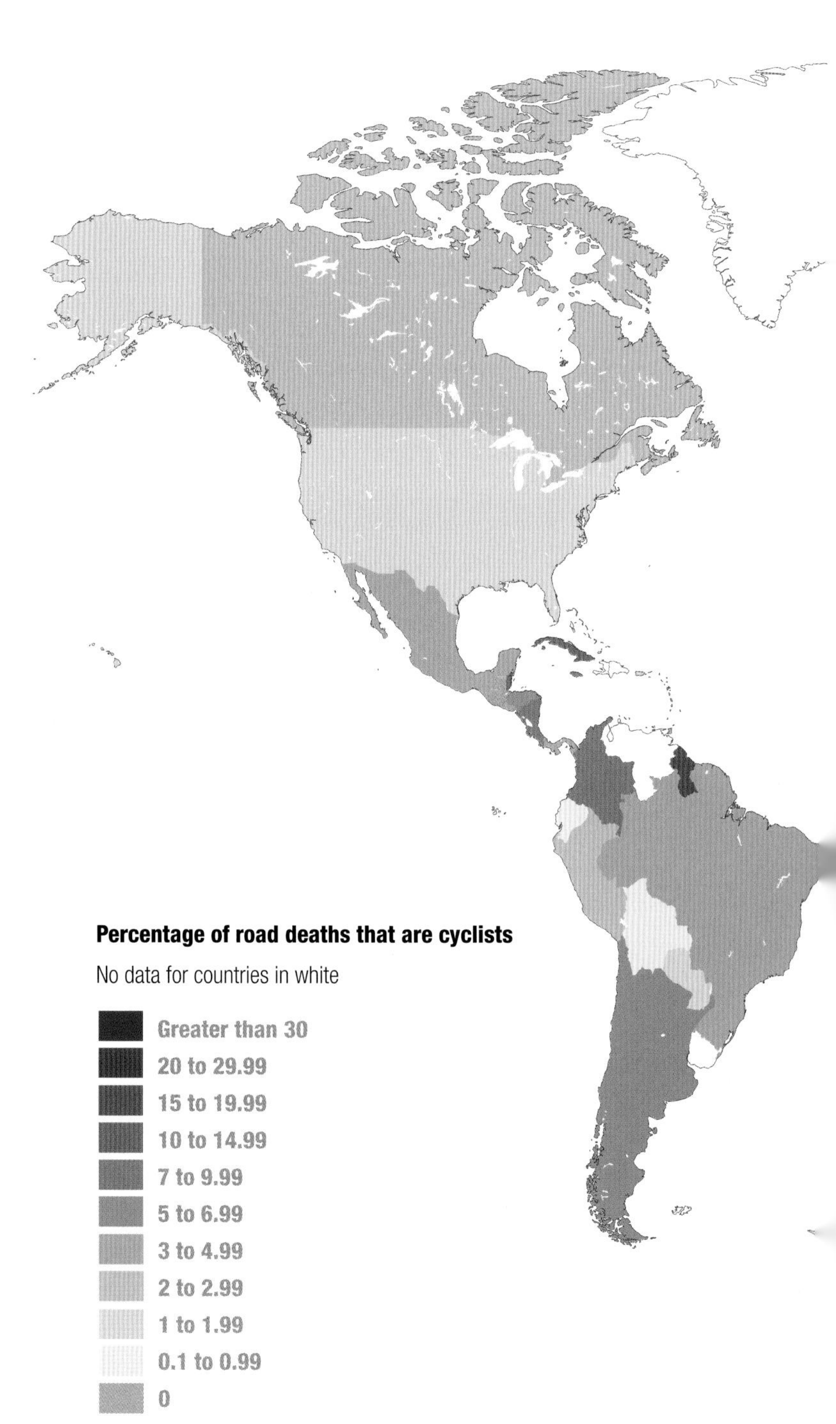

How safe is cycling around the world?

▼ **Road safety** *The World Health Organization's latest data reveals that cyclists comprised 24 percent of all road deaths in the Netherlands, the fourth-highest proportion in the world. However, studies have revealed that bicycle usage is far higher in the Netherlands than almost any other country for which data is available, which helps to explain why the proportion of road deaths is so high. Further analysis has revealed that the average distance ridden per person per year is approximately inversely proportional to the number of deaths per mile of travel—in other words, the more cyclists there are on the road, the safer it is to cycle. There may be a number of causes for this, from drivers being more aware of cyclists when there are more around, to better urban planning.*[25]

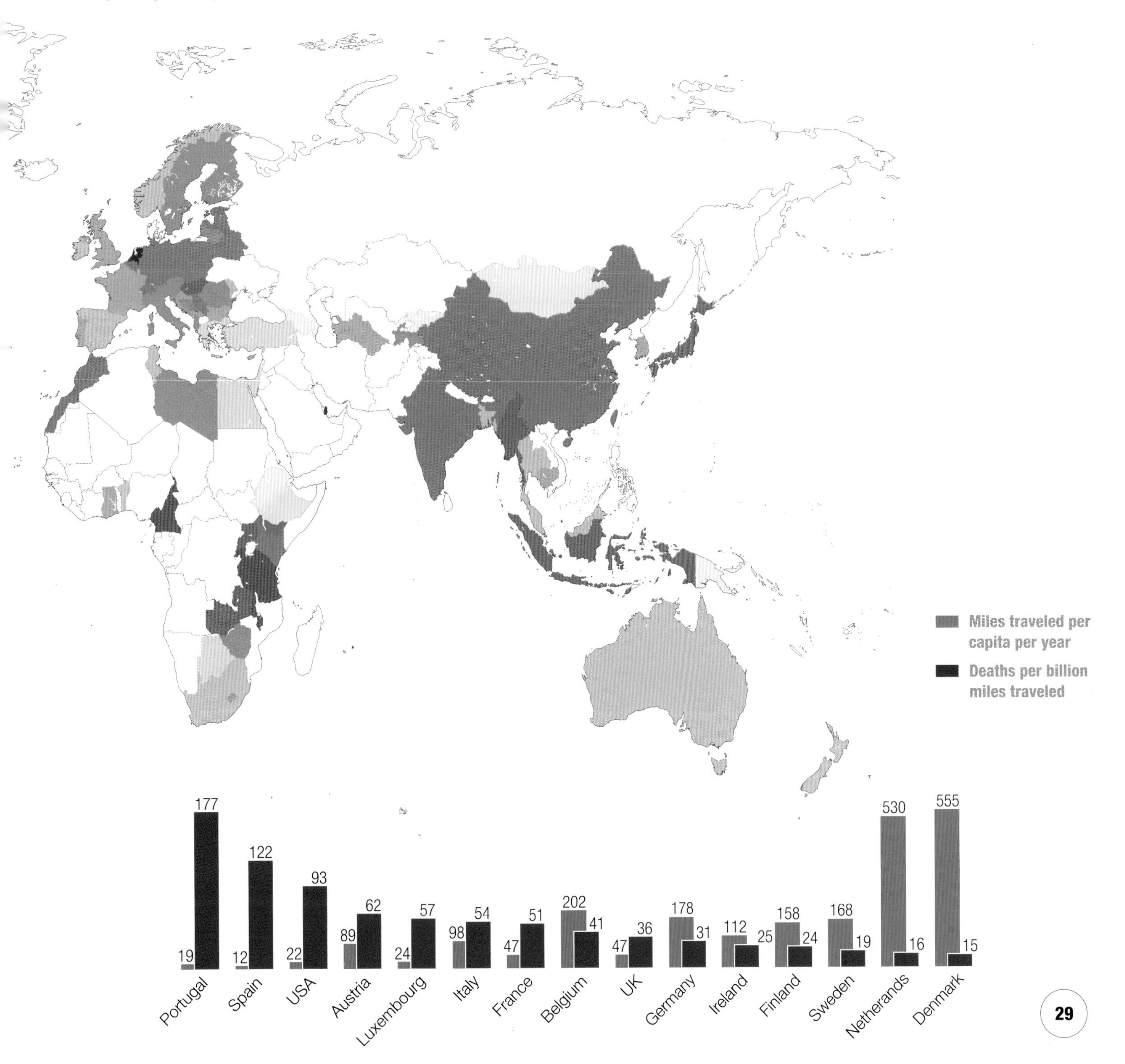

equipment: suspension

After the bicycle was invented, nobody sat comfortably for many years. The Draisine and Boneshaker put pressure on the most delicate parts and the Ordinary was little better with only its large wheel diameter to mitigate bumps. The safety bicycle frame design with sprung saddle, coupled with pneumatic tires, was a revelation. Even so, it's no surprise that cyclists were the first to demand smooth roads, years before cars were popular. Once slick asphalt had become ubiquitous, cyclists went off-road and demanded an equally comfy ride. To supply this and make handling easier, engineers first eased vibrations with seat pillars supported on coiled springs or resilient elastomers, but the big breakthrough was when front suspension was added to forks. Rear suspension derived from motorcycles soon followed, with increasingly diverse designs to suit different off-road disciplines. Today, even commuter bikes have suspension for bumpier roads.

▲ ***Becker full suspension*** *(ca. 1890)*
Front suspension is provided by a pivot at the head tube with a spring to the down tube, rear suspension by a pivot at the bottom bracket and a spring at the top of the fork to the seat tube.

▶ ***Sage telescopic fork*** *(ca. 1922)*
A coil spring inside the upper part of the fork allows the blade to move up and down as the wheel crosses rough ground. The same telescoping principle is used in most suspension forks today.

▼ ***Moulton Dry Cone*** *(1962)*
Rubber-cone technology invented for the original British Mini car was applied to the rear triangle of Moulton bikes to reduce road shock from its small wheels.

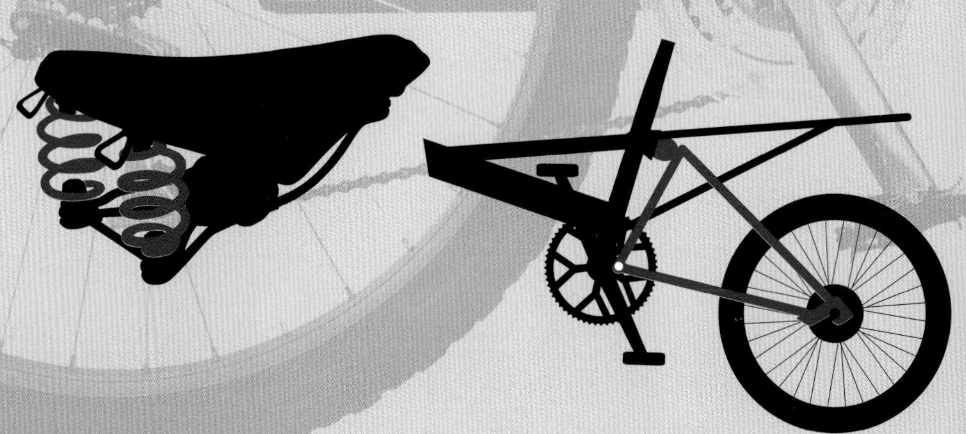

▶ ***Brooks B66*** *(1927)*
Until 50 years ago the vast majority of saddles were sprung with metal coils. The rise of lighter synthetic foams and padding has displaced most but this model is still available today.

Comfort is of fundamental importance to cycling, literally and metaphorically. With up to about 70 percent of a rider's load on the saddle, any technology that cushions sudden vertical acceleration forces from road bumps by storing it and then releasing it gradually is welcomed. Springs, shock absorbers, and linkages all add weight, but they provide an easier ride that reduces the toll on human energy.

Some nineteenth-century frames were designed to act like one large spring. This both reduced control and absorbed the rider's energy through "pedal bob," the tendency for the suspension to compress and expand with each crank rotation. A well-designed suspension system should, instead, improve a bike's handling, braking, steering, and safety—as well as comfort. The Pedersen, designed in 1893, did exactly this, with a hammock saddle suspended between frame tubes.

More commonly, coiled metal springs were used to suspend the saddle, but this left the frame unsprung. A century later, springs were put inside telescopic front forks of mountain bikes, and new designs use compressed air in a sealed unit that can be tuned by changing the pressure, although their reliability depends on seals that remain airtight under hostile conditions.

On a modern full-suspension bike, a swing arm holds the rear wheel, with a coil or air spring combined with a pneumatic or hydraulic shock absorber to dampen rebound. The pattern of pivots and linkages on these twenty-first-century designs is crucial to handling characteristics.

▼ ***Softride*** *(2007)*
A suspended flexing beam absorbed the shocks for on-road and off-road riders alike. First sold in 1991, the design was used to win several trophies but when the UCI, the sport's governing body, tightened its regulations, the frame with no seat tube or seat stays fell outside the rules and production ceased. Triathlons have more liberal rules so it can still be seen at some events.

▼ ***Orange 22x*** *(2012)*
Downhill mountain-bike racing has driven the need for more robust suspensions with longer travel and better control.

How much power can a cyclist generate?

How many cyclists does it take to charge a lightbulb?

The question of how many cyclists it takes to power a lightbulb is not entirely frivolous. In 2009, the lights on a Christmas tree in the main square of Copenhagen, Denmark, were lit by up to 20 pedaling volunteers on stationary bikes. In New York, gyms are adding generators to their exercise bikes, and a British theater company called Milk Presents lights its performances with tandem power.[26] The bicycle is a great way to make the most of human power potential.

Work is defined by physicists as the product of a force times the distance over which it acts, and has units of energy. On a bike, work is done in an arc or circle: the cyclist pushing the pedals to turn the cranks. For such a rotating system, work is more conveniently expressed as the product of a torque—a "circular force" or "twisting force"—times the angle through which it acts. Power is the amount of work done per unit of time, and the power output of a pedaling cyclist can be calculated by multiplying the torque applied to the cranks by the angular velocity of the cranks. Power can be expressed in ft·lb/s (or, in SI units, in watts, W).

Competitors in Tour de France mountain stages maintain an average of some 220 ft·lb/s (almost 300 W) of power. When the notorious Alpe d'Huez was used in 2004 for a time-trial stage, a 365 ft·lb/s (495 W) output was reported for Lance Armstrong.[27] Riders embarking on breakaways on the flatter stages generate as much as 200 ft·lb/s (275 W).

An ordinary cyclist does not approach such achievements. Most healthy adults can produce up to 70 ft·lb/s (100 W) continuously at speeds of up to 15 mph (24 km/h). Enthusiastic amateurs may reasonably expect to produce 90 to 130 ft·lb/s (125 to 175 W) for one or two hours. They could watch an ordinary television powered by their own endeavors, but would require the help of almost 50 of their neighbors if they wanted to do so in the comfort of an air-conditioned room.

Comparative power

▼ ***Pedal power*** *Using cyclists to generate power is an idea that has been around almost since cycling began. While it is possible to attach a bicycle to a generator to power an electrical appliance, the power output of a cyclist relative to the power consumption of an average household means that we would each need about 100 cyclists on constant standby to keep things running.*

To light a room with energy-saving lightbulbs: 11 W (8.1 ft·lb/s)

To power an average LCD television: 130 W (96 ft·lb/s)

To keep a device on standby and power a refrigerator: 240 W (180 ft·lb/s)

To power a large plasma television: 320 W (240 ft·lb/s)

To power a clothes dryer: 3,000 W (2,210 ft·lb/s)

To power a central air conditioner: 4,850 W (3,580 ft·lb/s)

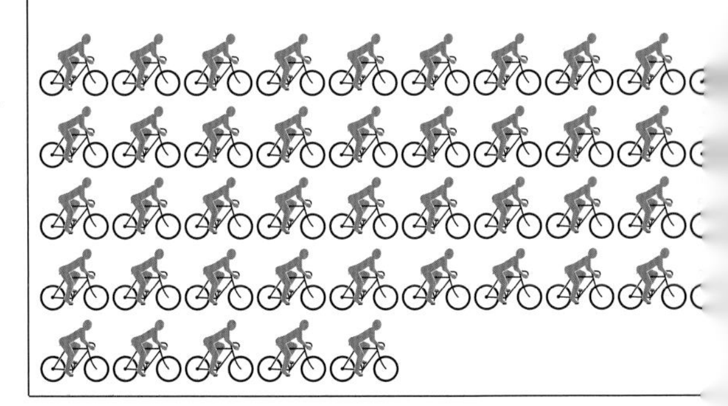

Torque

Distance (d)

Force (F)

Torque (τ)

▲ ***Torque talk*** *When cranks are turned, chainrings and sprockets spin and wheels revolve. They are all experiencing torque—work done in a circle. The magnitude of the resulting force is relative to the work input and radius of the circle described.*

Need to know

The "circular force" applied by a rider to make the axle and chainset rotate is known as torque:

$$\tau = Fd$$

where:
τ = torque (ft·lb)
F = force (lb)
d = distance (ft)

For rotating systems, the power output of the cyclist, in ft·lb/s, can be calculated by multiplying torque by angular velocity. Although angular velocity is commonly measured in revolutions per minute (rpm), a consistent set of units for everything else demands the use of radians per second (rad/s); a full circle corresponds to an angle of 2π radians.

Power and exhaustion

▼ ***Time to exhaustion versus human power*** *Manfred Nüscheler of Switzerland produced a massive 1,754 ft·lb/s (2,378 W) for barely five seconds, but this was in ideal conditions on a static roller bike. As this graph shows, power generation on a bike has a limited duration because of the capacity of the bike's battery: the rider.*[28]

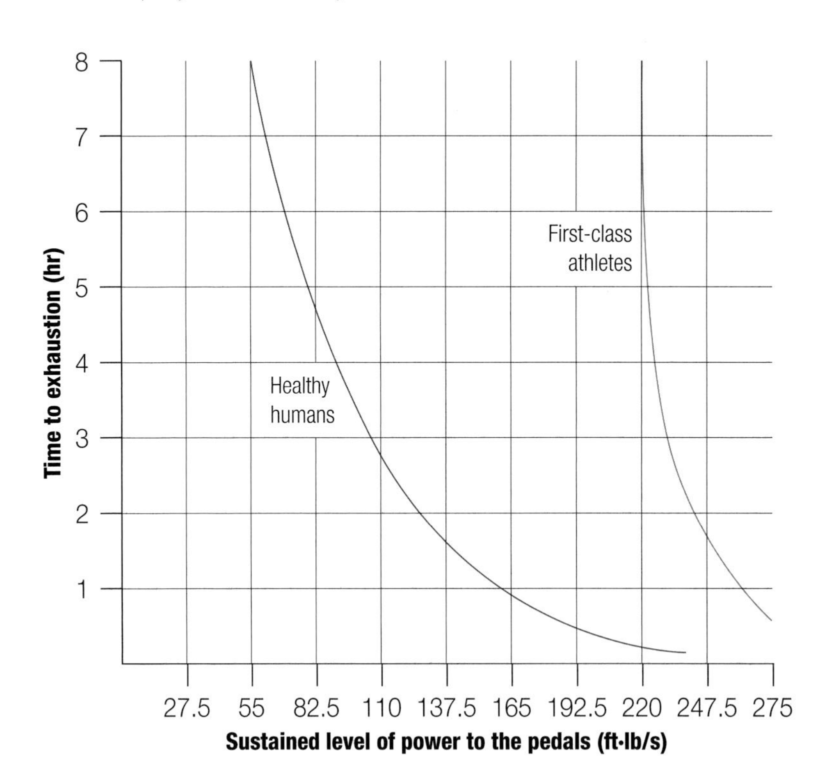

How can I compute the power output?

Is the force with me?

▼ ▶ ***Power mad*** *There are many places on a bicycle where instruments can be fitted to gather data used to calculate the rider's power output. Yet each device must be robust, lightweight, and able to communicate its measurements without affecting the cyclist's performance. Increased miniaturization is making the task easier and the equipment more affordable.*

The emergence of semiconductors and digital processing has enabled the design of accessories that are small, lightweight, reasonably affordable, and able to display power data in real time. This means that power meters are accessible to anyone who is interested in measuring how much work they do on their bike. Power is of some use as an indication of fitness, but it is perhaps better as a training aid, revealing the quality of performance and the intensity of exercise. Average speed over a regular route may show general improvements over time, but extrinsic factors, such as wind and temperature, have to be taken into account. Information about power is much more, well, powerful.

Getting the relevant data—particularly torque—out of a moving bike without putting a monkey wrench in the works is a small challenge to physicists. The power transmission train has to be intercepted and measured somewhere on its path from the rider to the road. The distortion in mechanical parts can be miniscule as the force passes through, and it could be measured at the shoe, the cleat, the pedal, the crank, the chainring/spider, the bottom bracket, the chain, the cassette, the rear hub, the rear spokes, or even the wheel rim. These options have given scope to manufacturers to develop several different proprietary systems.

Whichever location is chosen, the strain gauge is the most commonly used technology. A metallic conductor subjected to mechanical strain changes its electrical resistance by a small amount, and a strain gauge allows such a change to be detected to a high precision. For example, standing on a pedal at the bottom of its stroke distorts that crank ever so slightly, and this changes its electrical resistance. A strain gauge, calibrated under standard conditions, will read this resistance and indicate the applied force.

Quarq
The Quarq is a gauge that sits between the crank and the chainring.

Garmin Vector & Look Keo Power
Pedal gauges from Garmin and Look allow the power from each leg to be monitored.

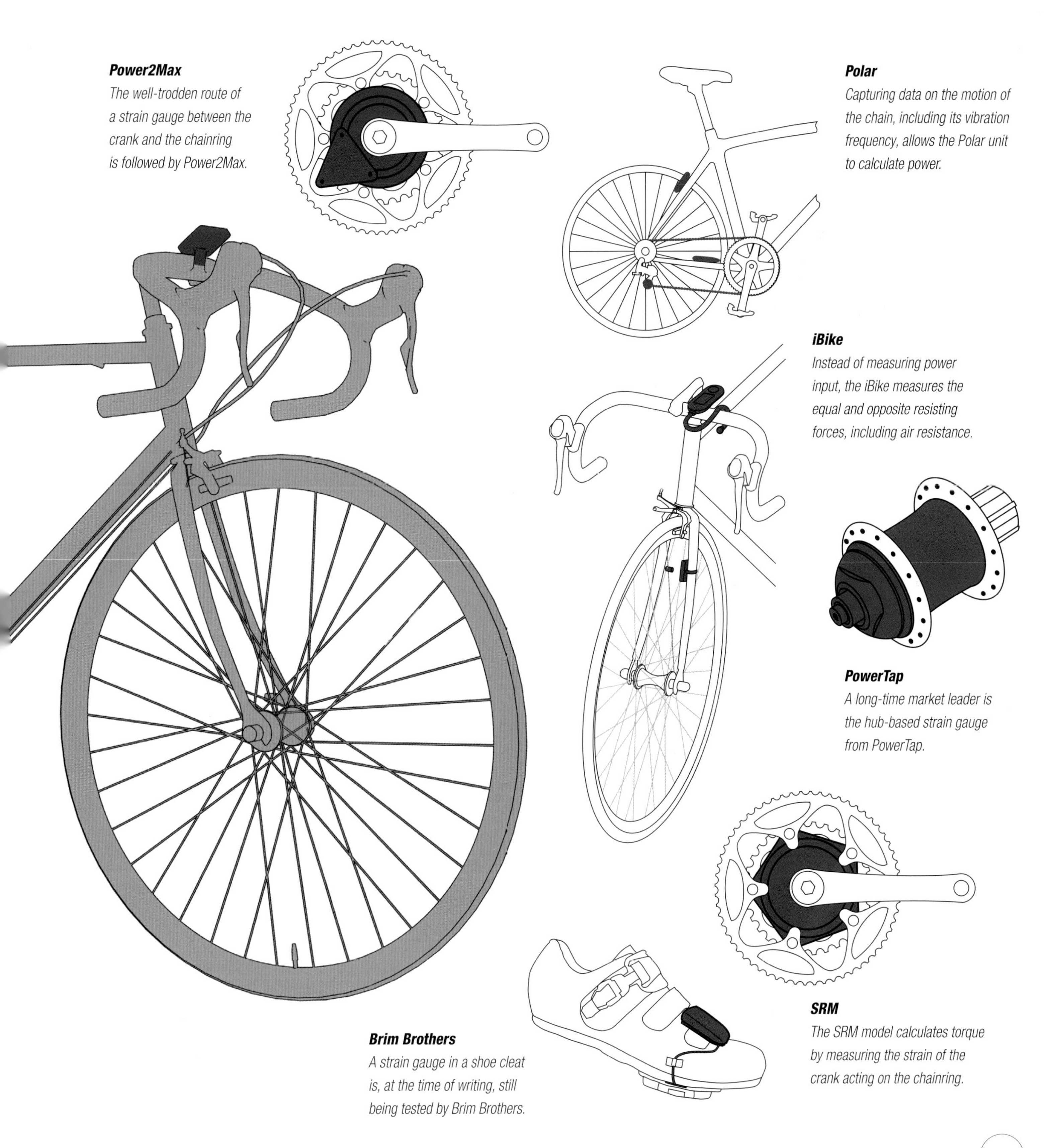

Power2Max
The well-trodden route of a strain gauge between the crank and the chainring is followed by Power2Max.

Polar
Capturing data on the motion of the chain, including its vibration frequency, allows the Polar unit to calculate power.

iBike
Instead of measuring power input, the iBike measures the equal and opposite resisting forces, including air resistance.

PowerTap
A long-time market leader is the hub-based strain gauge from PowerTap.

SRM
The SRM model calculates torque by measuring the strain of the crank acting on the chainring.

Brim Brothers
A strain gauge in a shoe cleat is, at the time of writing, still being tested by Brim Brothers.

Does a tandem have scientific advantages? → Are two heads better than one?

Tandem cycles might be considered to occupy the more light-hearted end of bicycle design, but they do offer an excellent framework for some serious cycling science. The tandem has one frame, two wheels, and two saddles, but standard two-seat tandems only weigh about 60 percent more than an equivalent solo bike. So with twice the number of riders and powerful legs turning the pedals, it makes sense that tandems are faster than solos.

Faster riding on a tandem is significantly enhanced by aerodynamics in comparison to a single cycle. According to pioneering American cycle aerodynamicist Chester R. Kyle, tandems have only half the air resistance of a pair of single cycles.[29] However, at low speeds, when air resistance isn't a factor, such as on a hill climb, their aerodynamic advantage disappears. To keep up with a solo cyclist on a hill climb, tandem riders have to be as strong and well-practiced in their technique as the solo rider.

However, cycling is not always about maximum speed. Riders pace themselves to make sure they cover their target distance without burning out. In this case, the weight savings and better aerodynamics of a tandem mean that the riders don't have to expend as much energy to produce sufficient power as they would if riding two solo bikes. Stokers, as the rear riders are known, have significantly lower physiological stress cycling on a tandem than when on a solo bicycle. Trials have showed that a stoker's heart rate is lower by 16 to 22 percent, while the level of lactic acid in their bloodstream is lower by 23 to 70 percent.[30]

Researchers note that, all other things being equal, riders can ride between 3 and 5 mph (5 to 8 km/h) faster on a tandem than when riding individually. However, tandems have longer and more complex transmission systems that lose slightly more energy than a single bike—but it's nothing compared to the friction of another kind that may arise between pilot and stoker if they do not work well together.

Tandem versus solo physiological effects

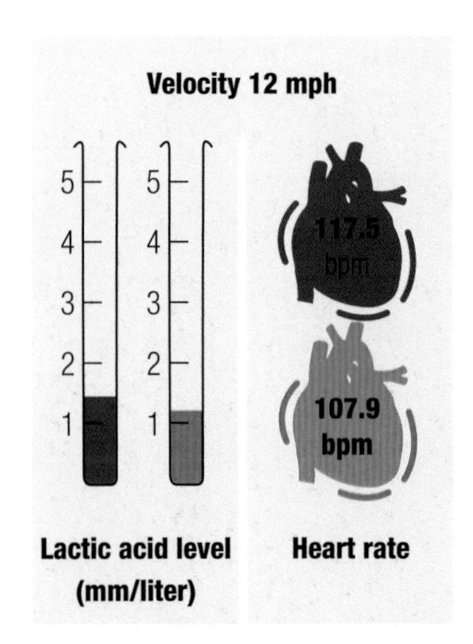

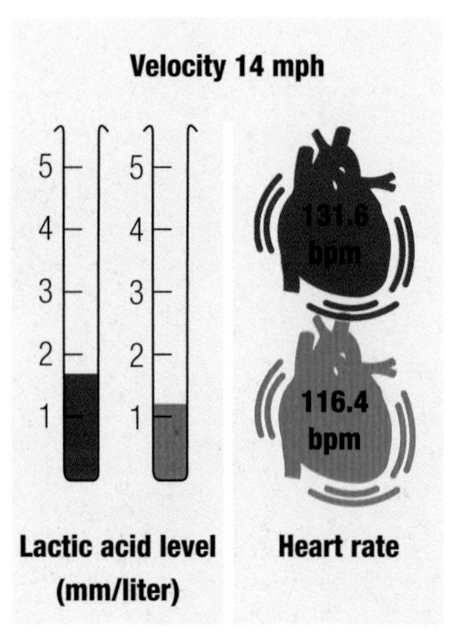

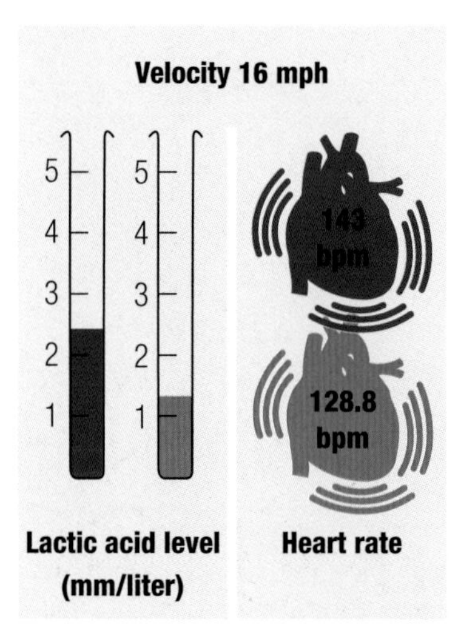

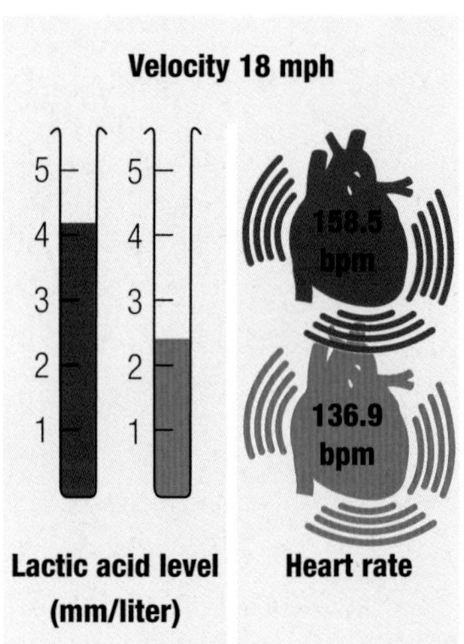

● Single bicycle ● Tandem bicycle

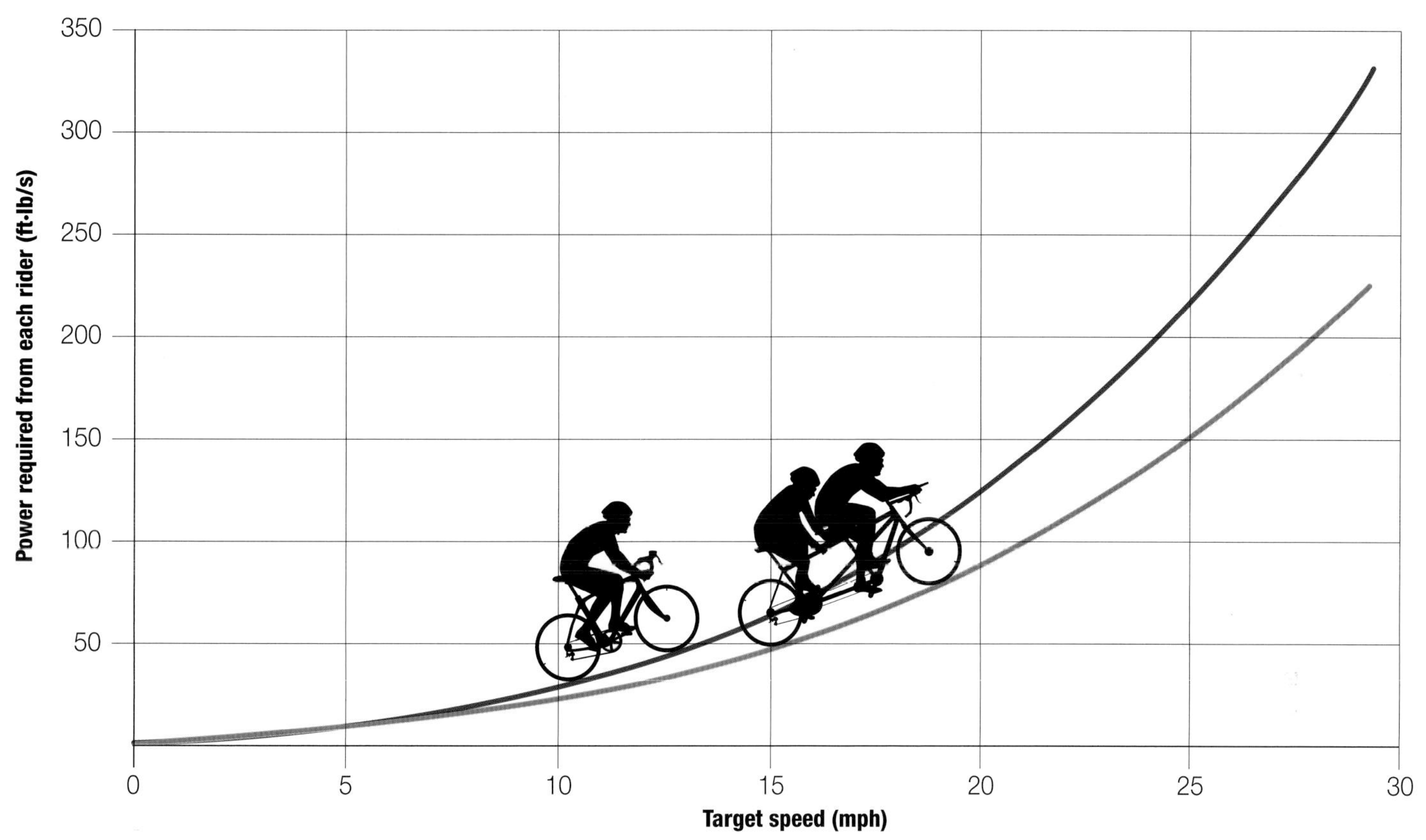

▲ ***Aerodynamic benefit of a tandem*** *A pair of identical twins, each outputting 120 ft·lb/s (160 W) on a tandem, could travel at 22.1 mph (35.5 km/h), while just one of them on a solo bike would achieve only 19.3 mph (31.1 km/h).*[31]

◀ ***Single and tandem heart rate and lactic acid*** *At a given speed, the rider on a tandem shows measurable signs of lower physical stress, including heart rate and bloodstream lactic acid levels. The figures are an average of the stokers' and pilots' results.*[32]

▶ ***The weight advantage*** *Two typical touring singles, at 29¼ lb (13.3 kg) each, have a mass of 59½ lb (26.6 kg), while a typical touring tandem weighs in at 47 lb (21.3 kg), a 20 percent weight reduction.*

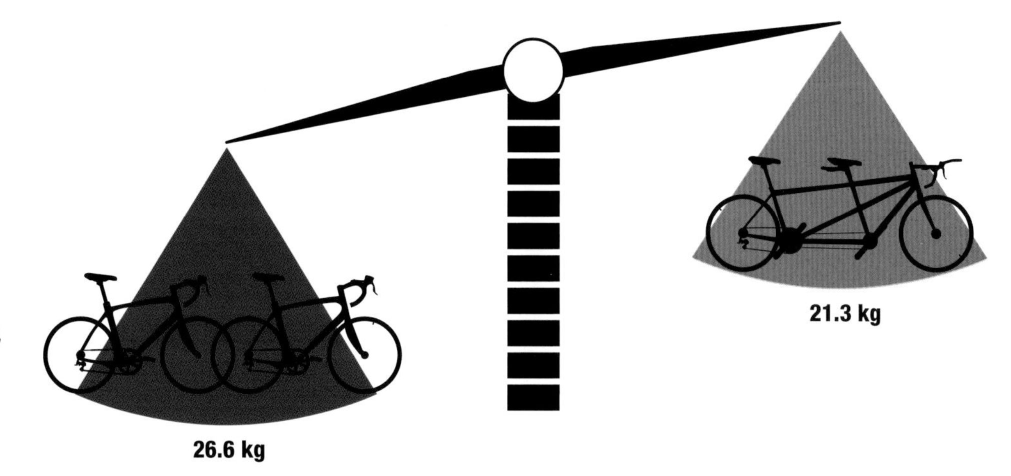

Bicycles fulfill a tough role. They must move, balance, and steer while supporting the cumbersome and often clumsy energy source—the human rider. The bike's wild collection of hundreds of diverse components must stay together while absorbing impacts and vibrations. Nothing must break under normal conditions and everything must continue to function through hundreds, thousands, and tens of thousands of cycles. Bikes can even be more reliable, predictable, and resilient than the person who is pushing the pedals and turning the steering. Bikes that didn't meet these criteria have swerved off the cycle path of history, and the models we have today are consciously designed to function as intended without superfluous engineering. To a large extent, they are successful because of the science of physics. Cycle makers have built up an extensive understanding of the properties of materials, structures, and dynamics, and they apply that knowledge to the bike so that it can give us the ride we want. This chapter describes the physics that makes your bike work so well.

chapter two

strength and stability

How can I calculate the stresses of riding?

What kind of load can my bike take?

Compared to the structures used in other kinds of transport, bicycles give the impression of being frail, with a lot of fresh air triangulated by slender tubes. This is often deliberate; one goal targeted by some racing bike designers is to make their machines look insubstantial. If a frame appears light, it can offer psychological appeal to riders anxious to shave seconds off their best time. Whether or not a faster ride is the result, a spare, sleek cycle does beg the question—how does it support a full-grown human adult, six times heavier than it is and pumping the pedals with legs like pistons?

The first factor to understand about bike frame strength is stress, which is a measure of how an applied load (force) is distributed over a section of material. When a material—this could be a frame tube, a spoke, a chain link—is subjected to an external force, it adjusts itself until this force is balanced out. A liquid or gas can flow until this happens, but, in fact, solids can respond in a similar way. The soft part of a saddle or a suspension spring changes shape until internal forces balance out the external ones. Components we think of as "rigid"—such as frame tubes—do much the same thing, but the changes they undergo before internal and external forces balance out are so microscopic that they are unnoticeable by us.

The stress each part undergoes is the average force per unit area of the component. To complicate matters, components such as frame tubes often do not have a consistent thickness or shape throughout—they are engineered to optimize a combination of specific characteristics such as weight, stiffness, and impact-resistance. Stresses also fluctuate rapidly in response to hard braking, bumps, and sharp turns, so a margin of safety also needs to be built in.

▼ ***Area code*** *Stress is the external force divided by the area to which it is being applied. The material of a flat pedal experiences stress in proportion to the area of the shoe in contact with it. Likewise, the stress on a saddle depends on the breadth of the rider's backside. The stress on the hollow tubes from which a bicycle frame are made is related to the cross-sectional area of their walls. Imagine a cross section of your frame tube. The ring of material that bears the stress lies between the outer radius (R) and the inner radius (r). Its thickness is* $t = R - r$. *Its area is equal to the outer area* ($A = \pi \times R \times R = \pi R^2$) *minus the inner area* ($a = \pi \times r \times r = \pi r^2$). *The constant* π *(pi) is about 3.1416. So, for example, the cross-sectional area of a tube with an outer diameter of 1 in* ($R = \frac{1}{2}$ *in) and thickness* $t = \frac{1}{8}$ *in is about 0.344* in^2 *(about 0.000222* m^2*).*

Calculating cross-sectional area

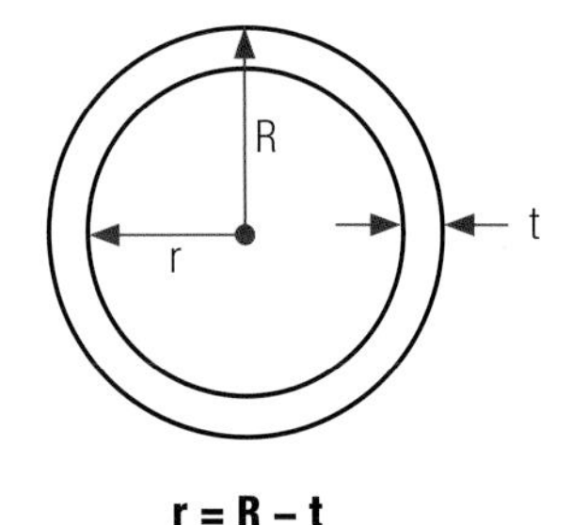

$r = R - t$

$A = \pi R^2$

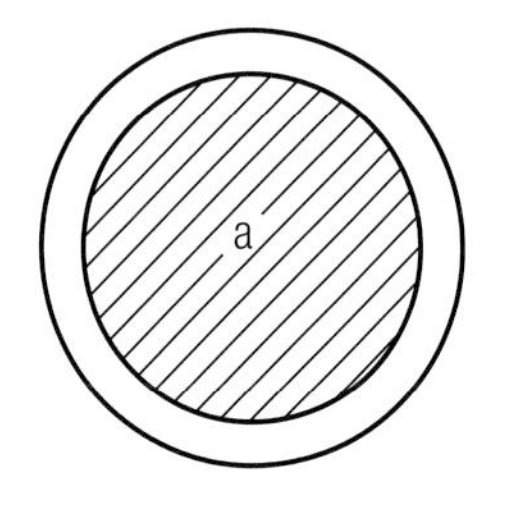

$a = \pi r^2$

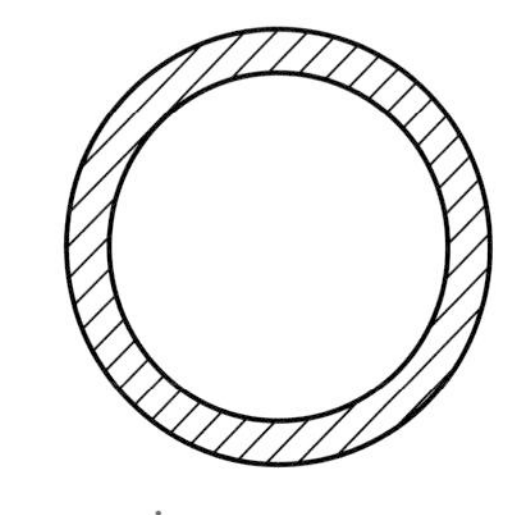

Cross-sectional area = A – a

Changes to stress

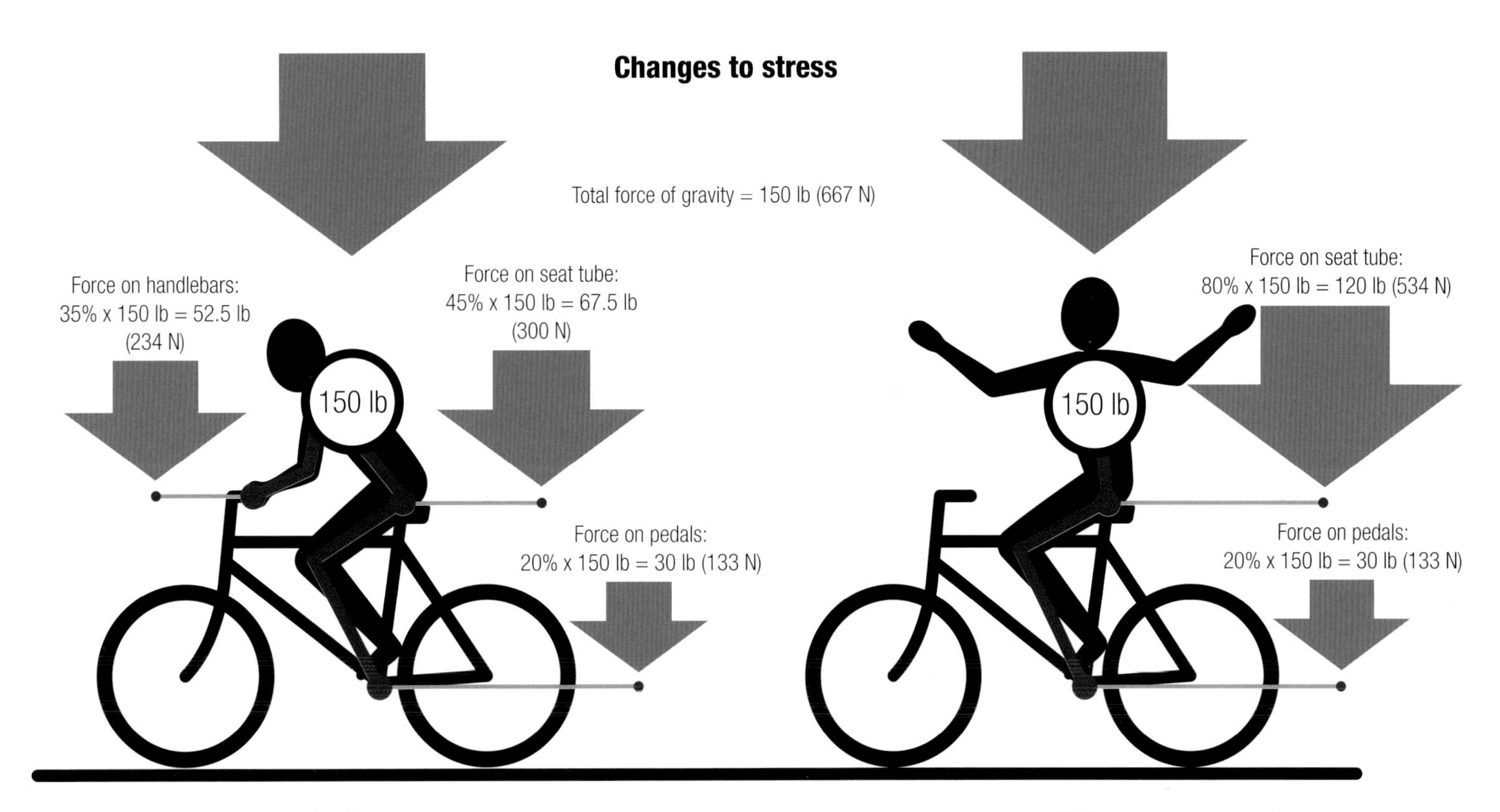

Stress on seat tube = $\frac{67.5 \text{ lb}}{0.344 \text{ in}^2}$ = 196 psi (1.35 MPa)

Stress on seat tube = $\frac{120 \text{ lb}}{0.344 \text{ in}^2}$ = 349 psi (2.41 MPa)

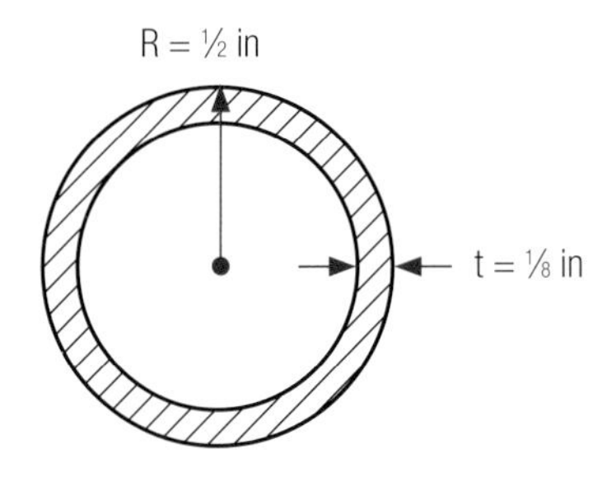

Cross-sectional area = 0.344 in²

▲ ***Stress raiser*** *The stress levels that the parts of a bicycle experience are always changing because the rider doesn't sit still for long. A cyclist changes position frequently to improve comfort, balance, steering, and pedaling efficiency. Each movement changes the distribution of loads on the three points of contact with the bicycle—the saddle, the pedals, and the handlebars. Every time the load changes, the stress on a component also changes. This diagram shows how the stress on a hypothetical seat tube can increase when a 150 lb (68 kg) rider takes their hands off the handlebars. Sudden impacts such as potholes can multiply stresses on key components such as rims, spokes, and pedals many times over.*

Need to know

This formula gives the so-called "normal stress":

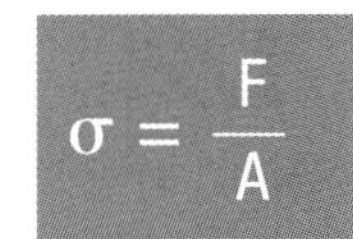

$$\sigma = \frac{F}{A}$$

where:

σ = stress (psi or Pa)

F = external force (lb or N)

A = area (in² or m²)

Stress is typically expressed in pounds per square inch (psi) or, in SI units, newtons per square meter (N/m²), also known as pascals (Pa). A 1 pound (1 lb) force applied over a 1 in² area results in 1 psi of stress (in SI units, a force of 1 N applied to an area of 1 m² produces a stress of 1 N/m², or 1 Pa; the unit megapascals, MPa, representing a million pascals, is frequently used). 1 psi is equal to about 6,900 Pa, or 0.0069 MPa. Stress has the same units as pressure.

How important is the elastic limit of materials?

Why doesn't my bicycle buckle?

When a component is put under stress, its dimensions change. This is clearly visible when you bend a spoke or stretch an inner tube. This is not so easy to see on a more substantial part such as a frame tube or crank. Nevertheless, the laws of physics cannot be broken, so when you apply a force to a pedal, the handlebars or a shifter, something has to give. This "give" can be torsion (twisting), tension (stretching), or compression (squashing) and is described by a quantity called strain. In all these cases, strain is the relative amount by which the shape of a part changes. Strain, commonly denoted by ε (Greek epsilon), has no units because it is a ratio of two measurements, and it is usually expressed as a fraction or percentage.

Almost all materials from which bicycles are made are "elastic"—that is, they can stretch and bend and absorb strain without breaking. For these materials, when the external applied load (the stress) is removed, the strain disappears and they return to their original dimensions. For example, when you stretch an inner tube and then let go, it snaps back into shape. The same thing happens with components made of metal, plastic, or carbon fiber, although the actual strains are very small.

However, there does come a critical point when the stress is so great that the strain is more than the material can take—beyond this point it can no longer return to its original shape. The level of stress at which this happens is called the material's elastic limit or yield point. If your frame tube or rim reaches its yield point, you would definitely call it "bent" or, in normal language, damaged. Like stress in general, the elastic limit has units of pounds per square inch (psi), or, in SI units, pascals (Pa) or megapascals (MPa). Bicycle designers use strain projections to make sure their components never get near their elastic limits, and, therefore, don't break in normal use.

Frame stress

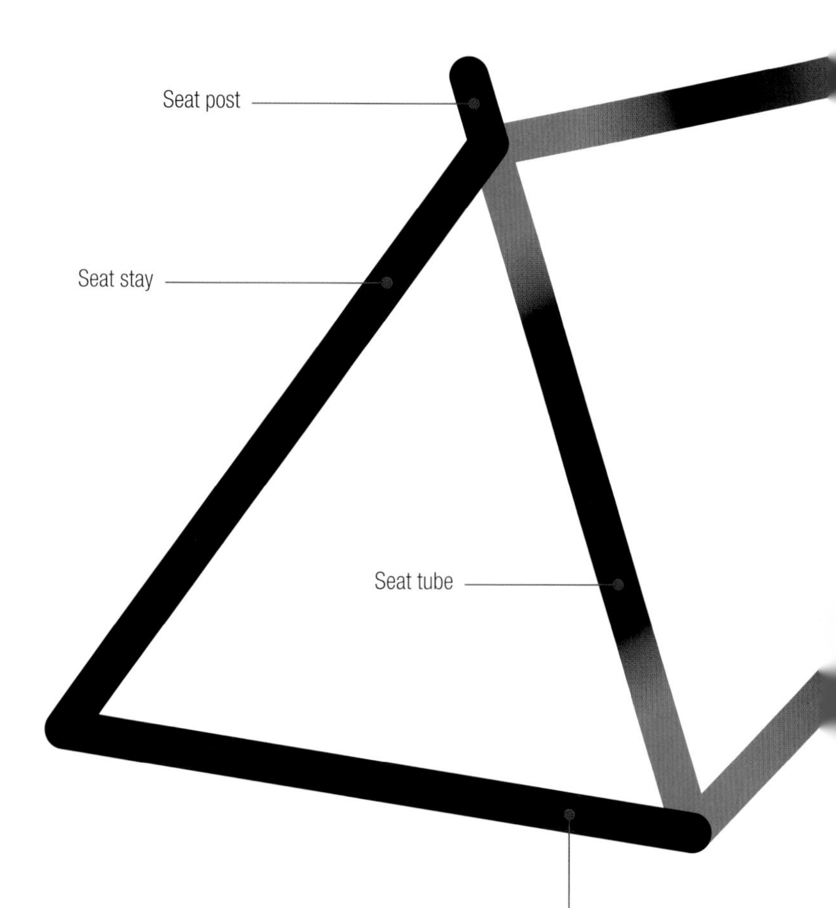

▲ ***Stress by color*** *Computer modeling of bicycle frames under various loads reveals the stress levels in the tubes. This color-coded image shows the stresses along a frame when a typical maximum load of 270 lb (1,200 N) is applied to the front fork. The load leads to bending in the top and down tubes, and the modeling shows that the greatest stress is where these adjoin the head tube. The stress here is about 50 percent of the maximum yield point so there is still a sizeable safety margin.*[1]

Head tube

Top tube

Down tube

70%

50%

30%

10%

0%

Stress as a percentage of the yield point (270 lb load on the front fork)

Handlebars stress

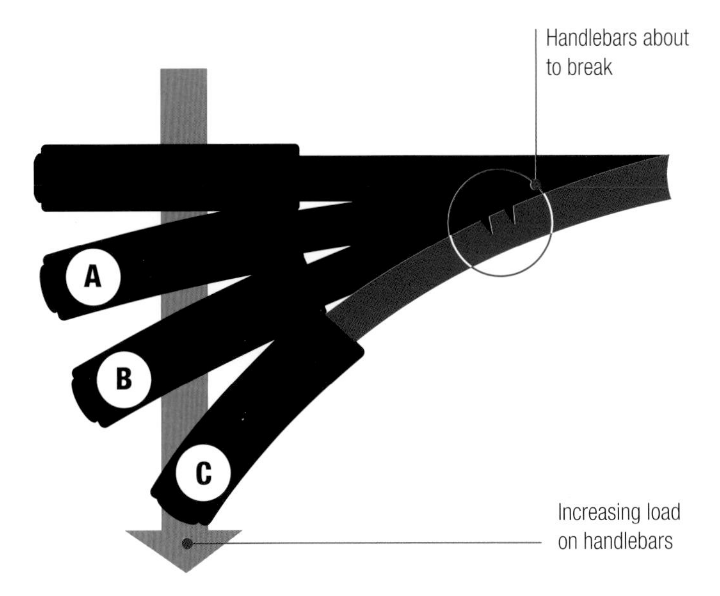

▲▼ ***Losing your grip*** *As the vertical load increases, the handlebars bend downward more and more. If the load is removed, the handlebars will return to their original shape unless the strain exceeds the material's elastic limit or yield point, at which point the handlebars are damaged and do not return to their original shape. The graph below plots progressively greater deformations, labeled A, B, and C, along the horizontal axis, against the applied force on the vertical axis. It shows how the dimensions of a component change as the load increases, and not in a purely linear way. Materials are put through deformation tests to determine strain and yield points as stresses increase, so that component manufacturers can specify the right material for the expected loads.*

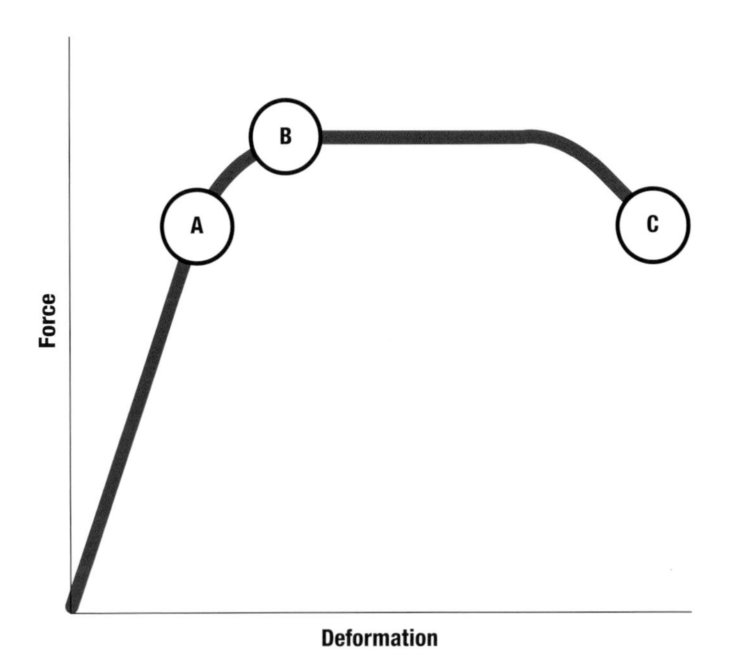

Need to know

Strain is a ratio of two measurements—the size of a deformation compared to the undeformed dimension. For example, in the case of a part whose length changes as the result of a load, a strain of $\varepsilon = 0.01$ represents a 1 percent increase in its length, while $\varepsilon = -0.23$ indicates a decrease of 23 percent.

Can a good Young's modulus help give a better ride?

Does stiffness make a bike feel better?

Engineering labs measure the stresses on the materials used to build bicycles to make sure they don't bend or give way in normal use. In addition, they are interested in the materials' stiffness—the ability to absorb shocks without flexing too much. The quality of any ride is subjective but designers need predictability above all. This helps give riders a sense of confidence, and it is why stiffness matters.

The scientific measure of a material's stiffness is its Young's modulus, named in honor of Thomas Young, a brilliant English polymath of the early nineteenth century. Young's modulus (also called the elasticity modulus) is denoted by the letter E and defined as the ratio of stress (σ) to the resulting strain (ε)—that is, how much a material distorts under a particular load. It has units of pounds per square inch (psi), the same units as stress.

If a designer specifies a material with a known Young's modulus, its stiffness, "bendability," or "springiness" can, therefore, be accurately predicted. If the designer also knows the dimensions of a part and the intended weight of the rider, the behavior of the part can be predicted. Broadly speaking, the higher the value for Young's modulus, the stiffer the component. This information is used by bike designers to adjust the comfort and rigidity of a bike frame. A stiff frame and an almost rigid crankset combined with seat post and bars with low Young's modulus to flex and absorb vibrations can help produce a more efficient overall machine.

Basic calculations are made with the simplest of frame materials—for example, metal tubes of consistent dimensions and thickness. In reality, bikes use butted tubes of variable thickness, tubes with aerodynamic profiles, curved tubes, hydroformed aluminum tubes, and carbon fiber reinforced plastic shapes that are far removed from plain tubes.

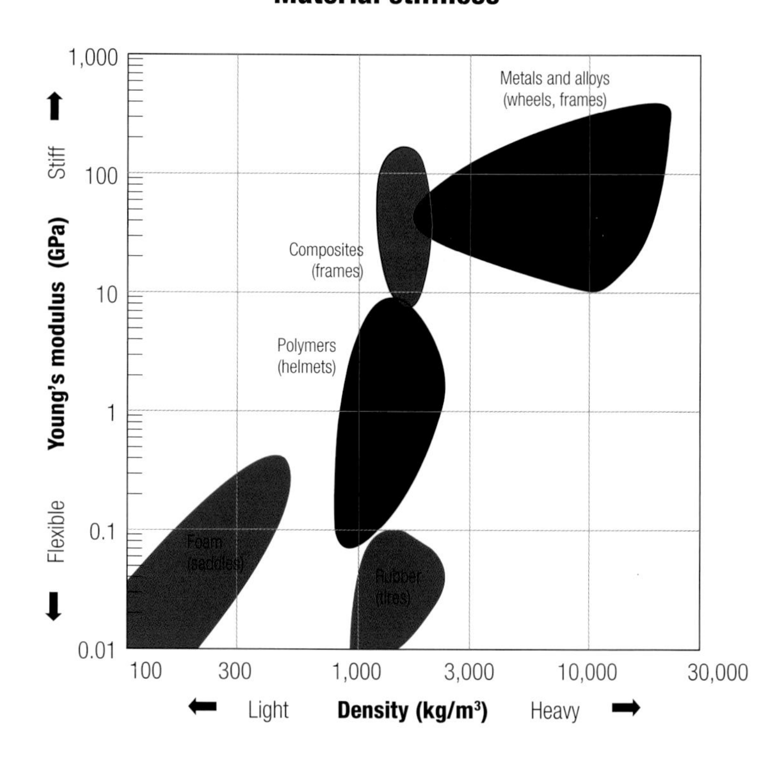

▲ ***Stiff or not?*** *This graph shows the range of stiffness and weight for materials used in bike construction. Nobody wants a saddle as rigid as a top tube, but pedal cranks should be as stiff as possible to transfer energy efficiently. It's a case of specifying material of the right density and stiffness for each application. An attempt in the 1980s to market a mass-produced plastic bike, called the Itera, was a commercial failure, partly because of its lack of stiffness.*[2] *The graph shows a logarithmic scale; 1,000 GPa is equivalent to 145 Mpsi and 30,000 kg/m³ is 1,873 lb/ft³.*

▶ ***Not all metals are alike*** *The greater the Young's modulus (E), the stiffer the material. For example, the medium carbon steel alloy used for most bike frames has a modulus of about 30 Mpsi (30 million psi), while 6061-T6 grade aluminum alloy has a modulus of 10 Mpsi. The fact that steel is stiffer than aluminum of the same thickness will not surprise anyone who has tried to bend both—it's harder to deform steel, and Young's modulus quantifies it. The diagram to the right shows Young's modulus for a variety of metals.*

Need to know

Young's modulus is calculated as follows:

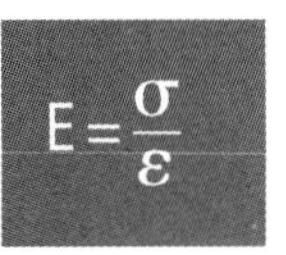

$$E = \frac{\sigma}{\varepsilon}$$

where:

E = Young's modulus (Pa)

σ = stress (Pa)

ε = strain

Need to know

There are two main reasons why most bikes feel OK. The first is that frame building is a mature industry—most of the mistakes were made many years ago and have long been eliminated through trial and error. Granted, some users of innovative materials have had to learn the hard way more recently, but even they have been helped by the second underlying reason for frame reliability—easily available modeling software and affordable computing power.

Frame metal stiffness

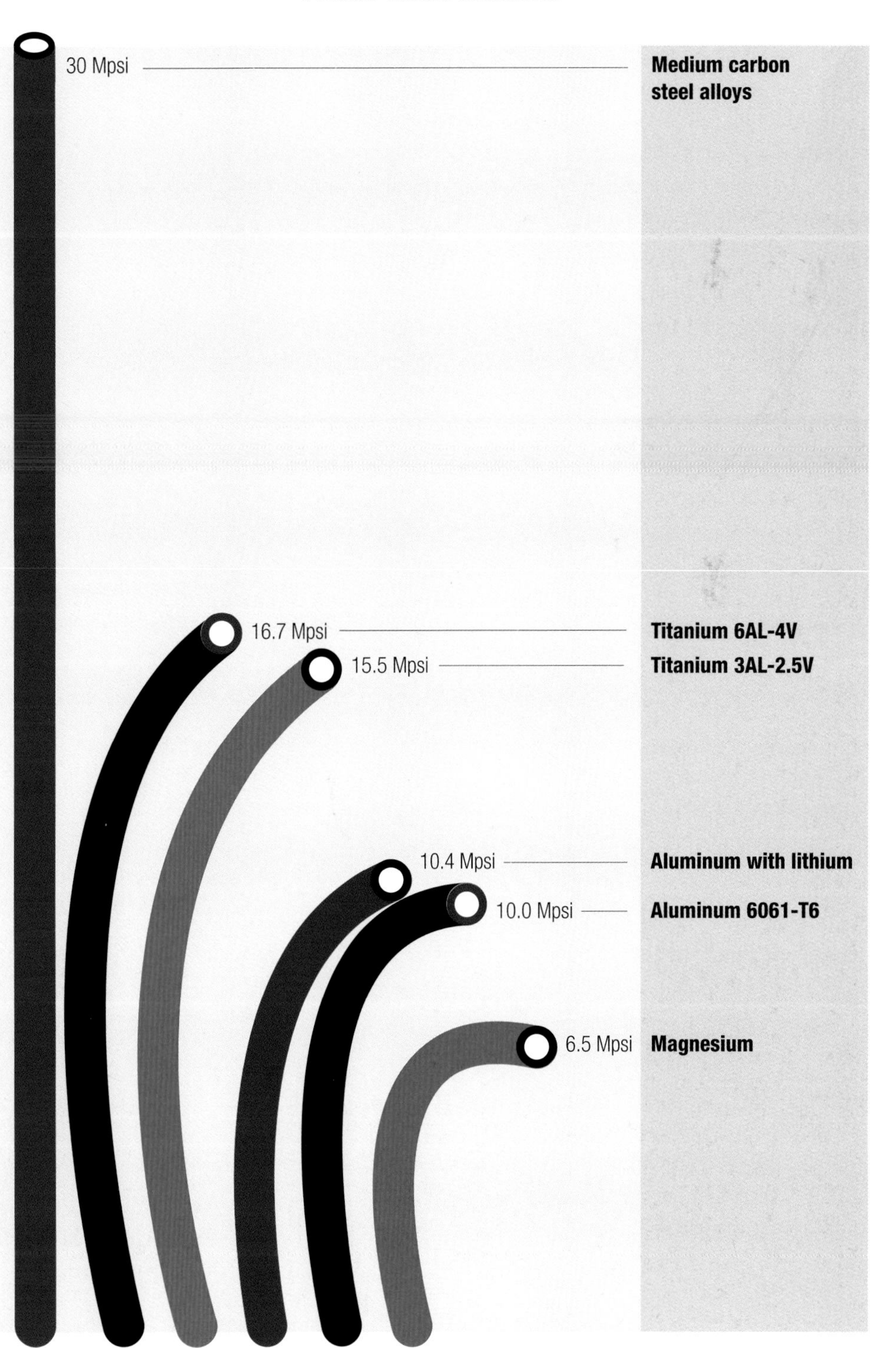

How important is ultimate tensile strength?

How near am I to breaking my frame?

Manufacturers don't want any part of their bikes to become bent or distorted in normal use, so they use materials with yield points far higher than the stresses and strains of normal riding. However, any bike will experience the occasional "unusual stress" moment—for example, when hitting a pothole. Bending a part so it is permanently deformed may be a nuisance, but breaking it is far worse, especially if it happens suddenly.

Breaking point is reached if the stress peaks way beyond the yield point, at a point called ultimate tensile strength (abbreviated to UTS, or just TS), resulting in a catastrophic failure. If a bike frame can be made from a material with a large difference between its yield point and its ultimate tensile strength, the rider should have an opportunity to spot a deformation before it turns into a complete material failure.

How do manufacturers establish these crucial, highly safety-dependent figures for the materials they use? Data is collected from destructive tests in the laboratory, in which samples are subjected to measured loads. The necessary data includes the cross-sectional area of the sample before the test begins and the maximum tensile (stretching) force that must be applied to break the material. The ultimate tensile strength (in psi) is the ratio of the breaking force (in lb) to the cross-sectional area (in in^2); a more useful unit is a million psi, or Mpsi.

Of course, thicker components can help to avoid catastrophic failure—for a particular material, the greater the cross-sectional area, the greater the force needed to cause damage—but this will make things heavier, which designers try to avoid. The engineers will have juggled all these conflicting requirements on their design pads to specify materials that will remain intact during their expected use. So, avoid those massive potholes and UTS should keep the bike in one piece.

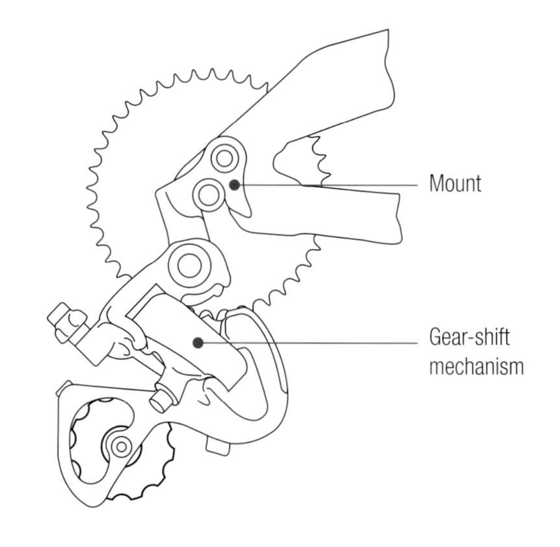

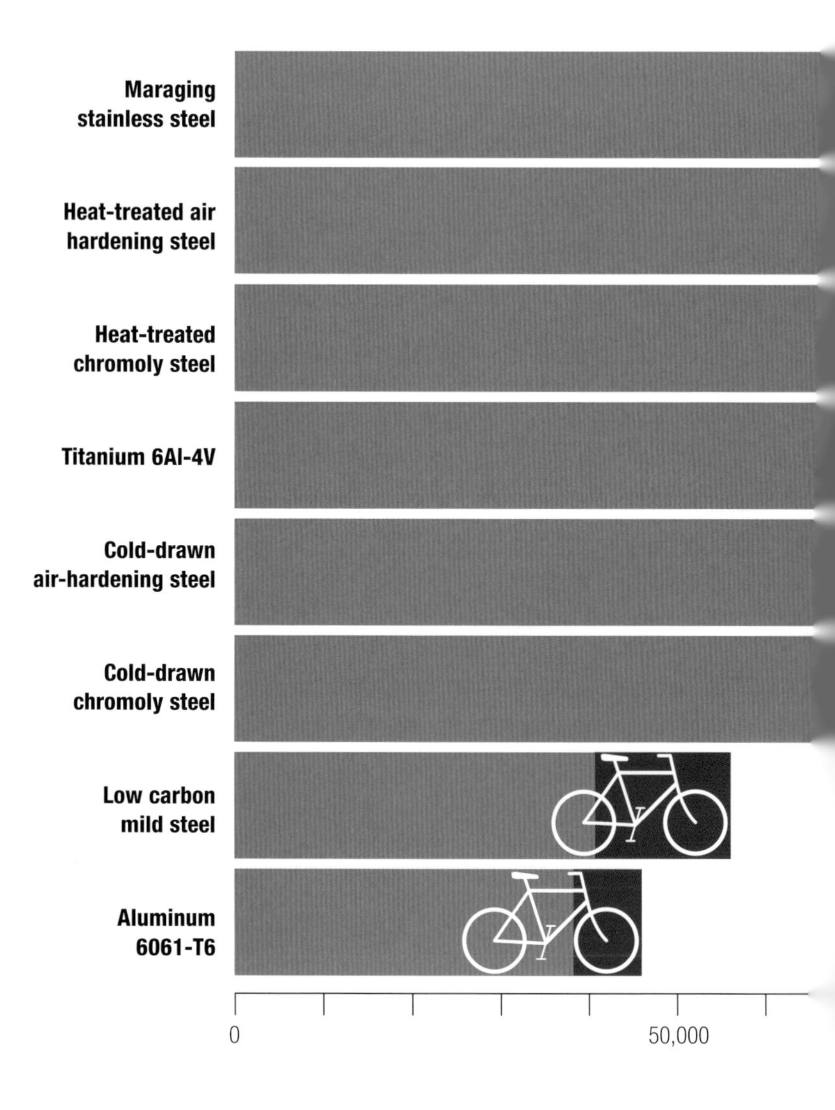

Designing with relative strength

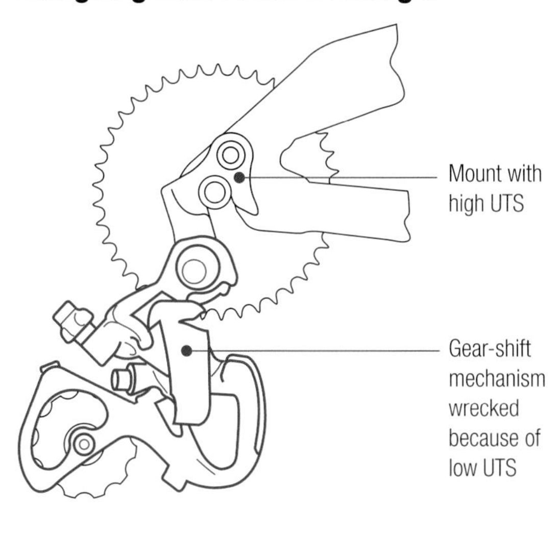

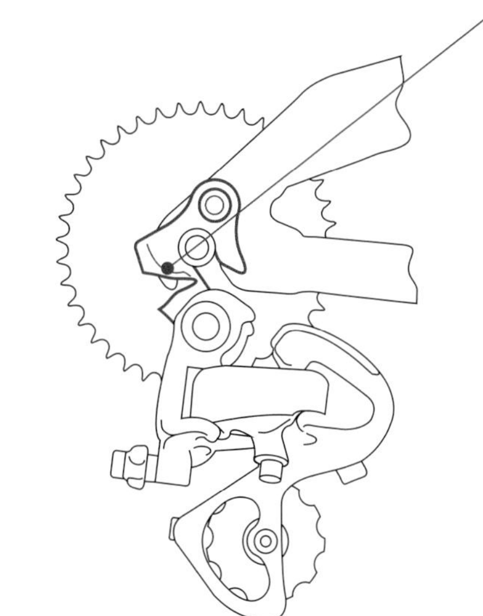

◀ ***Survival of the fittest*** *A bike frame is made up of many components, each with its own UTS. If one part is too strong, it could lead to failure in an adjacent part, so making sure they all work together is vital. Some manufacturers deliberately design in "sacrificial" elements to protect delicate areas such as rear derailleurs. In the early days of gear mechs, many were wrecked because the mounts were too strong. Today, mounts are designed to reach UTS and break before the expensive gear-shift mechanism does. The mounts are easily replaced.*

Deformation and failure

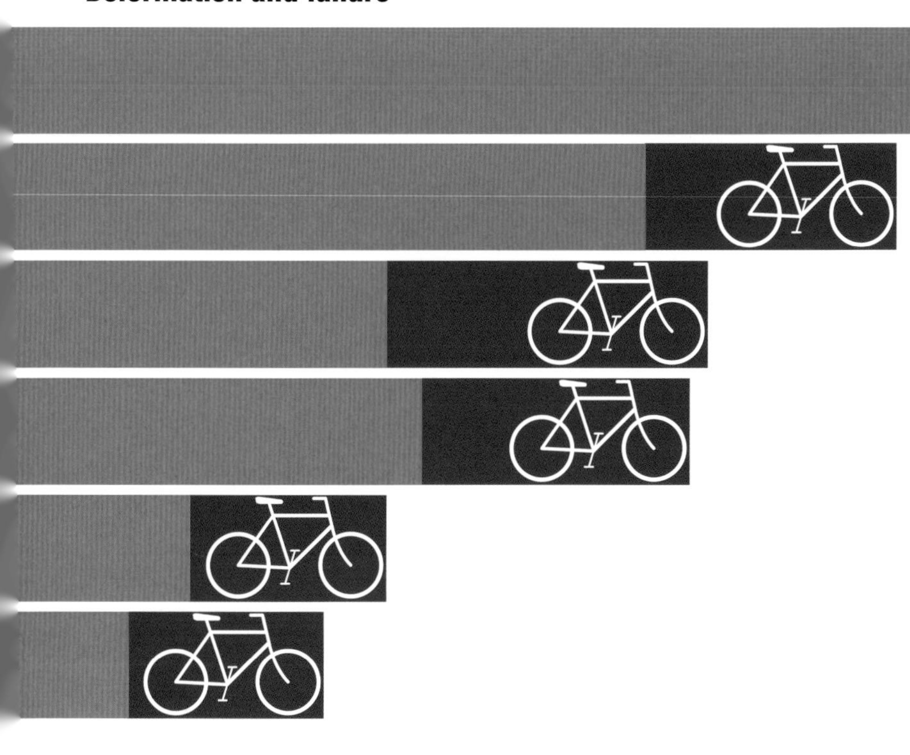

◀ ***Margin for error*** *The difference between yield point and ultimate tensile strength determines how much leeway there is between deformation and failure. The smaller the red zone, the less chance the rider has of noticing a problem before the part breaks.*[3]

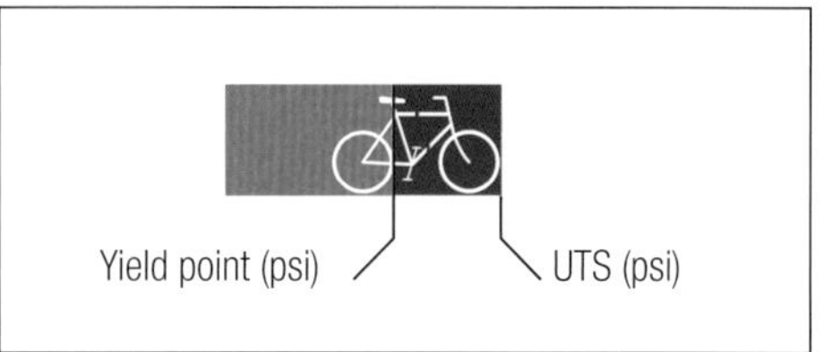

equipment: frame geometry

The vast majority of bike frames rely on structural engineering's favorite shape—the triangle. Why? Because it's the stablest of all simple two-dimensional geometric shapes. Once the lengths of its sides have been decided, all the angles are automatically determined and cannot take on any other values. It can only be deformed by changing the length of one of its sides or by breaking it. This means that if the corner of a frame is subject to significant stress (for example, in a hill climb or heavy braking, even a crash), the frame will tend to hold its shape. This is not true for other two-dimensional figures—such as a square—which readily change their angles even with fixed side lengths.

The main triangle on a bike consists of the top tube, down tube, and seat tube. It shares the seat tube with another triangle, comprised also of the seat stays and chain stays that run to the back wheel axle. The two triangles together are commonly described as a "diamond frame." Admittedly, the head tube may add a short fourth side to the main triangle, and not all frame elements are made in straight lines, but designers try to get as close as possible to triangles, because this gives the most rigid structure in a single plane, for a given mass of material. Even on full-suspension bikes that do away with seat stays, there are triangles holding the rear wheel.

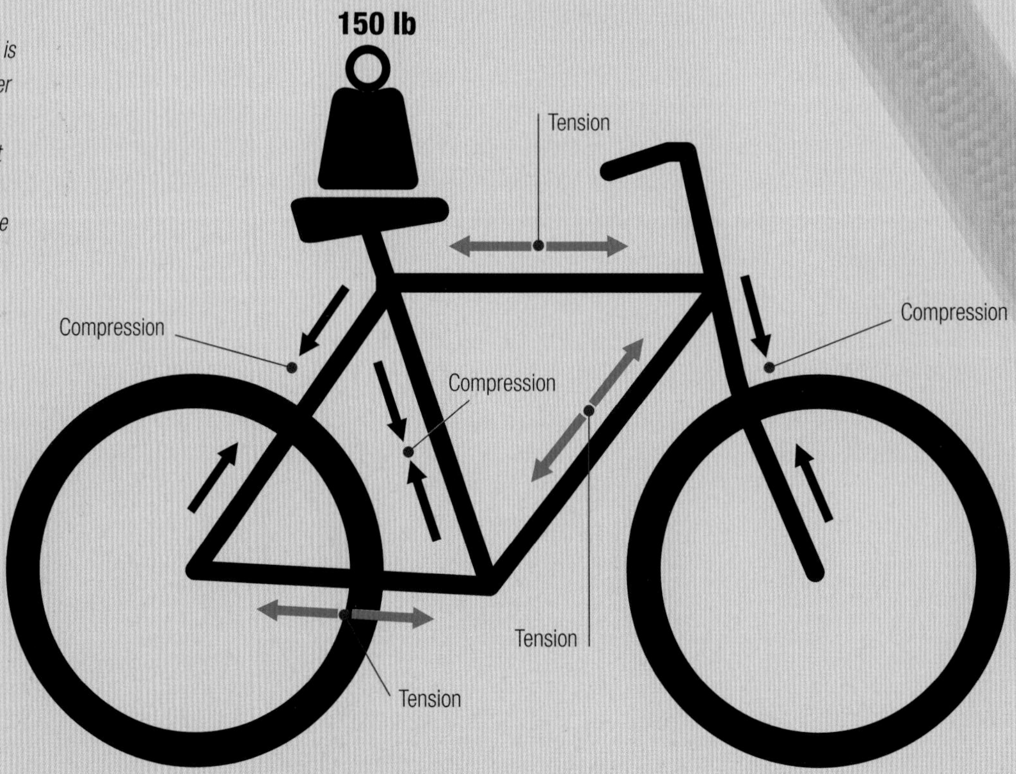

▶ ***How triangles distribute forces*** *When a force is applied at one corner of the frame, such as when a rider sits on the saddle, the load is distributed through the entire framework to the wheels. The seat tube and seat stays are under compression, while the top tube, chain stay, and downtube are under tension. The stable shape of the triangles means that the load is transmitted with minimal risk of distortion of the frame. The fork is also under compression.*

Distribution of forces

150 lb

Tension

Compression

Compression

Compression

Tension

Tension

The triangle is key to bike frame stability because the three fixed tubes share the transmission of forces between them when they're put under pressure or stretched, with minimal risk of tube distortion.

▼ ***How triangles work laterally*** *Triangles play their part laterally as well as vertically. Look at any bike from behind and it is apparent that the wheel axle and seat stay/supports form a third triangle, giving lateral stability. That's why the seat stays almost always go to a central point behind the seat tube. A fourth triangle can be seen by looking down on the rear wheel; the chain stays diverge from the bottom bracket and also use the rear axle as the third side. And the front forks? Another triangle—wheel axle to steering head—offers rigid three-sided configuration.*

Lateral triangles

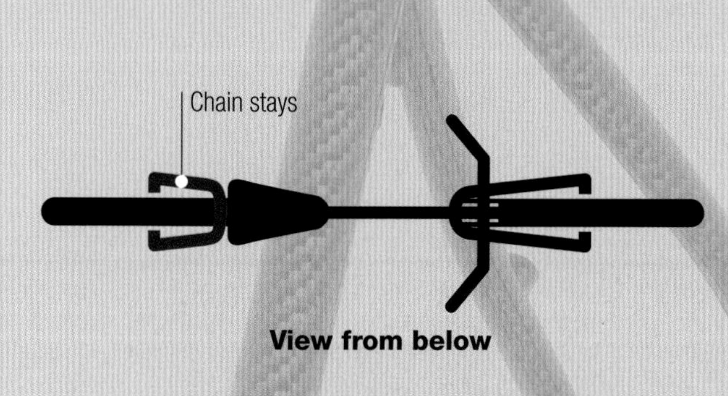

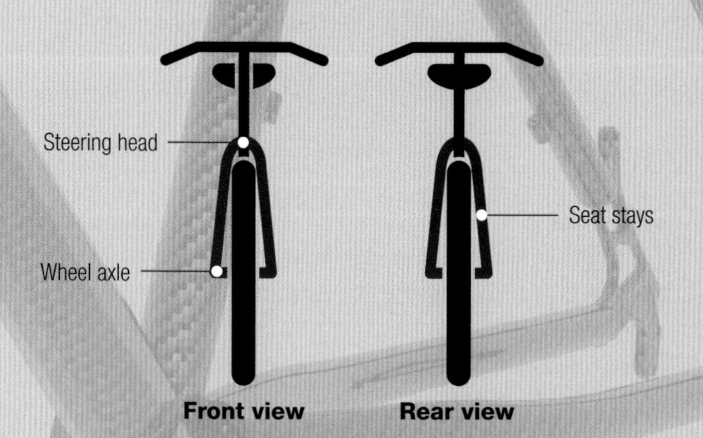

Triangles in a diamond frame

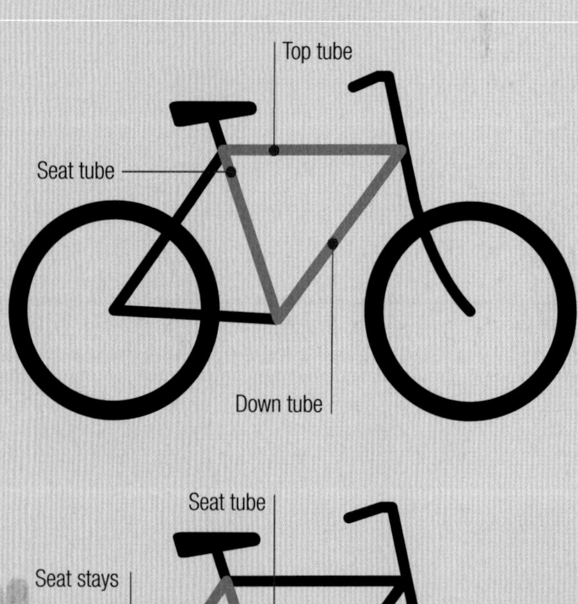

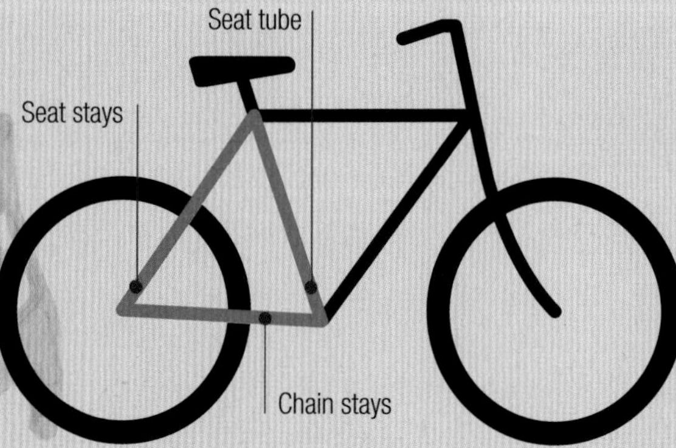

Is there an optimum frame shape? → Is there a perfect bike?

You may think your bike is perfect for you, but, in fact there are hundreds of thousands of perfect bikes. To choose the one that is perfect for you, it's necessary to learn a little about basic frame geometry. The lengths and angles of the tubes, and the geometry of the stays and the fork, are all crucial for the perfect fit and ride.

For bicycle frames, level ground is the reference plane so a vertical tube is said to be at 90 degrees and a horizontal tube at 0 degrees. At its simplest, the more vertical the seat and head tubes, the more the frame will be suited to fast racing, because the steering will tend to be more direct and sensitive. This is because the point where the front tire touches the ground is close behind the point where a line down through the head tube, called the steering axis, would touch it. This imaginary gap is called the trail, and its size matters. Conversely, a laid-back angle for the head tube tends to make steering less twitchy because the trail is larger. A relaxed angle also provides a smoother ride because the cantilevered forks more easily absorb road vibrations. So elite triathletes may opt for a frame with seat and head tubes angled as steeply as 76 degrees from the horizontal, while a sensible commuter will have angles as shallow as 65 degrees.[4] The angle of the steering axis isn't the only important factor in bike handling. Most forks are raked forward in a curve, placing the axle ahead of the steering axis and reducing the trail. The greater the rake, the more the steering wants to straighten itself, an effect known as caster, and the more effort is needed to turn the handlebars. Frame builders today seem to have converged on a rake of about 1¾ in (45 mm).[5]

The seat tube angle affects comfort and power. A steep tube angle puts the rider's thighs more directly over the crankset, so pedaling is more efficient. A shallow tube angle, however, absorbs more vibrations, so they are not transmitted to the saddle.

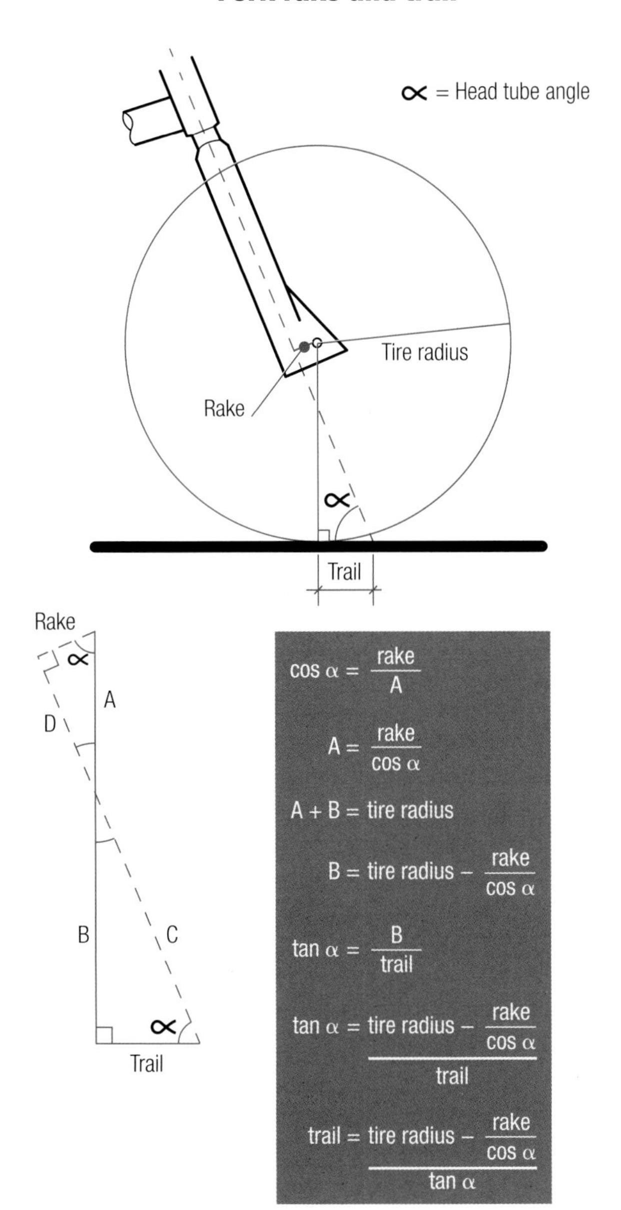

▲ ***A perfect fit*** *It is not only the head tube angle and length that are important. Other parameters that help the bike fit include the bottom bracket height, the angle of the top tube, and whether tubes and stays are straight or curved—the permutations are endless. For the perfect bike, you will need a frame built to fit you, but many off-the-peg bikes can be adjusted close to perfection if you are of normal build. Fork rake and trail play a significant part in the feel of the steering.*

Comfort verus efficiency

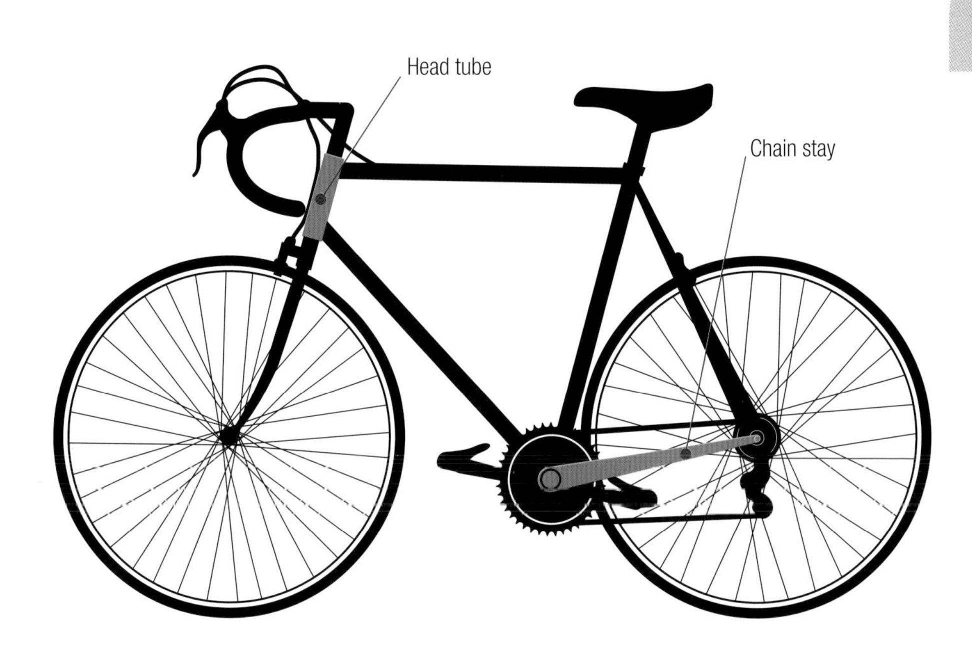

> “A relaxed angle also provides a smoother ride because the cantilevered forks more easily absorb road vibrations.

▲ ***How will you ride?*** *As well as their angles, the frame tube lengths decide the rider's comfort and efficiency. To be more aerodynamic, choose a short head tube and steep angles, a combination that will keep the handlebars low, so the rider tends to bend forward and down, slicing through the wind. To be more upright and relaxed, opt for a longer head tube, which raises the handlebars for a comfortable, more upright ride. Longer chain stays will provide more ride comfort by increasing the distance between the wheels, but the longer, heavier chain they require will be less mechanically efficient.*

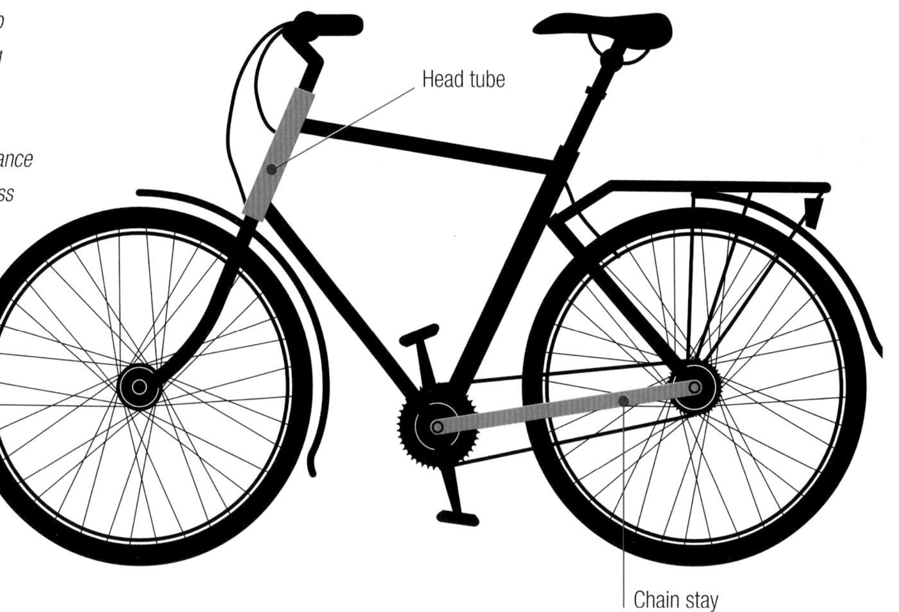

How can I calculate the perfect fit?

Have I got the right bike for my body?

Bike shops generally carry a range of frame sizes to fit only the most common dimensions—although riders come in a huge variety of shapes and sizes. It's important to get the right size to optimize power output, balance, comfort, and control.

There are almost as many theories about the best approach to fitting as there are bones in the human body. Several commercial sizing systems go into fine detail, but, for most riders, there are just a few measurements needed to get close to the ideal. The key measurement is your inside leg measurement (inseam), however, height, torso, and arm lengths are useful for refining the possibilities. Stand upright on a hard floor with your back against a wall and your bare feet 6 in (15 cm) apart. Slide a book between your legs and against the wall, then pull it up firmly but gently into your crotch. Measure from the top edge of the book to the floor. Repeat the process three times and calculate the average. Use this to choose the size of bike you want—most manufacturers supply a chart.

However carefully you measure yourself, the bike industry still leaves plenty of room for confusion. Mountain bike sizes are widely specified in inches and road bikes in centimeters. Some manufacturers simply describe frames as "small" or "large," so ask for specific dimensions or take a tape measure. Frame dimensions may be defined in different ways (see below), so always check. Saddle position is also critical, making a noticeable difference in comfort, stability, and power delivery. A rule of thumb is to have the steering head/bars an elbow's length from the front of the saddle. Don't worry if you suspect your frame isn't exactly the right size, because adjustments can always be made to seat post height, saddle position, handlebars, stem, and cranks.

Defining dimensions

▶ ***In the frame*** *There are a number of different ways of stating a frame's dimensions. For example, seat tube length can be described to be from the center of the bottom bracket to the center of the top tube (C–C), to the top of the seat tube (C–T) or to a virtual point where the two tubes would intersect if the top tube were horizontal.*

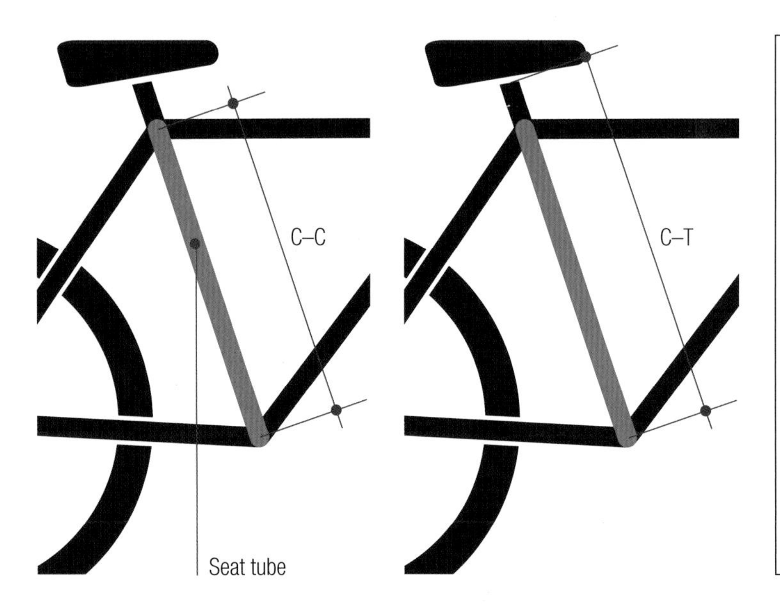

Need to know

Riders with uncommon measurements or with absolute perfection in mind may prefer to buy a made-to-measure machine from a craftsman frame builder. The bike will then fit better than a glove. Common formulae used to calculate frame dimensions are below (a tape measure and an experienced dealer can be invaluable).

Seat tube = inside leg x 0.64

Top tube = (arm length + crotch to sternum length) x 0.51

Frame size chart for 700c adult road bikes

Rider's height / **Rider's inseam**

overall	in	cm	25 in / 64 cm	25½ / 65	26 / 66	26½ / 67	27 / 69	27½ / 70	28 / 71	28½ / 72	29 / 74	29½ / 75	30 / 76	30½ / 77	31 / 79	31½ / 80	32 / 81	32½ / 83	33 / 84	33½ / 85	34 / 86	34½ / 88	35 / 89	35½ / 90	36 / 91	36½ / 93	37 / 94	37½ in / 95 cm
4' 8"	56	142	43	44	**45**	**46**	47	47																				
4' 8½"	56½	144	43	44	**45**	**46**	**47**	47	48																			
4' 9"	57	145		44	45	**46**	**47**	47	48																			
4' 9½"	57½	146		44	45	**46**	**47**	**47**	48	49																		
4' 10"	58	147			45	46	**47**	**47**	48	49																		
4' 10½"	58½	149			45	46	**47**	**47**	**48**	49	50																	
4' 11"	59	150			45	46	**47**	**47**	**48**	49	50																	
4' 11½"	59½	151				46	47	**47**	**48**	**49**	50	51																
5'	60	152				46	47	**47**	**48**	**49**	50	51																
5' ½"	60½	154					47	47	**48**	**49**	**50**	51	52															
5' 1"	61	155					47	47	**48**	**49**	**50**	51	52															
5' 1½"	61½	156						47	48	**49**	**50**	**51**	52	53														
5' 2"	62	157						47	48	**49**	**50**	**51**	52	53														
5' 2½"	62½	159							48	49	**50**	**51**	**52**	53	54													
5' 3"	63	160							48	49	**50**	**51**	**52**	53	54													
5' 3½"	63½	161							48	49	50	**51**	**52**	53	54	54												
5' 4"	64	163								49	50	**51**	**52**	**53**	54	54												
5' 4½"	64½	164								49	50	51	**52**	**53**	54	54	55											
5' 5"	65	165									50	51	**52**	**53**	**54**	54	55											
5' 5½"	65½	166									50	51	**52**	**53**	**54**	54	55	56										
5' 6"	66	168										51	52	**53**	**54**	**54**	55	56										
5' 6½"	66½	169										51	52	**53**	**54**	**54**	55	56	57									
5' 7"	67	170										51	52	53	**54**	**54**	**55**	56	57									
5' 7½"	67½	171											52	53	**54**	**54**	**55**	56	57	58								
5' 8"	68	173											52	53	54	**54**	**55**	**56**	57	58								
5' 8½"	68½	174												53	54	**54**	**55**	**56**	57	58	59							
5' 9"	69	175												53	54	54	**55**	**56**	**57**	58	59							
5' 9½"	69½	177													54	54	**55**	**56**	**57**	58	59	60						
5' 10"	70	178													54	54	55	**56**	**57**	**58**	59	60						
5' 10½"	70½	179														54	55	**56**	**57**	**58**	59	60	60					
5' 11"	71	180														54	55	**56**	**57**	**58**	**59**	60	60					
5' 11½"	71½	182														54	55	56	**57**	**58**	**59**	60	60	61				
6'	72	183															55	56	**57**	**58**	**59**	**60**	60	61				
6' ½"	72½	184															55	56	57	**58**	**59**	**60**	60	61	62			
6' 1"	73	186																56	57	**58**	**59**	**60**	**60**	61	62			
6' 1½"	73 ½	187																56	57	58	**59**	**60**	**60**	61	62	63		
6' 2"	74	188																	57	58	**59**	**60**	**60**	**61**	62	63		
6' 2½"	74½	189																	57	58	59	**60**	**60**	**61**	62	63	64	
6' 3"	75	191																		58	59	**60**	**60**	**61**	**62**	63	64	
6' 3½"	75½	192																		58	59	60	**60**	**61**	**62**	63	64	65
6' 4"	76	193																		58	59	60	**60**	**61**	**62**	63	64	65

Crank size

165 mm (6.5 in) 170 mm (6.7 in) 172.5 mm (6.8 in) 175 mm (6.9 in)

◀ ▼ ***Getting fit*** *This table is for average-sized riders and doesn't always account for people with long limbs and short bodies or vice versa. However, changing components such as the handlebars, stem, seat post, or cranks should help. There are different tables for road bikes, mountain bikes, and BMX bikes to reflect different riding requirements such as ground clearance and clearance between crotch and top tube.*[6]

Frame sizes are in cm, center-to-top (C–T); subtract ½ to ¾ in (1.5 to 2 cm) for center-to-center (C–C) measurement. Frame sizes and crank lengths are only starting points.

45 Common frame size for your height/inseam

45 Possible frame size for your height/inseam

How much energy does a frame absorb? → Am I wasting my pedal power?

Nobody likes waste, particularly when it's their own hard-won energy. That's one reason why there is a strong focus on stiff frames for keen riders—some cyclists fear that a frame that deflects can divert pedal power away from the drive train—and away from the rear wheel.

Pedaling is not a symmetrical action—one leg is pushing forward and down while the other is either resting or pulling backward and up. Good cranks are quite stiff, so this imbalance of forces could be transmitted to the bottom bracket. The bottom bracket twists from side to side, applying torque (twisting force) to the down tube, seat tube, and chain stays, and wasting energy. Some riders swear they can feel the movement, yet it is extremely difficult to find scientific research that proves that some of their kinetic energy is being used to twist the frame tubes and is then lost as heat.

The dominant view is that the frame acts like a spring. When the bottom bracket is twisted to one side, some energy is stored as potential energy (as in a stretched spring). This is then released as the frame returns past its neutral position, adding to the pedaling force as the other foot pushes forward and down. This action repeats with each pedal stroke—so the net result should be zero loss.

In experiments where some high-end racing frames were clamped to a rig and subjected to deflection forces simulating pedaling action, a force of at least 17.5 lb (78 N) was needed to deflect the frames at the position of the bottom brackets by just 0.04 in (1 mm); the stiffest frames didn't budge until the force was up to 27.4 lb (122 N). The experimenters judged that anything over about 18 lb (80 N) represented acceptable stiffness (and insignificant energy loss).[7] The upshot is that riders shouldn't worry about frames stealing their power—bicycles are one of the most efficient methods of transport on the Earth.

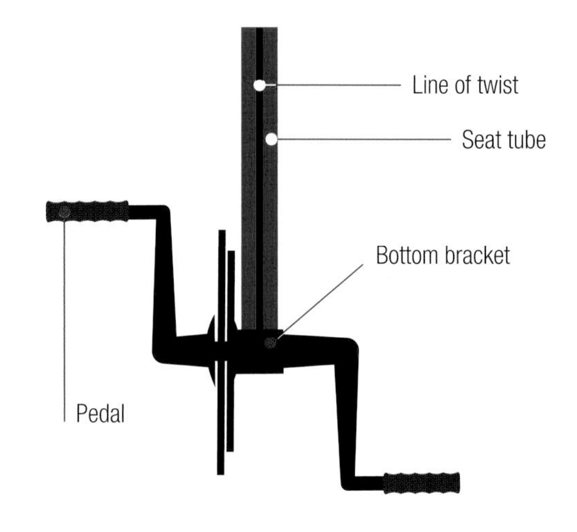

▲ ▼ ***How pedaling can twist the frame***

The dynamics of pedaling create forces of torque and tension in the frame tubes that are joined by the bottom bracket. These may twist the bracket and deflect the tubes, although they do spring back when the force is applied to the opposite pedal. The most likely result is that this flexing does not absorb much of the rider's energy.

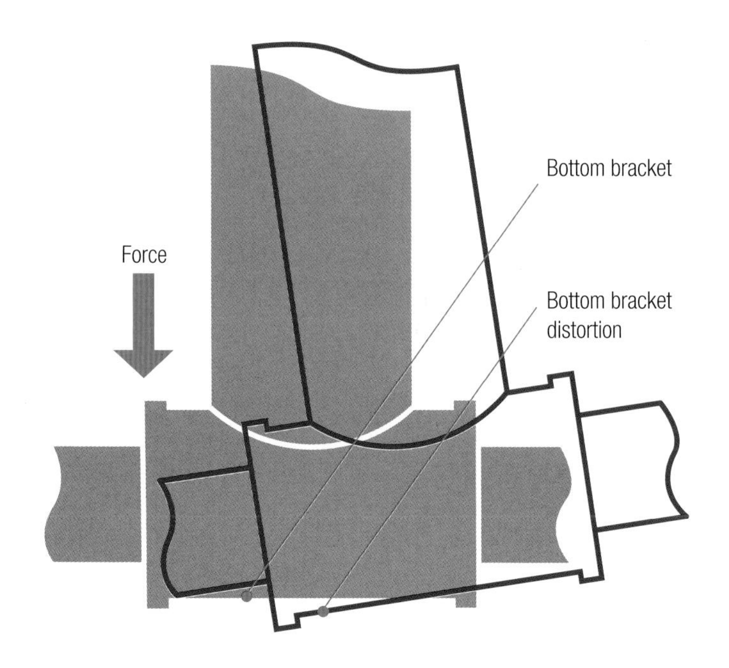

“The upshot is that riders shouldn’t worry about frames stealing their power.

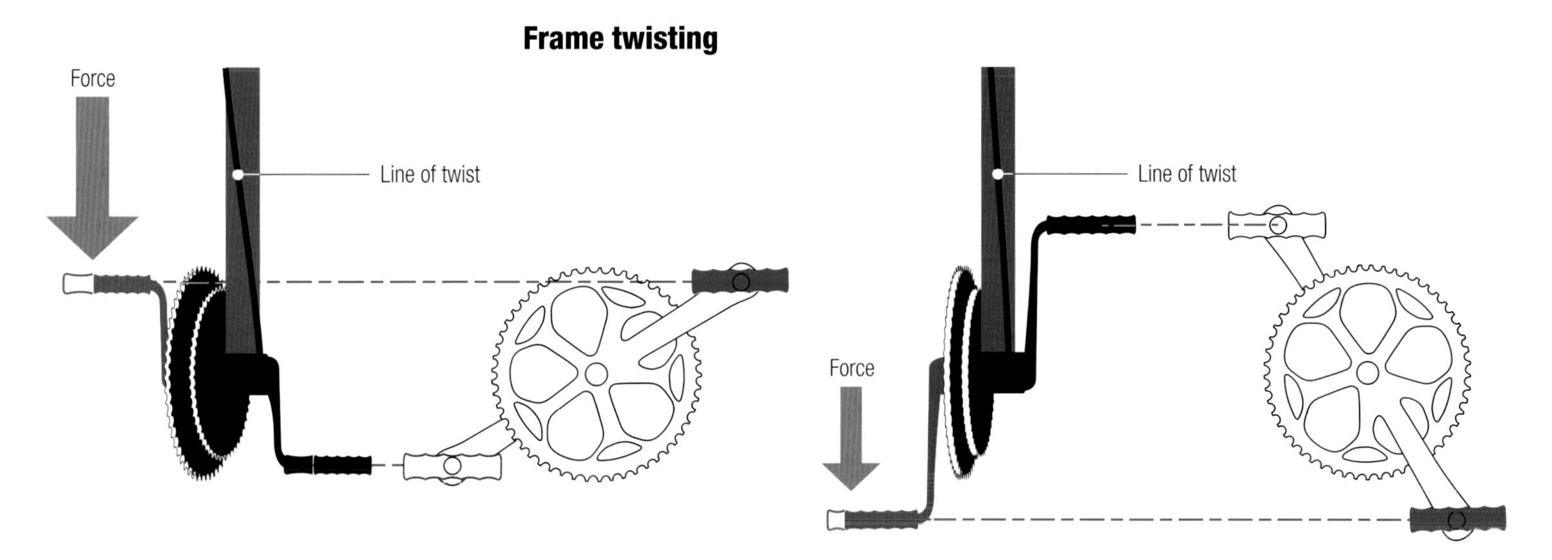

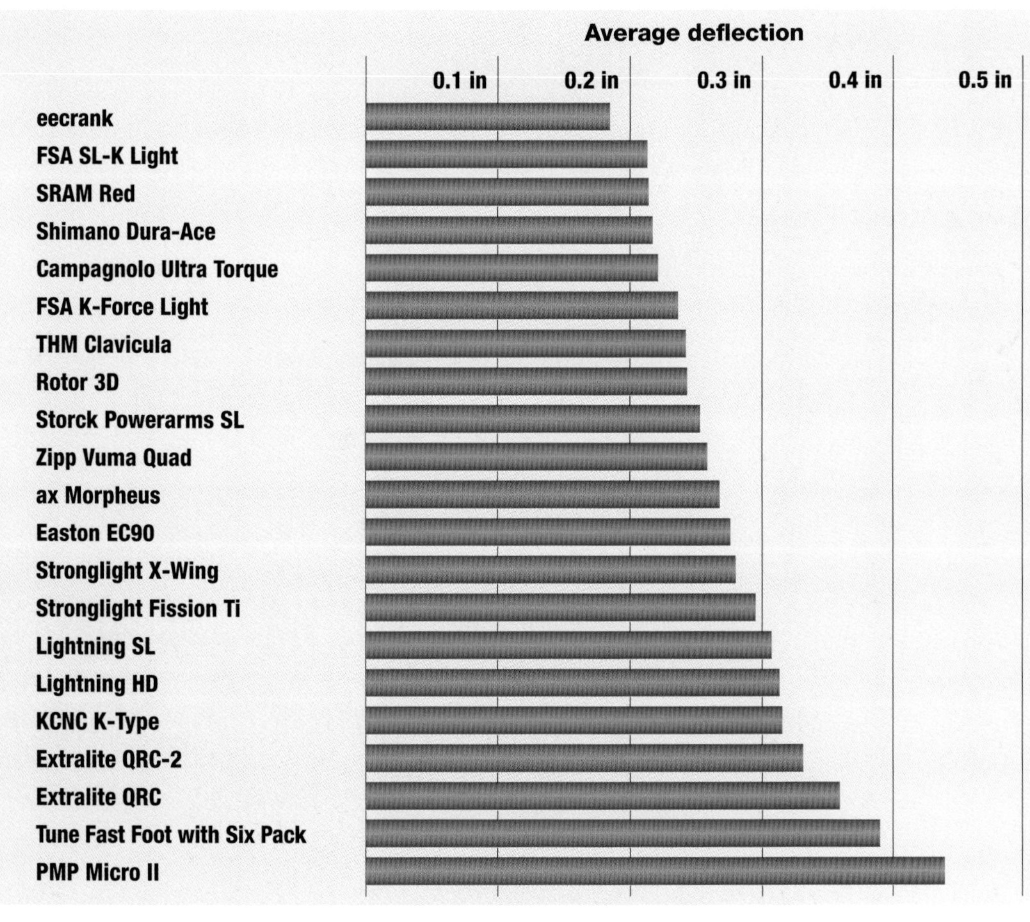

Crank deflection

◀ ***Stiff as a crank*** *In the end, the degree of stiffness is down to personal preference for handling and comfort. A very stiff bike can lose contact with the ground over bumps and reduce the rider’s ability to control it. Conversely, a very flexible frame may increase comfort but be less predictable in sudden maneuvers. Bicycle engineer Jason Woznic tested a range of popular cranks to calculate their average deflection. Each arm was loaded with 50 lb (222 N), then an additional 200 lb (890 N) was added and the change in deflection recorded. The experiment was repeated and the average deflection calculated.*[8]

science in action

controlling the forces

Cyclists use their strength and movement while riding not only to pedal and steer but also to attenuate forces that might otherwise damage their bike or threaten their stability. Just as a stuntman rolls to survive a fall, cyclists bend their arms, torso, and knees and shift their weight to mitigate a sharp change in velocity, sparing their bike from damage and helping it to stay upright. Experienced riders may believe these actions are instinctive, but humans have had to learn how to cycle, so it is only through practice that we know what to do to avert a fork-snapping jolt or wheel-buckling landing. We don't want overengineered, heavy bikes, so we deploy our bodies to moderate the combination of velocity and gravity safely.

BMX competitions are among the sharpest tests of these cycling abilities. Riders must pedal fast while reacting swiftly and strongly to avoid pitfalls, and they utilize the terrain to their advantage, curbing the potential likelihood of catastrophic failures at all times. On an Olympic supercross track, peak pedal rate (cadence) can exceed 170 revolutions per minute (rpm) within 6½ yd (6 m) and 1.6 seconds of the start. Elite men finish their time trials in under 40 seconds, of which only 12 seconds are spent pedaling, 10 seconds jumping, and 18 seconds pumping. Women riding a modified track can spend 15 seconds pedaling, 7 seconds jumping, and 18 seconds pumping.[9]

Switching from one mode to another during rapid steering demands powerful controlling movements. Video analysis by researchers at the Millennium Institute of Sport & Health in Auckland, New Zealand, has revealed that during the jumping and pumping periods, the dominant pattern is the horizontal abduction and adduction of the arms—moving them away and toward the body to maximize control. Considering the high demands of the sub-minute event, it's no surprise that it is described as the closest to a pure strength and power sport of any cycling discipline. This necessitates a training program, which includes squat, deadlift, and bench press exercises.

▶ ***Airborne relaxation*** *There's no point in pedaling when airborne, so it's a chance to rest the leg muscles while shifting arm positions to stay well balanced. Then it's a matter of preparing the body to absorb some of the landing forces. Once back on terra firma, it's time to turn those pedals in bursts of just a few seconds. BMX supercross Olympians have been shown to make best use of jump time to relax, which means they are less fatigued when they enter the final sprint.*[10]

Does suspension make a significant difference to efficiency?

Why doesn't every bike have springs?

Suspension is meant to improve comfort, handling, and efficiency, but it incurs a cost: added complexity, price, and weight. It may also need more effort to ride, with the sprung system absorbing some of the rider's energy that might otherwise be used to propel the bike. Suspension systems were popular in the 1800s—mostly because the pneumatic tire had not arrived. Once tires were readily available, bicycle suspension almost completely vanished. Now it's in fashion again, thanks to mountain biking. Engineers designed mountain bikes first with front-wheel suspension (hardtail) and then with both front and rear suspension (full suspension) to take on ever rougher terrain.

The benefits and disadvantages of riding with a rigid frame, hardtail, or full suspension have been assessed in scientific experiments. One lab has shown that, when riding over a 2½ in (6 cm) bump at a typical riding speed, front suspension reduces vertical forces on a bike by 37 percent and horizontal forces by 28 percent.[11] Those horizontal forces slow you down, so if you can reduce them by over one-quarter, it takes significantly less energy to propel the bike forward. That is, you need less energy to ride a hardtail bike with front suspension than to ride a rigid-framed mountain bike.

Another test compared a hardtail bike with a fully suspended bike. The full-suspension system halved the vertical movement of the saddle as the bicycle went over a bump, providing a more comfortable ride. More interestingly, it revealed a difference in rider effort. The results showed that the full-suspension rider needed 30 percent less oxygen than the hardtail rider over the same track. Heart rate was also slower, by 20 to 50 beats per minute. This correlated with a reduction of 30 to 60 percent in the power that had to be transmitted through the cranks to maintain the same speed as the hardtail.[12] In other words, despite the added weight, it was significantly easier to ride a fully suspended bike.

Suspension and oxygen intake

▼ ***Gimme air*** *Suspension brings a particular advantage to the rough-terrain rider—reduced oxygen usage. The oxygen demanded by the rider is an indication of the effort needed to ride a bike. In a trial with eight riders, comparative oxygen intake (values in ml/kg/min) showed the hardtail to be harder to ride on rough terrain than the fully suspended bike, but marginally easier on a smooth surface.*[13] *The hardtail has also been shown to be easier to ride than a rigid-frame bike.*[14]

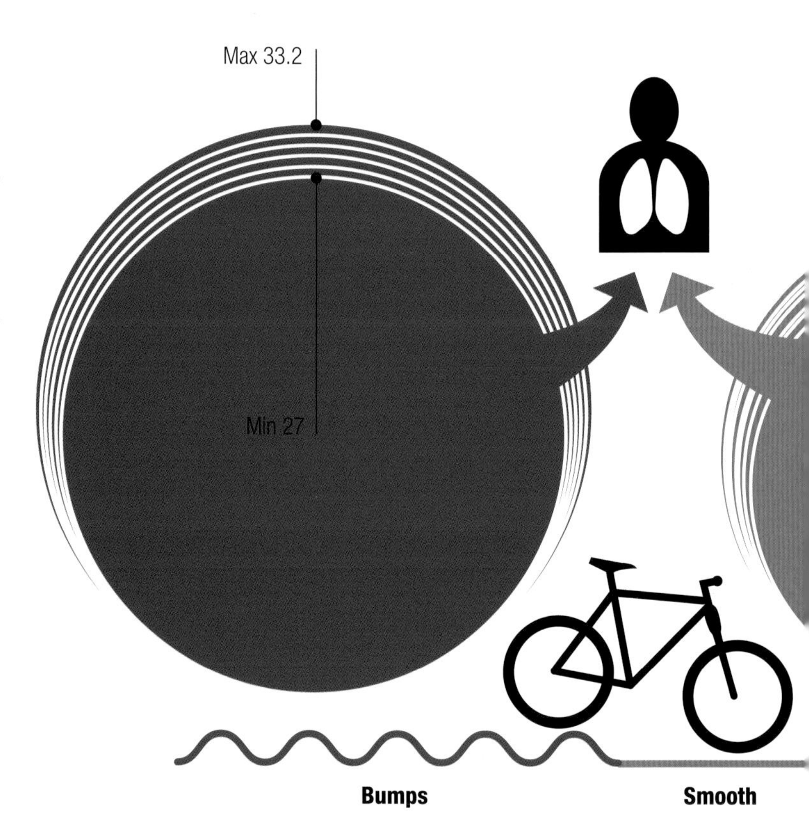

▶ ***Smoothing out the bumps*** *On the rough stuff, suspension wins out. The blue line shows the vertical forces on the front tire of a mountain bike riding over a single bump. The red line shows the same bike but with front suspension added. It doesn't vary as much or peak as high because the suspension absorbs some of the vertical forces resulting from the bump.*[15] *(This is all fine for rough terrain, but it's a very different situation on smooth asphalt. The extra effort required to propel the extra weight of the suspension over the course of a long stage race cancels the advantages of reduced impact load on frame and rider. This means road racers are unlikely to be using sprung bikes any time soon.)*

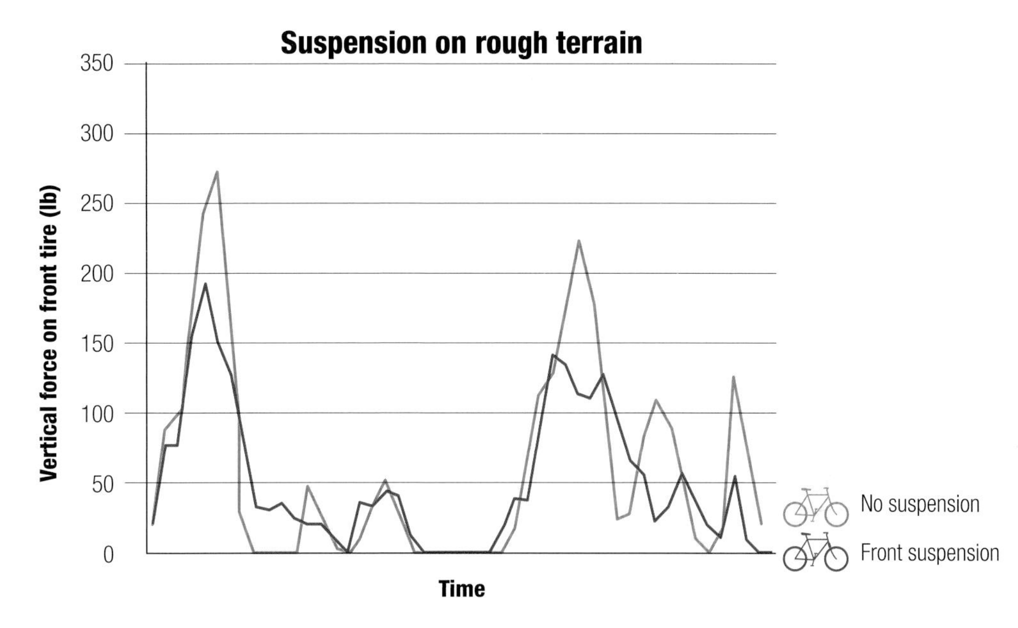

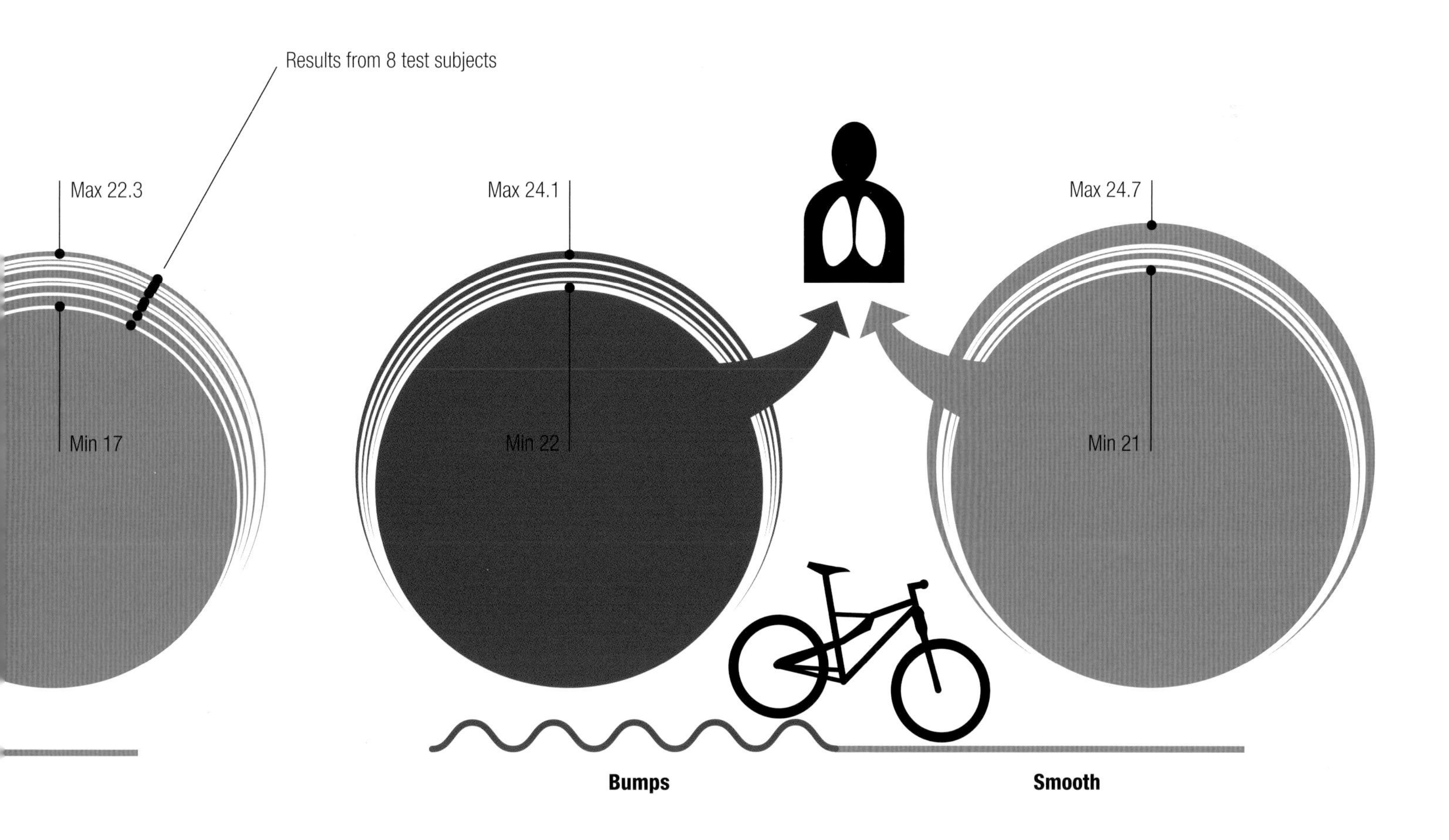

What are self-stabilizing dynamics?

How does a bicycle keep itself upright?

This is the greatest unsolved mystery of cycling. Astonishingly, a bicycle can stay upright without a rider as long as it's moving at a speed of about 8.7 mph (14 km/h). Scientists can say with confidence only that there are several factors at work.

To stay upright, a bicycle first needs a front wheel that is freely steerable. This requirement is familiar to every cyclist who has had the misfortune to ride their front wheel into a narrow rut—when the front wheel is trapped, balancing is virtually impossible. Second, it needs a caster effect for the front wheel—the more relaxed the angle of the fork, the more stable the bike. Third, the distribution of the handlebars and fork mass has an effect on how the steering reacts to a change in verticality (wobble). For example, a bike with a handlebar basket full of bricks will be less stable than one whose low-rider front panniers carry the same heavy load.

Put these three properties together in the right proportion and the result is "self-stabilizing dynamics." One explanation is that when the moving bike begins to lean to one side, gravitational torque rotates the front wheel away from straight ahead and the bicycle starts to describe a circle. In reaction, the road surface applies a centripetal force, which restores the wheel to pointing straight forward. The centripetal force also exerts a torque on the entire bicycle, which pushes it out of the leaning stance; then the bicycle centerline regains the same vertical plane as its center of mass.

That's the theory. The problem is that nobody has yet been able to quantify the relationships among all of the forces active during self-stabilization, and how they result in the self-stabilizing dynamic. In fact, experiments have shown that the gyroscopic effect of the wheels of a bicycle without a rider is negligible, and that the caster effect isn't vital either. Like the bicycle itself, the question appears simple but the reality is extremely complex.

How self-stabilizing works

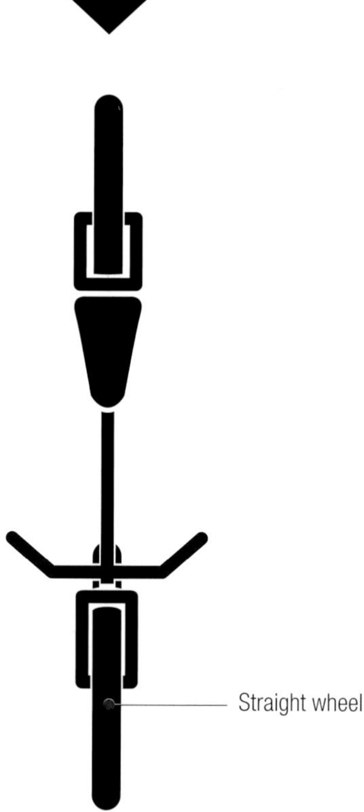

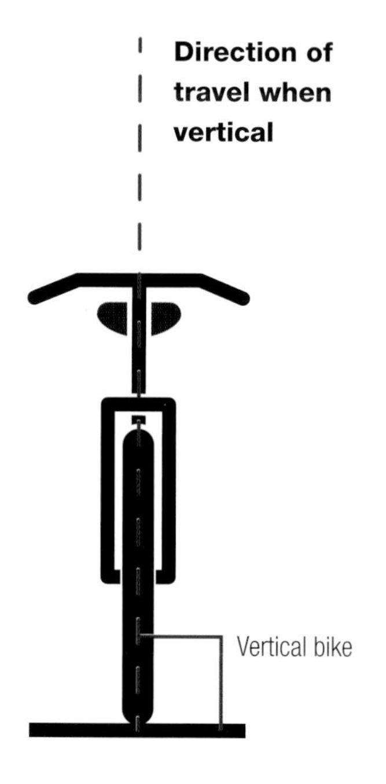

▶ ***Balancing act*** *Release a bike at speed across level ground and watch it lean and steer itself to maintain its balance. Centripetal force and gravity combine with the motion of the articulated frame and the mass distribution to keep the machine upright while it is traveling above a threshold speed. If the frame starts to lean, the wheel turns in the same direction, helping to right the bicycle.*

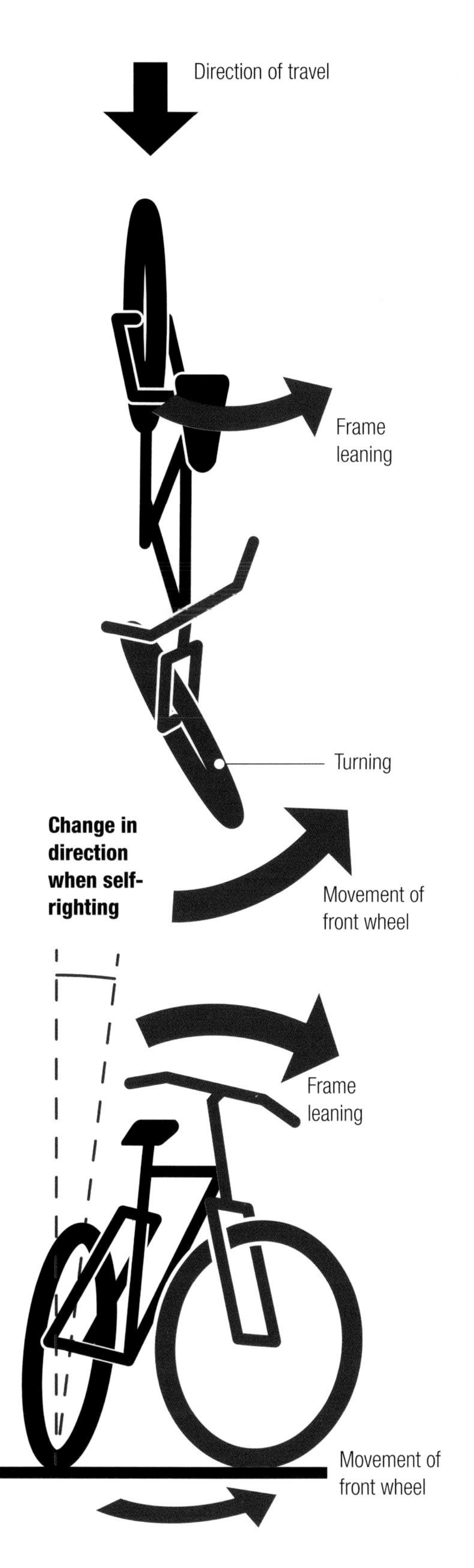

Caster effect and gyroscopic forces

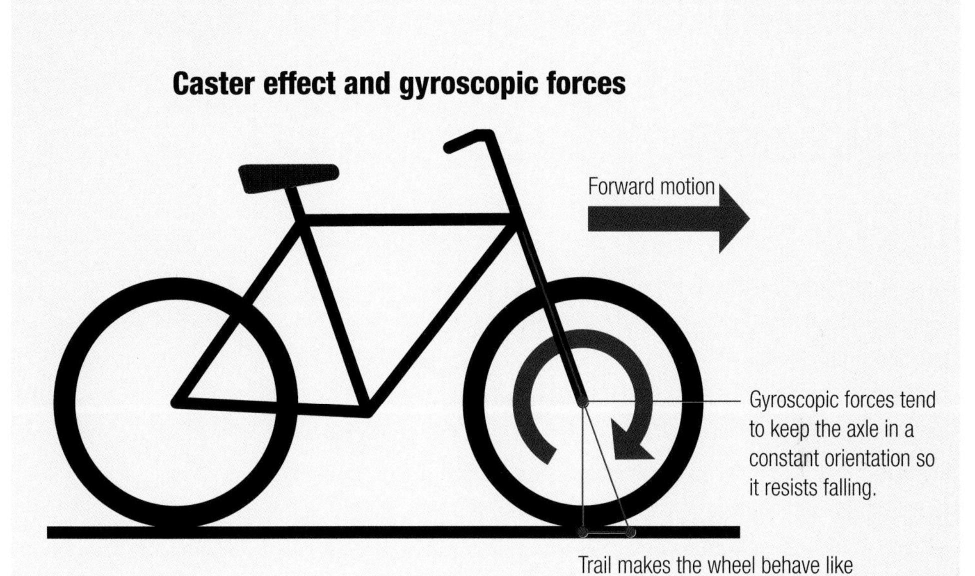

Torque effect

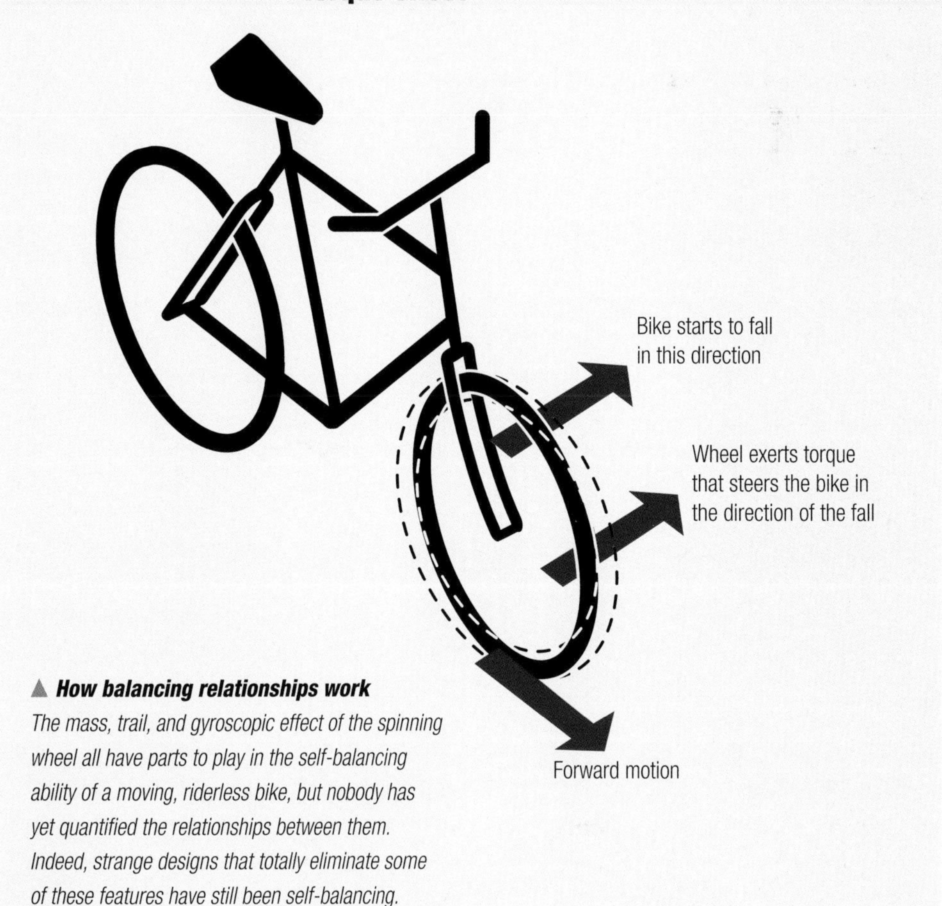

▲ ***How balancing relationships work***
The mass, trail, and gyroscopic effect of the spinning wheel all have parts to play in the self-balancing ability of a moving, riderless bike, but nobody has yet quantified the relationships between them. Indeed, strange designs that totally eliminate some of these features have still been self-balancing.

What are the rider input forces that affect cornering? → How does countersteering work?

Put a person on a bicycle and its dynamics change. The combined center of mass is higher than that of the bike alone. What's more, the center of mass shifts as the rider shifts on the bike. The scientific challenge is to identify how those movements influence the directional behavior of the bike. The basic principle for changing direction for all single-track vehicles is countersteering. To turn to the right, it's necessary to first turn to the left momentarily. This small, counter-intuitive maneuver, performed without thinking by all but novice cyclists, harnesses the power of gravity.

Here's what happens. Imagine a vertical plane through the top tube and the points where the tires contact the ground. If the combined center of mass of rider and bicycle lies in this plane, then the bicycle is balanced. It will travel in a straight line. If the handlebars are turned fractionally and briefly to the left (countersteer), the center of mass shifts to the right of this vertical plane. So the moving bicycle now starts to lean to the right under the influence of gravity. To "catch" it before it falls to the ground, the cyclist turns the handlebars to the right, taking the bicycle along the rightward curve the rider wants to travel.

Once the front wheel is turned, physics says an inward-pulling ("centripetal") force is needed to keep the bike moving in a circle—much as gravity pulls inward on a satellite to keep it in an orbit. This force is the friction between the tires and the ground. To convince yourself of this, imagine what would happen on ice or an oil slick, where friction almost vanishes: no turn, with the bike carrying on straight ahead.

That takes care of the tires, but how to keep the rest from carrying on straight ahead? A rider knows the answer instinctively: lean inward. The physics is complicated, but the angle of lean in an idealized situation—a steady turn in a fixed circle, no braking or speeding up, no wind or aerodynamic drag, and narrow tires—can be calculated (see opposite).

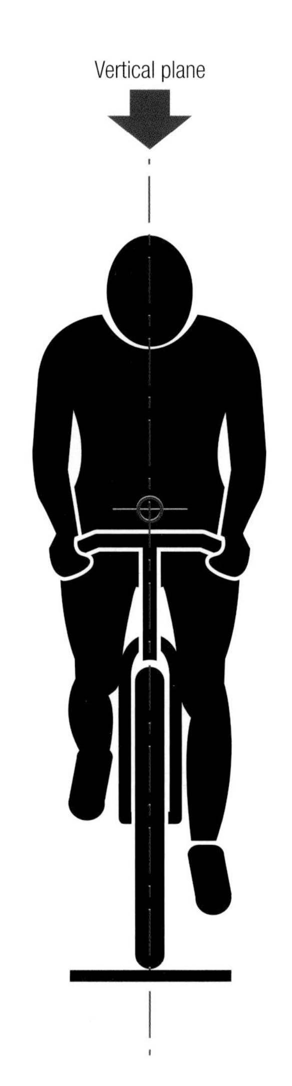

▶ ***The forces in cornering*** *A 150 lb (68 kg) rider on a 20 lb (9 kg) bike racing at 20 mph around a curve with a radius of 50 ft (15.2 m) experiences a centripetal force of over 90 lb (406 N). This is the force that keeps bike and rider on the curve instead of shooting off the road in a straight line.*

How countersteering works

▶ ***Steer and countersteer*** *The key point is that it is necessary to unbalance a moving bicycle to change its direction in a controlled manner. To start a right-hand turn, the rider needs to momentarily steer to the left so the bicycle becomes unstable and begins to fall to the right. By then turning to the right, stability is regained in the new direction.*[16]

Stability when upright and when cornering

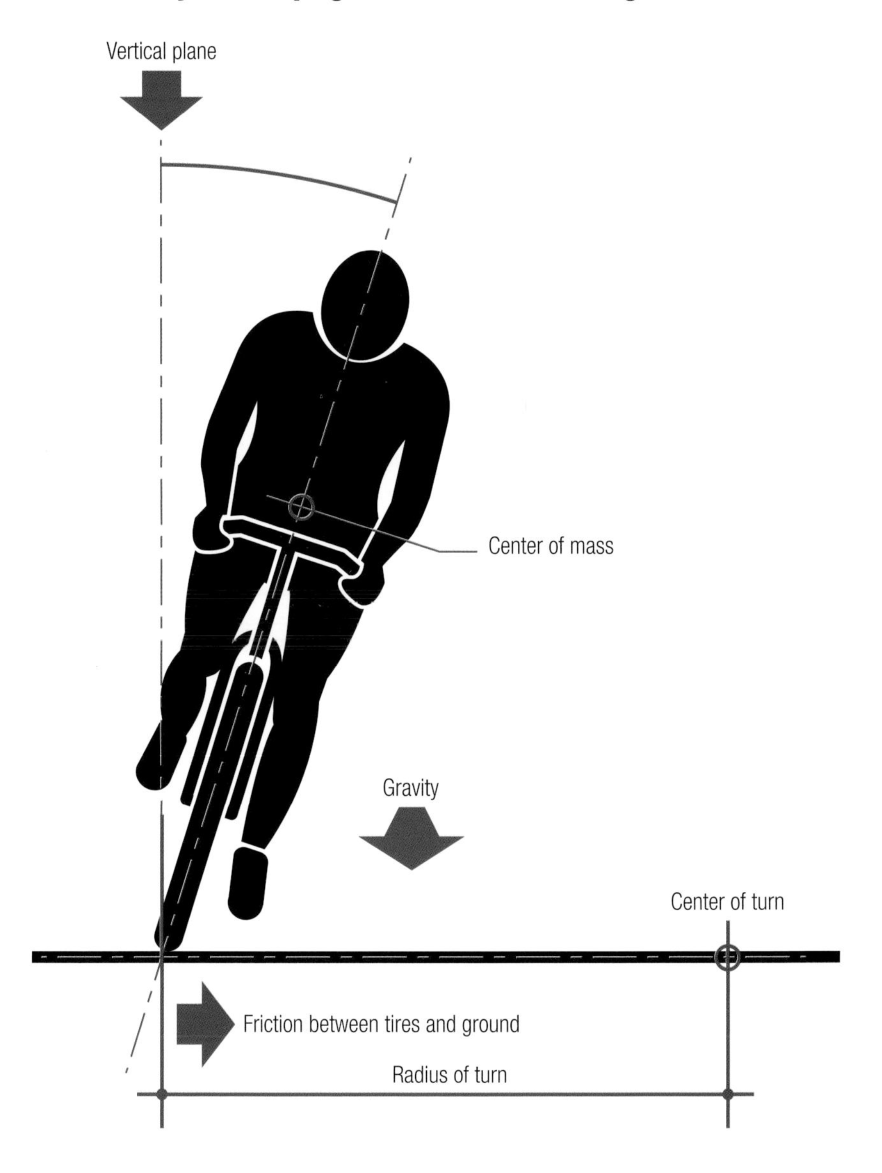

Need to know

Centripetal force:

The centripetal force acting on a bike and rider can be calculated using the following formula:

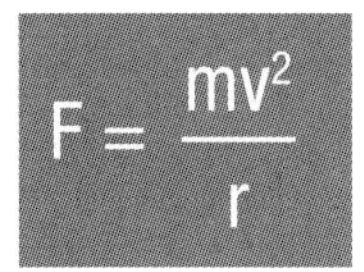

$$F = \frac{mv^2}{r}$$

where:

F = centripetal force (lb)

m = mass of bike and rider (slugs)

v = velocity (ft/s)

r = radius of the corner (ft)

A slug is equal to weight on the Earth divided by gravitational acceleration (32.174 ft/s²).

Angle of lean:

In an idealized situation, the angle of lean is given by:

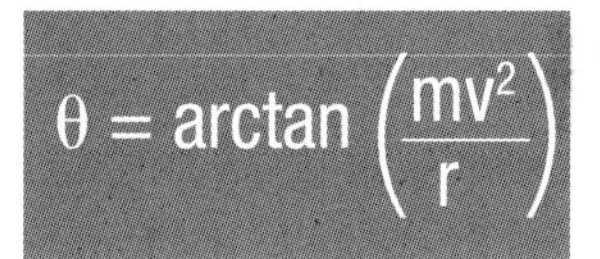

$$\theta = \arctan\left(\frac{mv^2}{r}\right)$$

where:

θ = angle of lean (°)

m = mass of bike and rider (slugs)

v = forward speed (ft/s)

r = radius of turn (ft)

equipment: **folding bikes**

▲ ***Earliest patent*** *(1888)*
No examples of Emmit Latta's patented folding bike survive, but the near-horizontal front fork would have produced an unusual steering action.

Cyclists who don't have space to store or transport their bikes buy folding versions. They are happy to forsake qualities such as handling, comfort, and weight for convenience. Emmit G. Latta of Friendship, Pennsylvania, was granted in 1888 the earliest known patent for a folding bike, and military demands rapidly became a major stimulus for its innovation. Inventors sought to design bikes that could be transported compactly by the armies of the United States, Japan, and European nations. Hinged frames, handlebars, and even wheels; separable frames and flip-out pedals: clever solutions were innumerable. After a slump in popularity in the 1970s, the success of small-wheeled bicycles coupled with a growth in multimodal travel has led to a more recent revival in folders.

▲ ***Columbia Compax Paratrooper*** *(1942)*
The latest in a long line of stowable bikes designed for paratroopers during World War II. The separable frame and folding handlebars were released with wing nuts.

▲ ***Brompton prototype*** *(1975)*
The simplified steel frame permits an unusually swift folding action that produces an extremely compact portable package.

Designing a bike that folds involves compromise. By deliberately introducing break points in the frame, the structure's ability to withstand the loads during cycling is weakened, so hinge mechanisms and frame tubes must be strengthened to compensate. Unfortunately, this usually adds weight.

This presents a dilemma for the designer—one reason why riders want a folder is so that it is portable. This is one reason why small wheels are frequently used for folders—they are lighter and make the bike easier to carry. The reduced diameter also cuts the stowage space needed. Another option is to use lighter, or less, material. Bickertons used aluminum long before it became generally popular for bike frames; another folder, a prototype called the Switch, uses carbon fiber; the Strida and the Brompton have fewer tubes—all in the pursuit of lower weight.

Another key issue is the ease of folding and unfolding. The solutions are diverse and the best are protected fiercely with patents. An alternative to folding is separable frames. Modern separables such as Moultons have small-diameter triangulated tubing to keep weight down, and suspension to ride as smoothly as full-sized wheels. Montague designed a kind of stealth mountain bike that is invisible to radar, with a near-silent electric motor and full-sized wheels, in perhaps the most sophisticated example—one funded by the United States' Defense Advanced Research Project Agency.

▲ ***Strida*** *(1987)*
With just three aluminum frame tubes and very small wheels, this folder keeps weight to a minimum. Ride quality and handling are traded for simple and compact folding

▲ ***Montague Paratrooper*** *(1999)*
The civilian version of the full-sized stealth folding mountain bike commissioned by the Department of Defense. A substantial top tube adds stiffness, lost with the removal of the down tube. The front wheel has to be removed for stowing.

How is cornering affected by an extra wheel?

Two wheels good, are three wheels better?

Riding on a tricycle or on a bicycle with small training wheels offers obvious benefits to stability, but the extra wheels make riding a tricycle quite different to riding a bicycle. While a racing bike can weigh under 15 lb (7 kg), even the lightest tricycle is one-third heavier at more than 20 lb (9 kg). A quadracycle is rarely less than 80 lb (36 kg). On an upright tricycle or quadracycle, the extra weight requires more effort to achieve the same speed or cover the same distance, although recumbent tricycles can achieve much higher speeds (see pages 132–33). The other issue is the handling: riding a multiwheeler is counterintuitive for those who learned to ride on a bicycle.

Steering a bicycle is learned unconsciously. Shifting the center of mass to one side of the vertical causes the bike to turn to that side. However, a lifetime of using this routine on two wheels is challenged immediately on a three- or four-wheeler. A tricycle or quadracycle simply goes in the direction that the handlebars are turned, so steering has to be relearned. When the handlebars are turned at a corner, the centripetal force that pulls it into a curving path is friction between the tires and the ground, just as with a bicycle. And as with a bike, the tendency for everything to carry on in a straight line leads to a sense of imbalance (from inside a turning vehicle this feels like an actual force pulling outward, referred to as "centrifugal force"). The rider can lean inward but—unlike a bicycle—the tricycle itself cannot. The center of mass moves toward the outside of the curve, and on an upright tricycle, if a corner is taken too fast, the inner wheel can lift and the whole thing begins to tip over.

Again the proper response goes against a bicyclist's instinct, which would be to raise the inner pedal. Instead, the inner pedal should be pushed down to the bottom, with the rider leaning onto it and hanging toward the center of the curve, like a sailor trying to keep a yacht upright in a strong wind. It's straightforward, but if you've only ridden bicycles, it can take a little time to adjust.

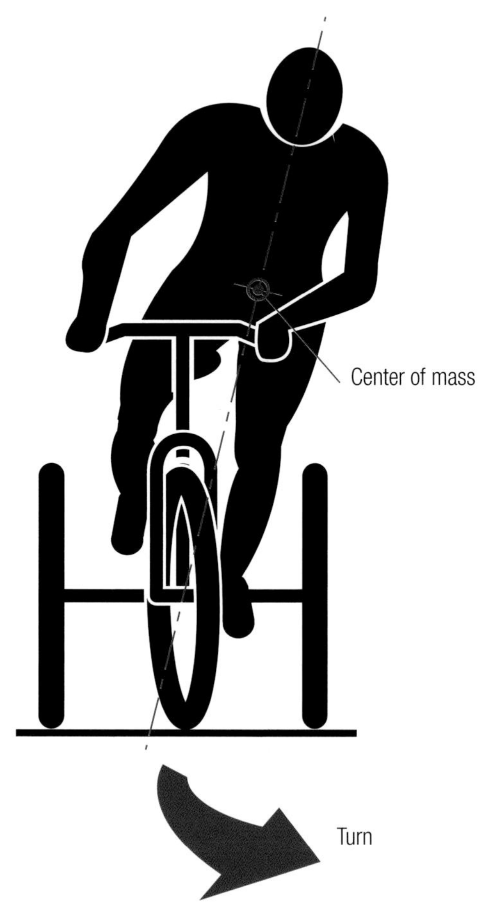

▲ ***Tricycle turn*** *A rider turning a tricycle adopts a position that is completely different to a bicyclist making the same maneuver. The tricyclist stands on the inside pedal and leans in because the machine itself stays upright. This keeps the combined center of mass of the tricyclist and tricycle inside the curve. The tricycle will only corner while the handlebars are turned.*

Turning on three wheels and two

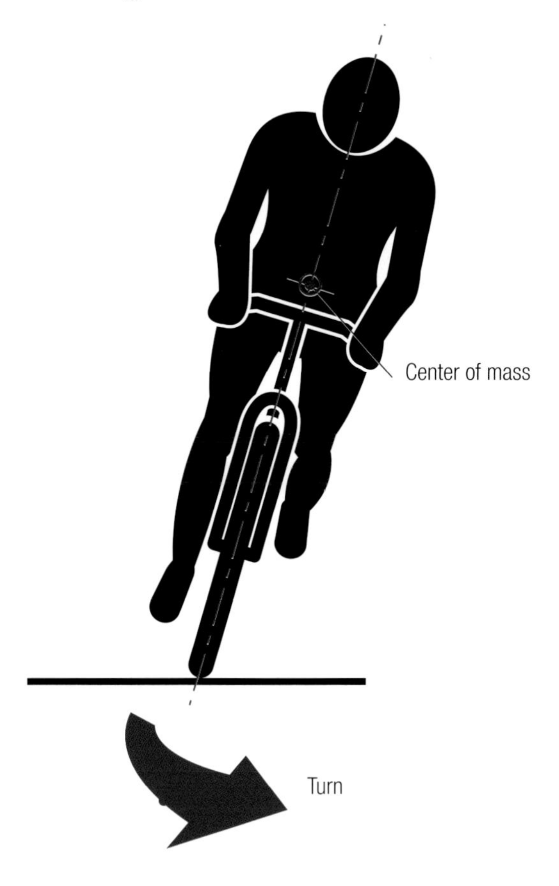

▲ ***Bicycle turn*** *With a bicycle, the lean maintains the turn (see pages 62–63); the cyclist does not need to steer with the handlebars and the front wheel stays in the plane of the bike frame.*

Recumbent tricycle

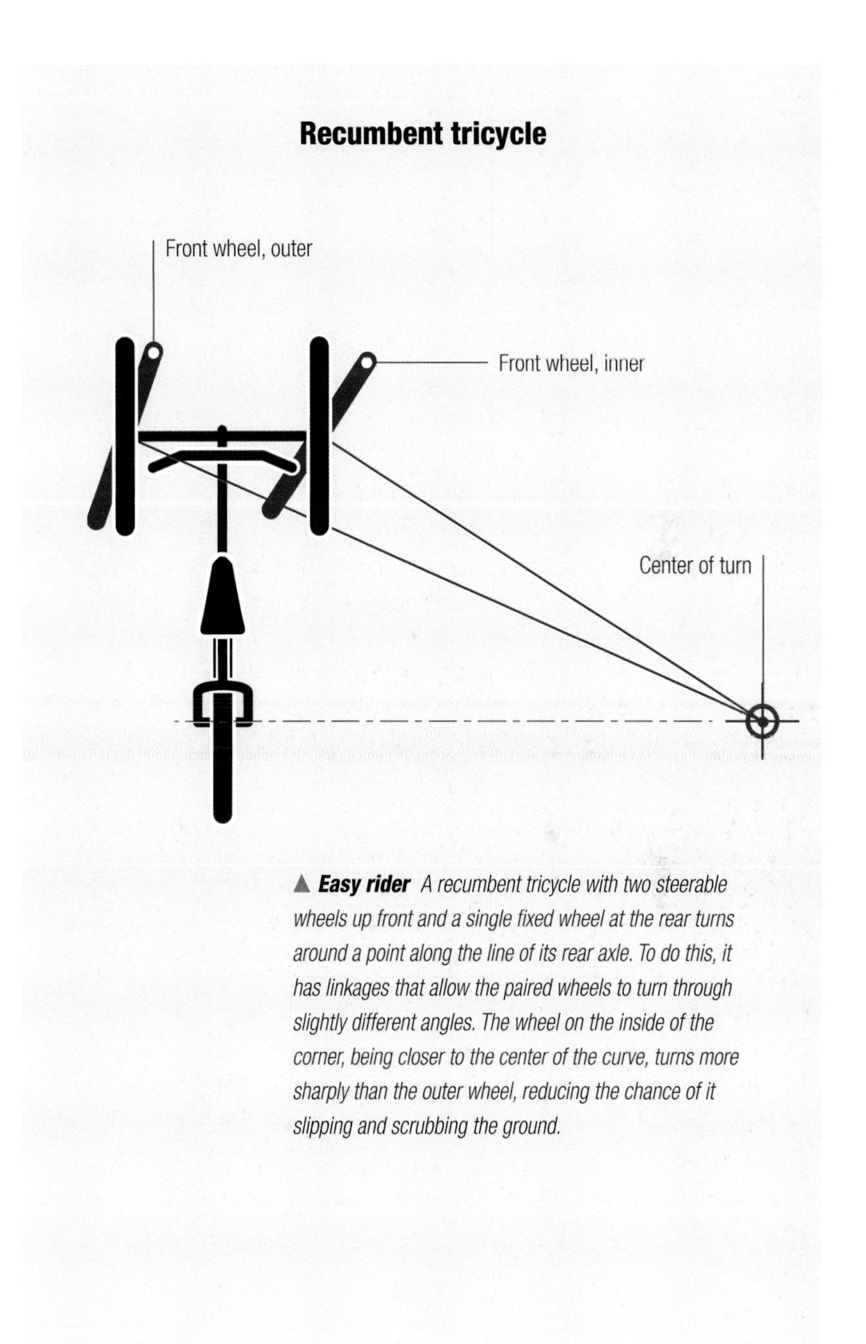

▲ ***Easy rider*** *A recumbent tricycle with two steerable wheels up front and a single fixed wheel at the rear turns around a point along the line of its rear axle. To do this, it has linkages that allow the paired wheels to turn through slightly different angles. The wheel on the inside of the corner, being closer to the center of the curve, turns more sharply than the outer wheel, reducing the chance of it slipping and scrubbing the ground.*

“Riding a multiwheeler is counterintuitive for those who learned to ride on a bicycle.

The dream bike is light, comfortable, durable, and costs nothing. Of course, it doesn't exist. The people who identify the best materials for the parts of a bike, create the materials, and manipulate them into the optimum shapes and sizes like to be rewarded for their efforts and expertise. They have studied the behavior of materials and applied what has been discovered by scientists to perform engineering feats that have made bicycles among the most efficient of human inventions. Engineers create bicycles by understanding the structure, potential, and limitations of materials. They cheat nature by creating materials that have never existed without human intervention and form them into parts that we can rely on. Their's is not a simple job, although it is made easier by the experiences of others. The engineers who have invented better ways to make an existing part or used new materials to introduce design innovations push the boundaries of bike making. This chapter provides the material evidence of their ingenuity.

chapter three

materials

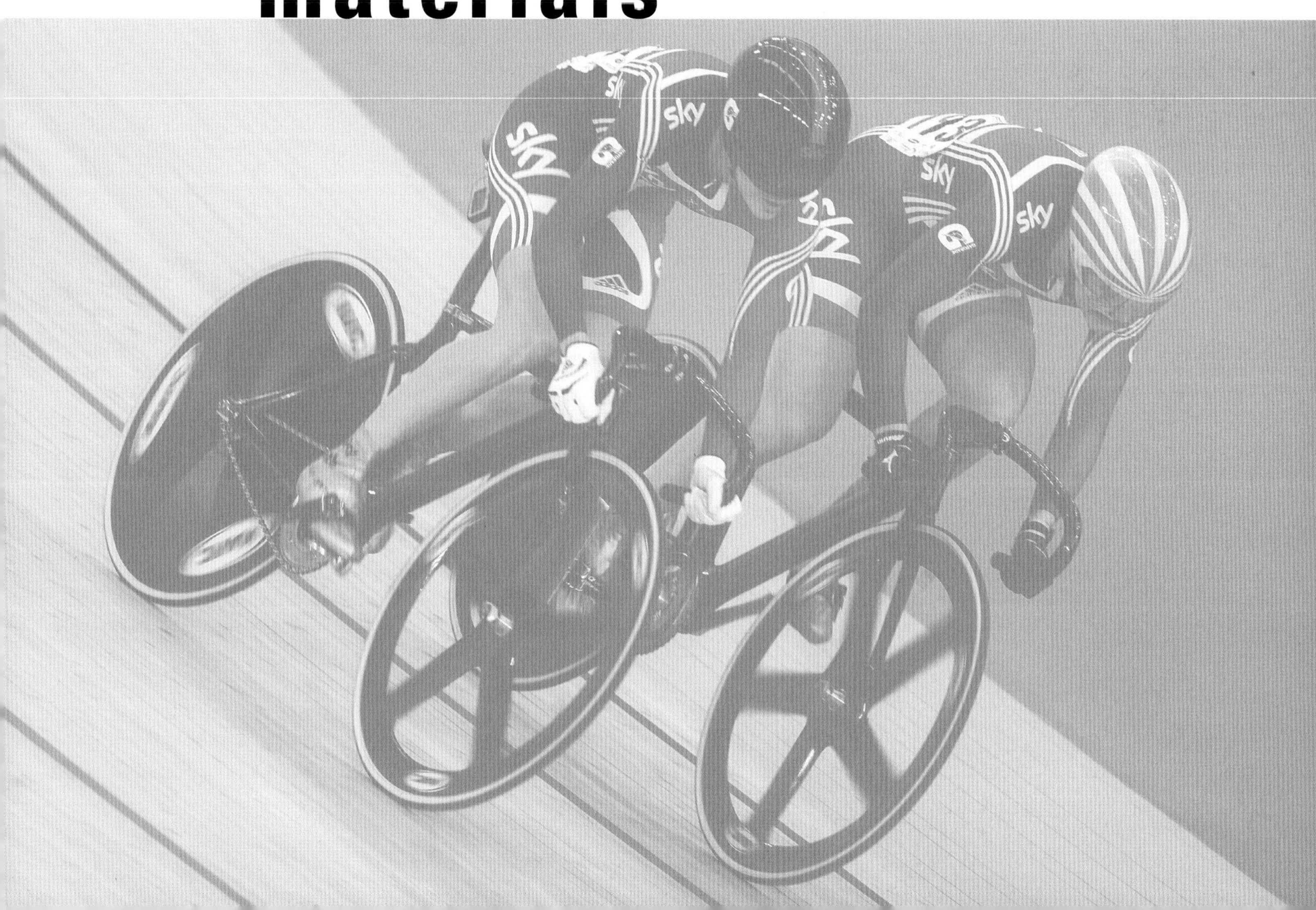

What matter makes up a bicycle? → Isn't a bike just made out of stuff?

Whichever kind of bicycle you have, it is made of matter. Matter is what makes up the observable universe. Current theories suggest that nearly all of the universe consists of so-called dark energy and dark matter—neither has been directly observed, but their existence has been predicted from looking at how the remaining fraction behaves. It's this remaining fraction that we know as matter, and it is categorized into five main states.

As far as cycling is concerned, the most obvious state is solid matter, such as metal, rubber, and plastic. Even though parts such as saddle covers and brake cables can be squashed, twisted, and bent, they are referred to as solid. The second state of matter, liquid, has become associated with bicycles only recently, with the workings of hydraulic brakes. The third state of matter, gas, has been an essential ingredient for comfortable and efficient riding ever since John Boyd Dunlop made the pneumatic tire popular in the nineteenth century. Suspension forks can also be pneumatic (which means that they use the properties of a compressed gas such as air). The fourth state of matter is plasma. While plasma makes up nearly all of the observable universe as part of stars and the space between them, it has not yet been harnessed for bicycles. Don't rule it out, however, because it may soon have an influence on cycling aerodynamics. The fifth state of matter, the Bose-Einstein condensate, was not created on Earth until 1995 and hasn't yet been applied to anything outside the laboratory, let alone to cycling.

All matter is made of atoms, or the chemical combinations of them called molecules, and has mass. The different states of matter are a result of how those atoms or molecules interact with one another and how they move about. Matter isn't constant—it can change from one state (or phase) to another when conditions alter. This chapter explains how differences in the structure of matter determine the properties of everything your bicycle is made from.

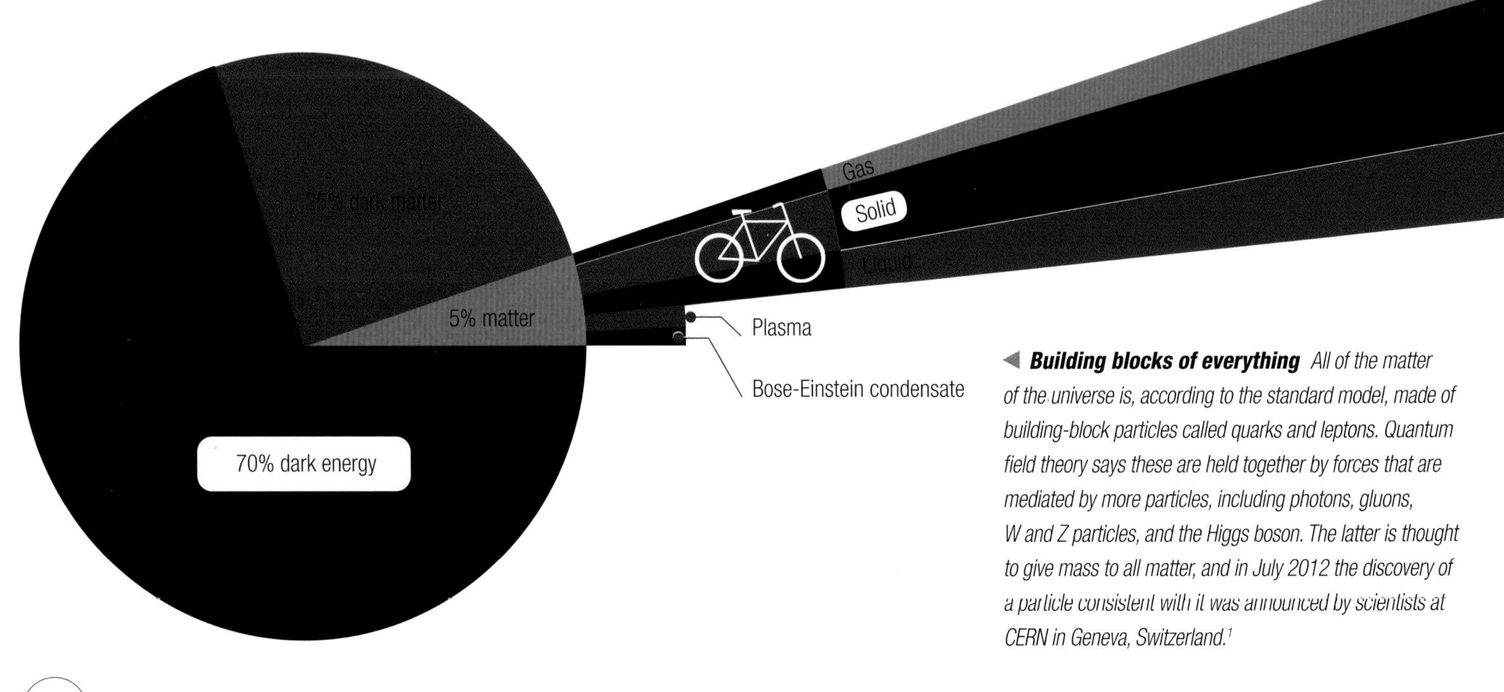

◀ ***Building blocks of everything*** *All of the matter of the universe is, according to the standard model, made of building-block particles called quarks and leptons. Quantum field theory says these are held together by forces that are mediated by more particles, including photons, gluons, W and Z particles, and the Higgs boson. The latter is thought to give mass to all matter, and in July 2012 the discovery of a particle consistent with it was announced by scientists at CERN in Geneva, Switzerland.*[1]

Bike matter

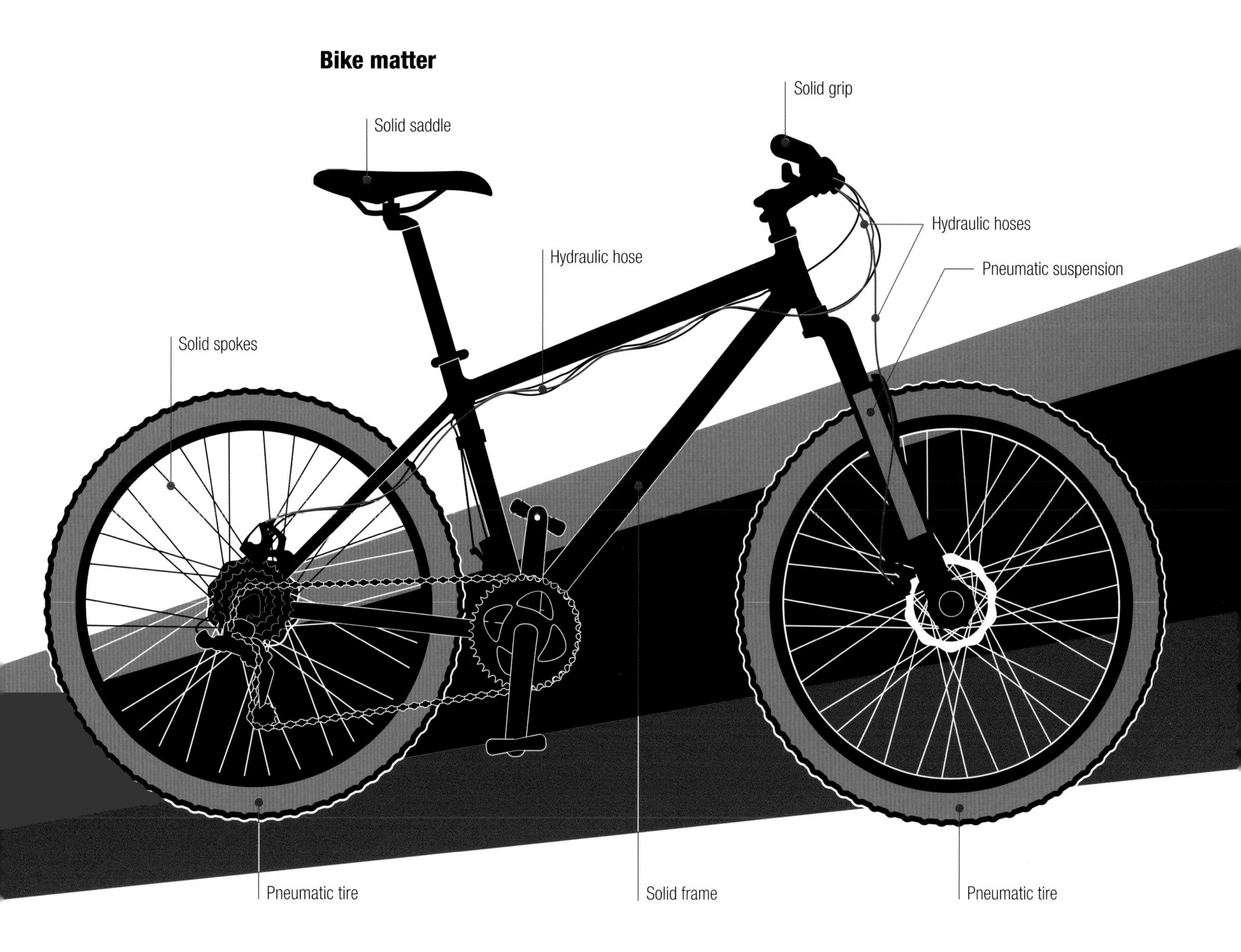

▲ ***What's my bike made of?*** *The matter that is used to make bicycles is stable under normal conditions on the Earth's surface, in that it does not change its state or phase. The first steerable two-wheelers harnessed matter only in its solid state, which is one reason why Draisines, Hobby Horses, and Ordinaries proved quite uncomfortable. It was only when two states of matter were put to work in combination that cycling became widespread. A flexible solid—rubber—was formed into a tube to contain a quantity of a gas—air—and wrapped around the rim. Liquids are of increasing interest to bicycle designers. They are used already to transmit braking forces from the levers to the pads, and there is research into how they could replace the conventional chain drive.*

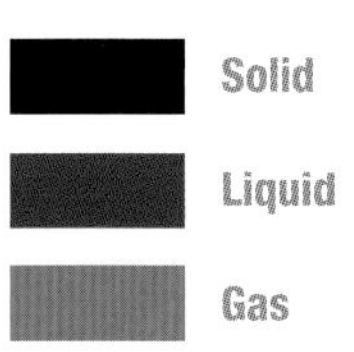

Why are metals strong? → Why aren't frames made of lightweight polystyrene?

The solid matter used for many bicycle parts is metal. Metals aren't the lightest of materials, but they are relatively strong, thanks to their atomic structure. The classical way to visualize an atom is as a swarm of negatively charged electrons orbiting around a nucleus consisting of an equal number of positively charged protons (plus, in all but a few elements, some uncharged neutrons). In metals, a number of electrons from each atom are so loosely bound to "their" nucleus that they are essentially free to drift among all the nuclei. This widespread sharing of electrons gives rise to electromagnetic forces that bind the metal atoms tightly together.

In general, the more of an atom's electrons that are free to join this "sea" of electrons, the stronger the metal. A titanium atom has 22 electrons, while aluminum has only 13, but, more importantly, titanium's atomic structure allows more of its electrons to join the "sea." For this and other reasons, aluminum is weaker than titanium and many other metals and is not suitable for hard-working bike components such as ball bearings and chain links.

The most common metal element used for bicycles is iron (26 electrons, 26 protons, 30 neutrons), the major ingredient in steel alloy. It pops up everywhere, even on bikes with frames made of other metals or of carbon fiber-reinforced plastic. This is because, as a constituent of steel, iron is strong, hard, durable, and ductile. It can be rolled into tubes, machined with a thread, or drawn into wire. These characteristics are determined by its atomic structure. Trial and error have taught bicycle mechanics which metal is best for each part of a bike. Quantum mechanics explains why.

The bonds between metal atoms pull them into a repeating pattern (much like a collection of ball bearings settles together). Such a long-range pattern forms a crystalline structure. This tight packing makes the metal resilient, because it allows atoms to slide over each other, just as it's easier to ride over closely packed cobblestones than ones with big gaps. The cobblestones on the Paris–Roubaix race are not just a test for riders; the infamous *pavé* also hammers the very atoms of their bikes. The right metals endure the punishment.

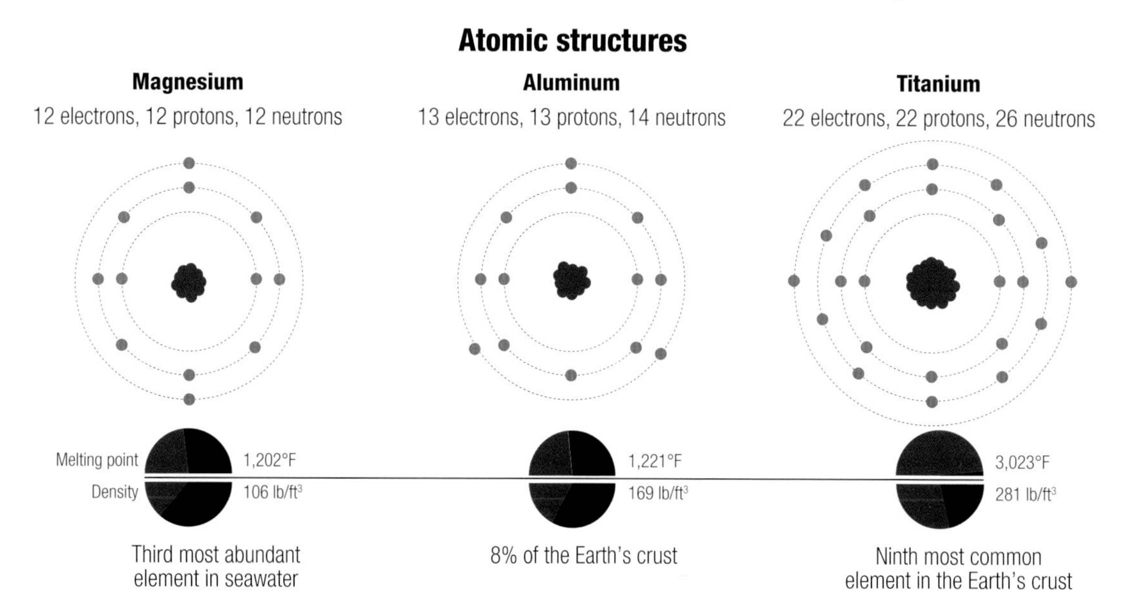

◀ ***What is your bike made of?*** *The four major metals used for bicycles—magnesium, aluminum, titanium, and iron—have different atomic structures, which lead to different properties, including crystal structures. They are rarely used in pure form because their properties can be improved by mixing them with other metals, as alloys. The diagrams represent the forms of each metal that are most abundant on Earth.*

The atomic properties of iron

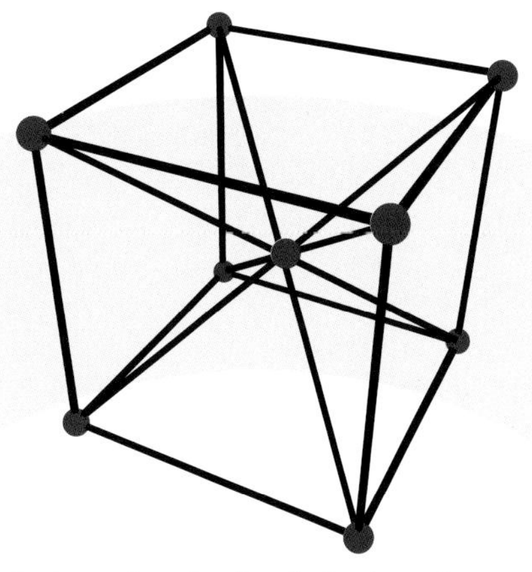

Body-centered cube of nine iron atoms

Iron

26 electrons, 26 protons, 30 neutrons

Structure

26 protons, 30 neutrons, 26 electrons

Abundance

Fourth most common element in Earth's crust

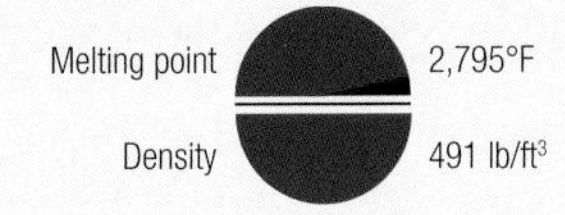

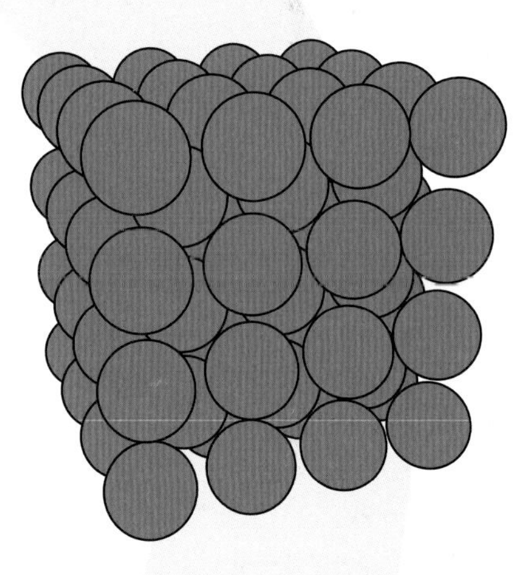

Iron crystal packed with body-centered cubes of atoms

▲ ***Hard but bendy*** *Iron atoms bond strongly because they share some of their electrons. This attraction pulls the atoms close together in a so-called body-centered cubic pattern, with an atom at each corner of a cube and another in the center; this repeated pattern builds up a crystal of iron. When a force is applied to the crystal, the "sea" of delocalized electrons allows the atoms to slip past one another without experiencing the strong repulsive forces that can cause other materials to break. This gives iron good ductility—it can be bent and stretched without fracturing.*

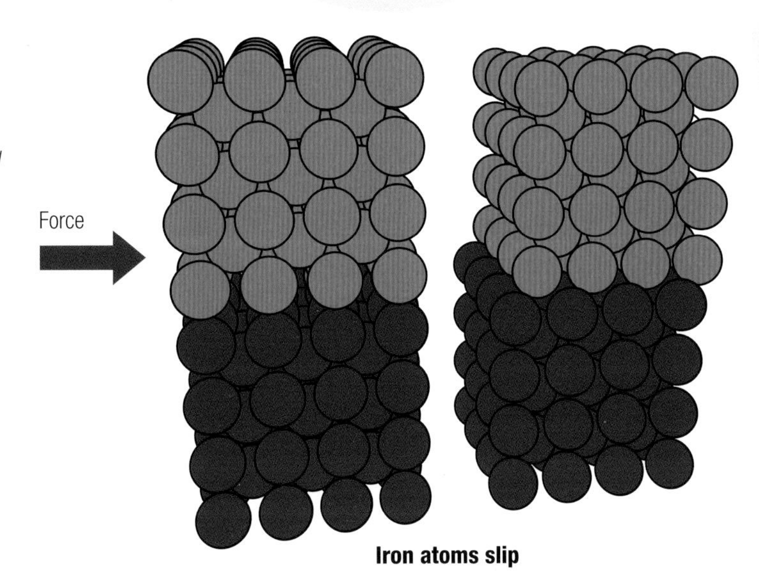

Iron atoms slip

What difference does alloying metals make? → What are the virtues of being impure?

Pure metal elements are not usually used for bicycle parts. Most often they are enhanced to make metallic alloys that are stronger, lighter, and more durable. These properties are determined by the atomic structure of the metal crystals. A common way to change this structure is to heat the metal until it melts, then add small amounts of other elements. When the resulting alloy cools and solidifies, the extra ingredients affect the crystal structure. Sometimes the additive atoms replace the host metal atoms, and at other times they squeeze in between them. The speed and frequency of the heating and cooling process and the thickness of the material also influence the characteristics of the resulting alloy.

The most common alloy in a bike is steel; it's the most popular material for the vast majority of bicycles and their components. Metallurgists have been alloying iron into steel for more than 160 years and there are well over 50 different recipes. Plain steel is made by adding carbon to iron. As the liquid mixture cools, a crystal compound called iron carbide aggregates in microscopic plates between crystals of pure iron. These make the steel stronger and lighter than iron, and other additions, such as chromium and molybdenum, can boost strength, reduce brittleness, and improve resistance to corrosion. Not all of the ingredients in a metal alloy may actually make any perceptible improvement to the performance of a bicycle, but they may help maintain the integrity of the material while it is being manufactured and then fabricated.

Iron isn't the only metal commonly alloyed. Pure aluminum is very soft and is alloyed with magnesium, silicon, copper, and chromium to make it harder. A rival to the high-performance aluminum frame, one made from titanium alloy actually uses aluminum as a key ingredient, along with vanadium. The few magnesium frames on the market have zinc and manganese included in the alloys used.

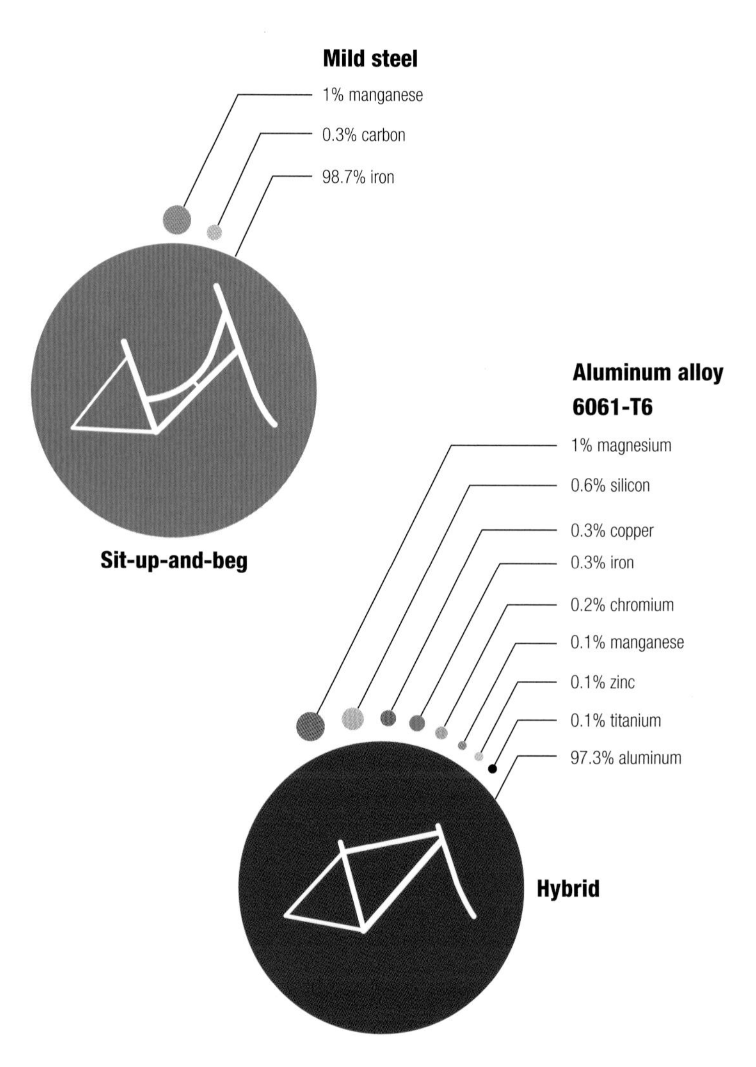

▲ ▶ ***Here's one of my favorite recipes*** *Different alloys suit different bikes—it's always a trade-off between price, weight, stiffness, and durability. The recipes for the various alloys have been perfected over decades, and metallurgists continually seek improvements by changing the mixes and the manufacturing and postproduction processes. These diagrams show the makeup of some typical alloys used for bike frames, with their ingredients listed by weight.*

Frame alloys

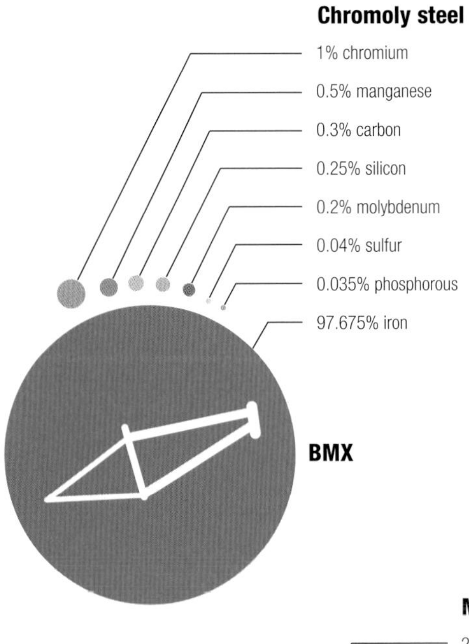

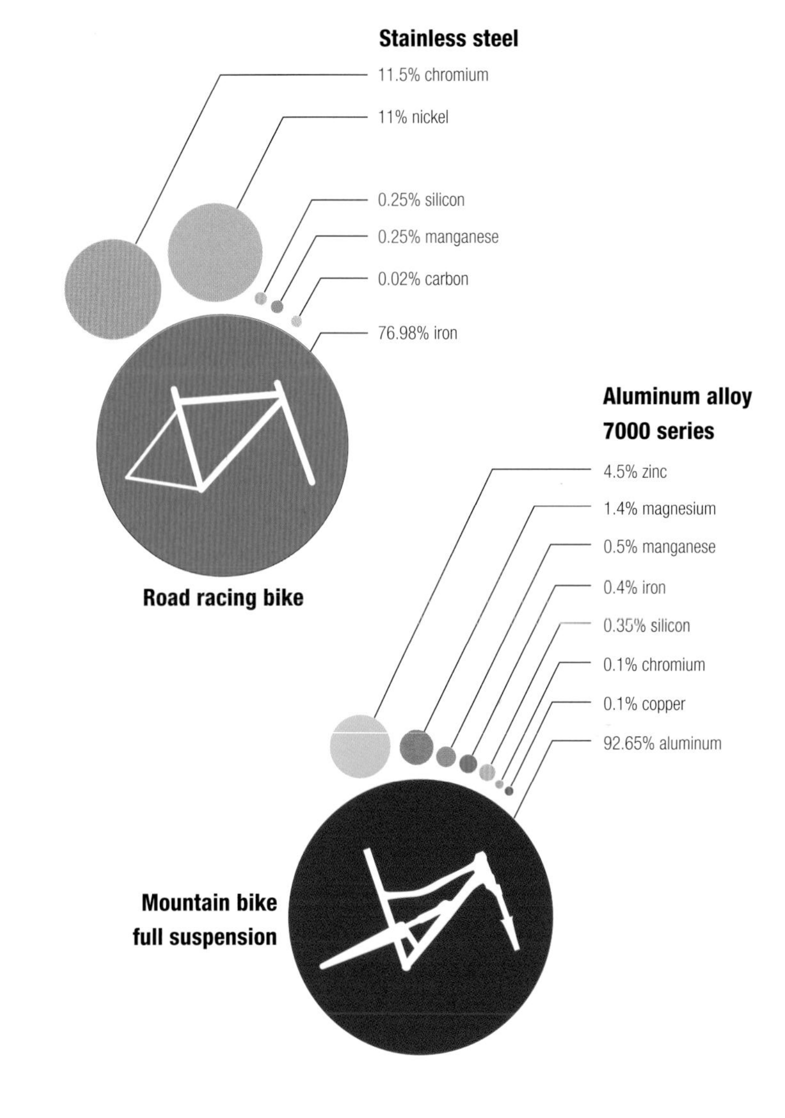

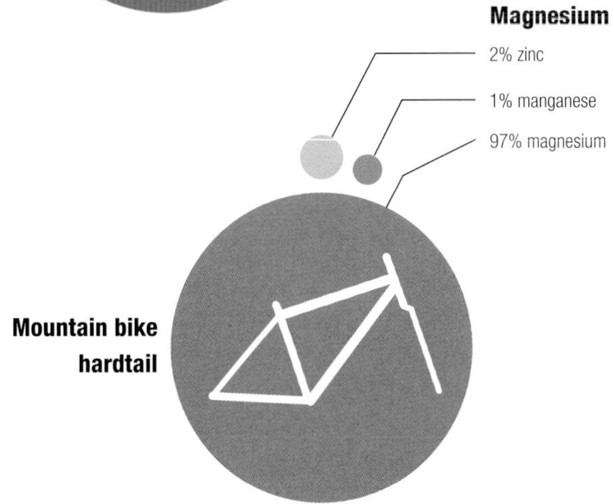

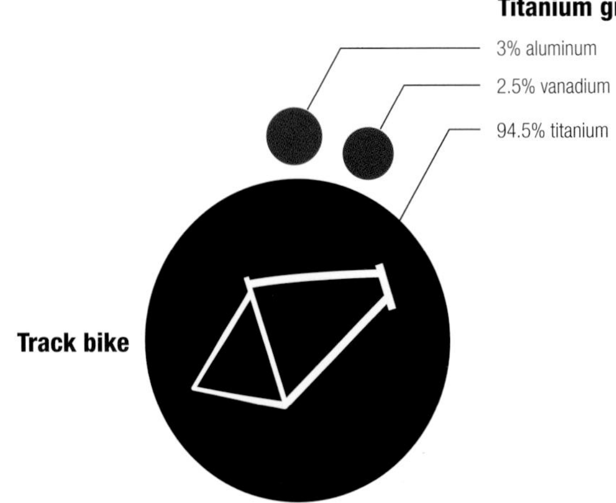

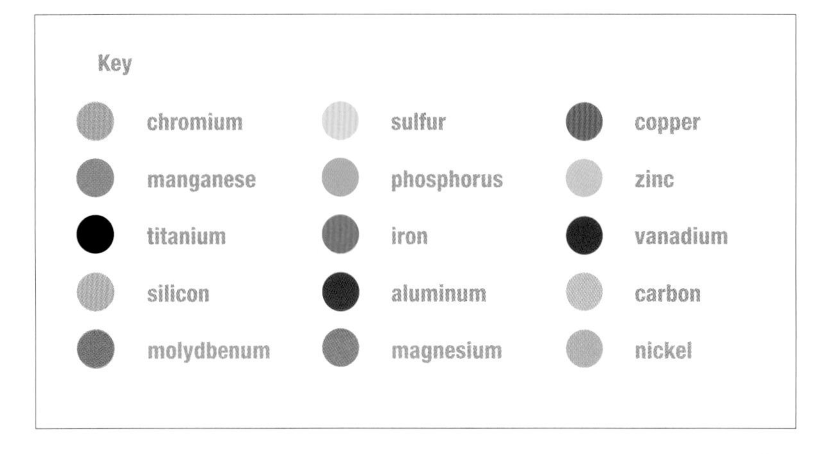

equipment: the frame

Hobby Horse *(ca. 1818)*
This was simply a beam with spoked wooden wheels attached, and the rider walked along with his legs on either side.

When Karl Drais invented the Hobby Horse in Mannheim, Germany, in 1817, he used wood from the European ash tree (*Fraxinus*), which was popular among coach builders because its grain is straight, easily shaped and joined, and has good overall strength relative to its weight. The material may have been familiar, but the bicycle's design was a technological breakthrough—its two wheels were in line and, crucially, it had steering. Since then, frame design has been propelled by advances in materials and technology. With the right combination, the optimum bike frames are now sufficiently strong, light, durable, and stiff to race down alpine roads at 50 mph (80 km/h), cross fields of boulders, or land in one piece after a triple back flip.

High Wheeler *(ca. 1870)*
The giant front wheel of this bicycle allowed the rider to cover a large distance with each rotation of the pedals.

Safety machine *(ca. 1885)*
By the end of the nineteenth century, the modern bicycle had tangentially spoked, same-size wheels.

Road bike *(ca. 1935)*
Fierce rivalry between manufacturers led to lighter, stronger bikes with straight frame tubes, cross-laced spokes, and aerodynamic dropped handlebars.

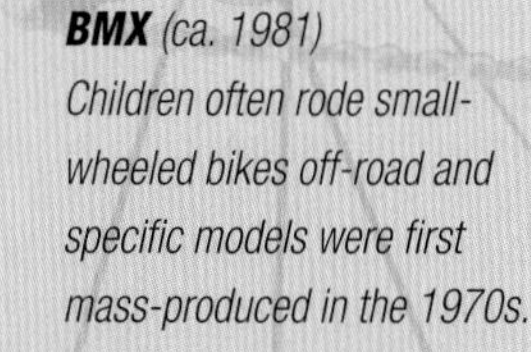

BMX *(ca. 1981)*
Children often rode small-wheeled bikes off-road and specific models were first mass-produced in the 1970s.

The first Hobby Horse weighed about 45 lb (20 kg). Today, the Union Cycliste Internationale (UCI), the sport's governing body, prohibits racing bikes lighter than 15 lb (6.8 kg). In the intervening two centuries frame builders have fashioned metals, polymers, and man-made composites to allow riders to travel farther, faster, and for longer. Improved manufacturing methods have been adopted, such as extruding tubes from blocks of hot metal, varying the internal diameter of tubes to cut weight yet retain strength, and hydroforming aluminum into computer-designed profiles. Joining and assembling now uses TIG welding for titanium and totally automating fabrication with composites.

Computer-based engineering techniques such as finite element analysis, computational fluid dynamics, and genetic design algorithms have opened entirely new routes to frame designs that work better for the human rider. At the same time, the study of human kinetics has revealed how frame dimensions should be refined to maximize the cyclist's potential.

Although the bicycle is a mature concept, the pace of change in frame design during the most recent three decades has driven the UCI to lay down strict, conservative rules to standardize the shape of a competitor's bike. The World Hour Records set in the 1980s and 1990s by Francesco Moser, Graeme Obree, and Chris Boardman wouldn't be recognized today because they were achieved on frames whose designs would not be legal. Inevitably, such rules will change. Not long ago, mountain bikes were unacceptable, and yet they're now being ridden in the Olympics. Bicycles don't have a reverse gear and neither does science.

Long wheelbase recumbent *(ca. 2003)*
Low riding improves aerodynamics, so speeds far exceed those of diamond-frame bikes.

Full suspension mountain bike *(ca. 2000)*
Full suspension absorbs the knocks of rough tracks, disc brakes aid control, and multiple gears ease pedaling.

Lotus Type 108 *(1992)*
Engineer Mike Burrows perfected a carbon fiber monocoque so Chris Boardman could become the fastest cyclist ever.

How is mechanical stiffness related to tube diameter?

Why do some frames have porky tubes?

Bicycle manufacturers optimize the weight and rigidity of a bike by choosing frame tubes that use varying cross sections and dimensions at different places. Most frames are fabricated from metal alloys, but the tubes must be formed to the right shape to make the best of a material's properties. Most makers still use hollow cylinders because this geometry minimizes the quantity of material required to produce the most resilient shape for a given weight—one that can withstand the many loads from different directions that are found in a hard-working bike. A square or rectangular cross-sectional tube shape cannot provide equal deformation or bending resistance in most directions unless more metal is used to reinforce it—and that imposes a weight penalty.

The diameter of a cylinder has a direct effect on its stiffness, and it is this dimension that has a significant influence on the rigidity of a frame. The rigidity of a material is defined by its Young's modulus (or modulus of elasticity). Lab tests show that steel is about three times stiffer than aluminum, so it would seem logical that a piece of aluminum needs to be three times thicker than an identical steel piece to be equally rigid. However, cylinders are not flat pieces of metal, and their mechanical properties provide an important benefit. When the diameter of a hollow cylinder is increased, its stiffness increases in proportion to the third power of the diameter. In other words, when the diameter is doubled, stiffness increases eight-fold (2^3).

At the same time, weight only increases somewhere between two- and three-fold, depending on the thickness of the cylinder wall. So an aluminum frame tube needs to be roughly 1.44 times ($1.4425^3 = 3$) greater in diameter than a steel tube to fully compensate for its lower Young's modulus, assuming there's no difference in wall thickness. One crucial limit on tube stiffness is the functionality of the frame—there's no point in having extremely stiff tubes if they are so fat that they get in the way of the cyclist's limbs and the moving components.

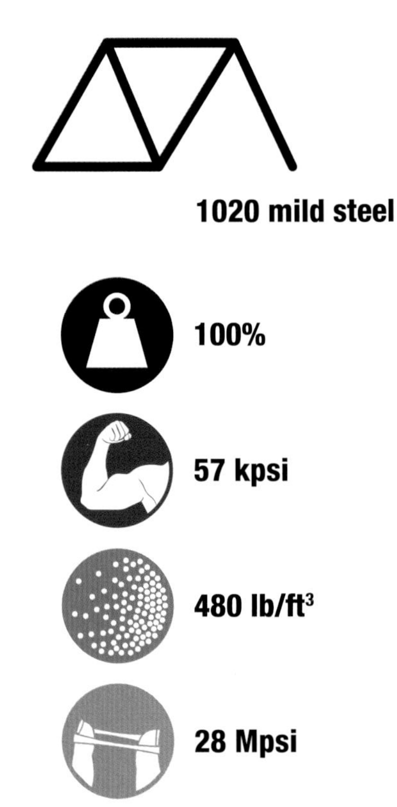

▶ ***To each his own*** *Every metal has its own properties and characteristics. The art of the frame builder is to balance these with the desired result—some cyclists want ultralight, stiff frames and others want rugged, more comfortable designs. These four alloys are the most commonly used for bike frames. To keep things simple, the data refers to straight gauge framesets with tube walls of constant thickness. The price of each material will vary according to market conditions. Nevertheless, based on historical data, it's likely that relative prices will not change, increasing in the order from mild steel through to titanium.*
(1 kpsi = 1,000 psi, 1 Mpsi = 1,000,000 psi.)

Key

Weight (as a percentage of a mild steel frame of the same size)

UTS, minimum

Density, minimum

Young's modulus, minimum

Yield strength

Mechanical properties of bike-frame materials

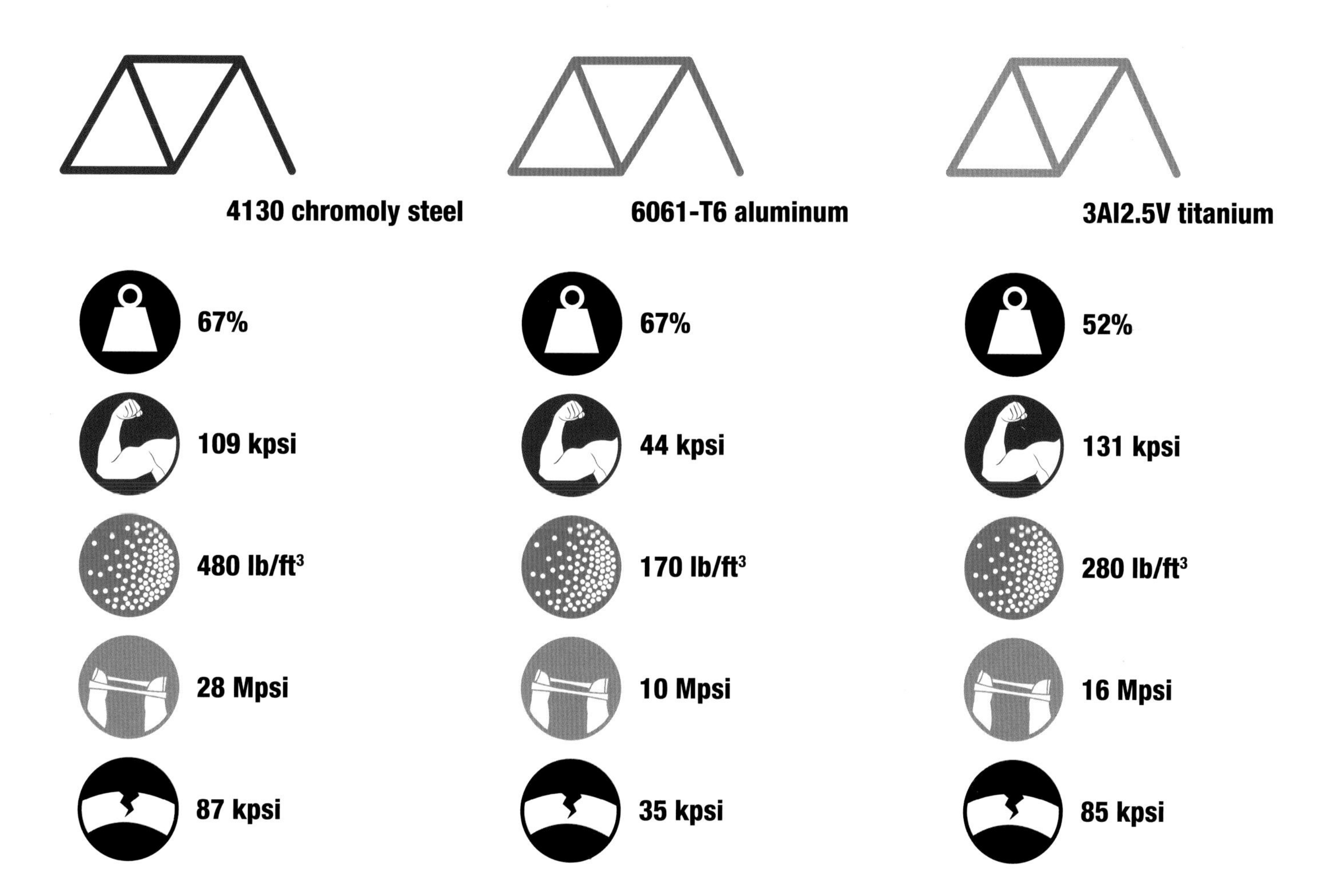

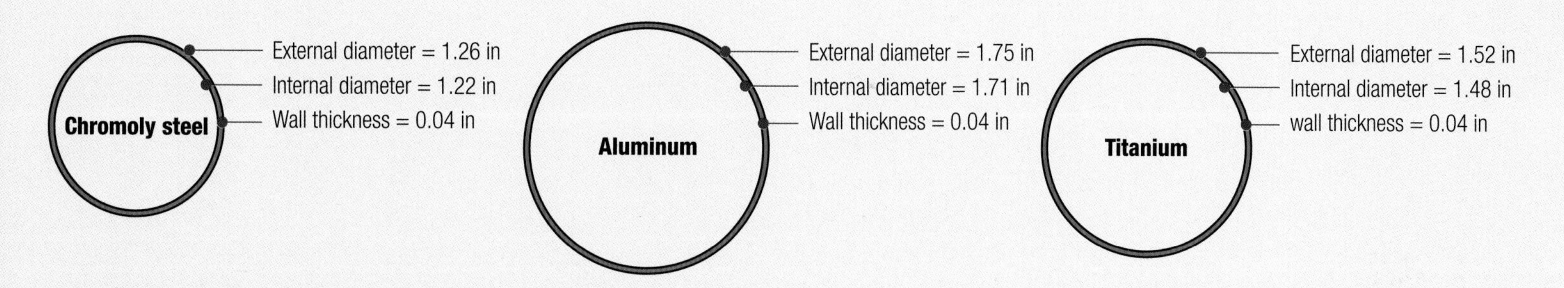

▲ ***How fat to function?*** *The stiffness of a frame tube changes according to the material and its dimensions—the diameter and wall thickness. If a steel head tube 1.26 in (31.7 mm) in diameter is to be replaced with another of aluminum or titanium but with the same wall thickness of 0.04 in (1 mm), the graphic reveals how much bigger the external diameters must be to compensate for the greater elasticity of the substitute metals.*

How are solid metals joined? What sticks my frame together?

Nearly every metal frame is assembled by either welding or brazing the tubes together, and both methods involve turning solid metal into liquid and back again. The joints must be at least as strong as the tubes, and the frame builder achieves this by harnessing the complex chemistry of hot metal.

Welding involves heating two adjacent areas of metal until they melt and fuse, and then letting them cool and solidify. It's complicated by the fact that molten metal may absorb gas molecules, which can weaken a joint considerably. Aluminum has a low melting point so it is easily welded, and an inert gas is used to shield it from potential contamination by gases in the air. For this process, gas tungsten arc welding (GTAW), also known as tungsten inert gas (TIG) welding, is used. Welding can reduce the strength of aluminum by 80 percent but this weakening can be reversed by heat treating the frame. Titanium especially needs a shield of inert gas to prevent oxygen, nitrogen, and hydrogen from diffusing into the hot alloy and making it brittle.

Brazing refers to a technique in which a small amount of a different metal, known as a filler, is used to hold the parts together. The filler is commonly brass or a silver/tin alloy of copper. The joint is heated, usually with an oxyacetylene gas torch, to a temperature hot enough to melt the filler but not the tube. When the filler melts, capillary action takes over and its molecules are drawn into the gap between the parts by a combination of adhesion forces (between filler and tube material) and surface tension (between filler molecules). This fills the gap and, as the brazing material cools, it solidifies into metal crystals that bind to the two parts of the joint. The advantage of brazing over welding is that it introduces a little more ductility (resilience) into a joint than welding, making it less prone to breakage.

Surface tension

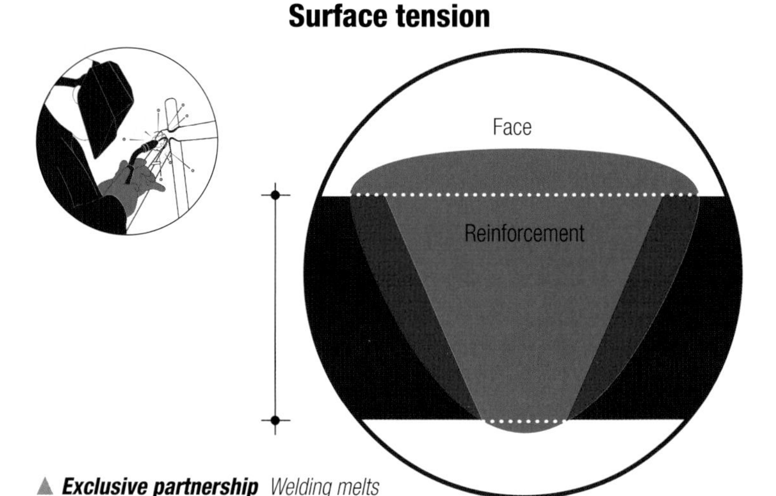

▲ ***Exclusive partnership*** *Welding melts metal with energy from a controlled electric arc and shields the working zone with an inert gas to exclude atmospheric gases that could contaminate and weaken the joint. A rod of material similar to the two tubes being joined is introduced to reinforce the junction. The fusion zone is created where material from the rod and the tube melt together and harden. The external surface of the weld is characterized by ripples, and the quality of the joint can be judged by their regularity.*

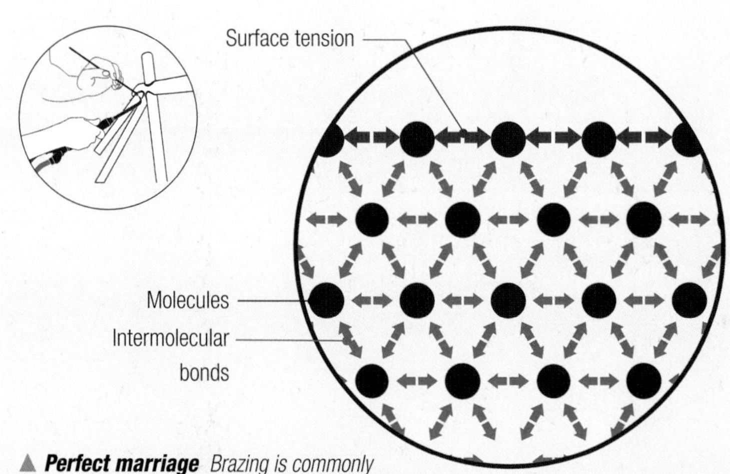

▲ ***Perfect marriage*** *Brazing is commonly used by the custom frame builder to marry steel tubes. A rod of filler metal that liquefies at a temperature lower than the melting point of steel is heated with a torch, and capillary action draws it into the gap between the tubes. Traditionally, close-fitting steel sleeves or sockets, called lugs, are placed around the ends of the frame tubes to strengthen the junction, but fillet brazing doesn't necessarily require them.*

Solid/liquid cycles: welding

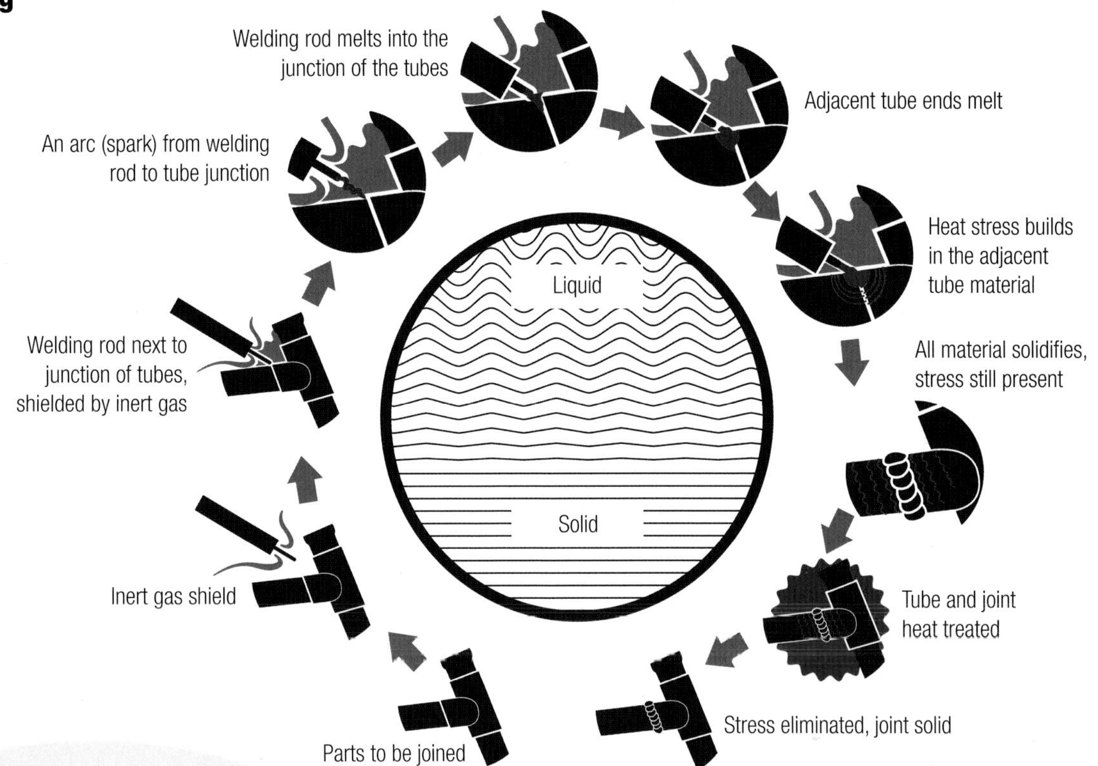

▶ ***Stress reduction*** *Structural components important to the strength of the bicycle melt during welding and some materials, such as aluminum, are susceptible to being weakened during the process. Heat treatment after the weld has cooled allows the molecules to restore their integrity.*

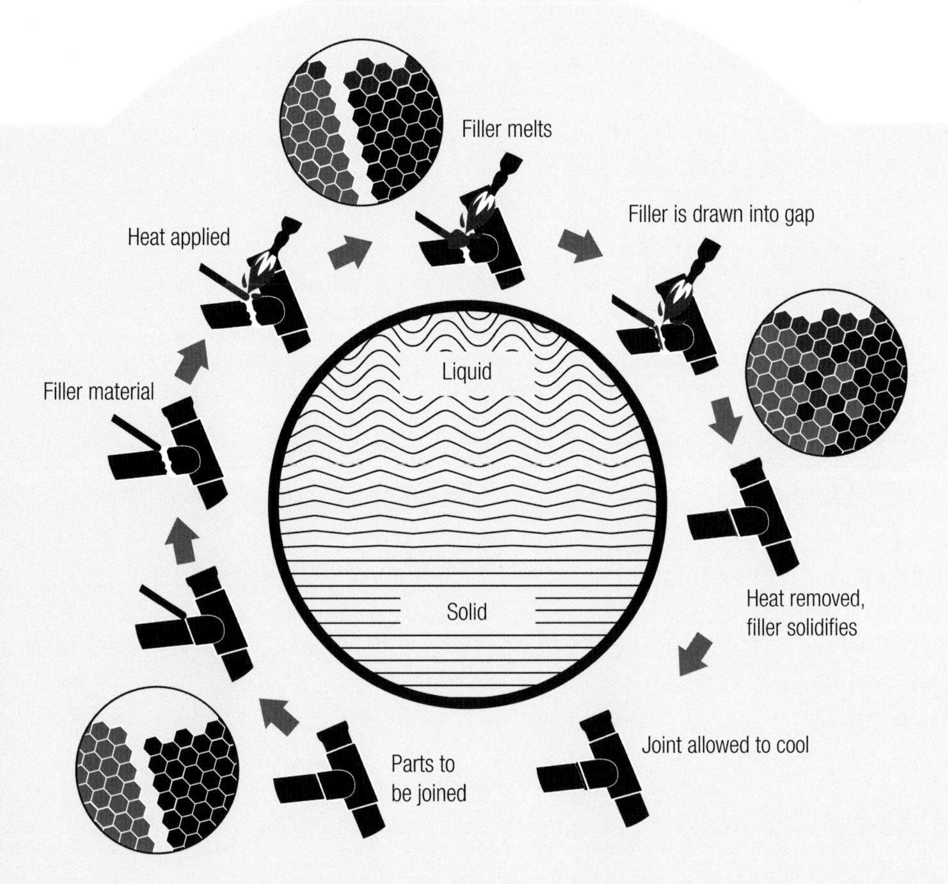

Solid/liquid cycles: brazing

◀ ***Joint venture*** *The filler is the material that melts, flows into the crack between the tubes, and fuses with the solid metal on each side, frequently creating a bond that is actually more resilient than the frame tubes themselves.*

equipment: frame tubes

The frame tubes of early bicycles were made of iron until steel alloys became available and affordable. Since that happened, in the second half of the nineteenth century, steel and its alloys have been used for the vast majority of bikes because they are relatively cheap and tolerate a wide range of fabrication processes. Aluminum became more popular in the late twentieth century, at the same time that titanium was first used by a few pioneers. Magnesium was tried in the 1980s but is still only offered by two manufacturers in the whole world. Away from the metal scene, composites have been growing for nearly 50 years and, in a more literal sense, so have bamboo frames.

Both the cheapest and the most expensive steel tubes are made from flat sheets rolled into cylinders and welded along their lengths. The weld is a weakness but is strengthened through heat treatment. Most other hollow metal frame tubing is seamless. For steel, a solid bar about 10 in (25 cm) in diameter and 35 in (90 cm) long is heated red hot, to more than 1,800°F (1,000°C), so it is compliant and can be rolled. To turn the hot, solid cylinder into a hollow tube, its end is pierced and drawn over a pointed steel rod called a mandrel. At the same time it passes through a conical hole, which reduces the external diameter. The process is repeated until a tube of the desired dimensions and a constant wall thickness—known as plain gauge—is obtained.

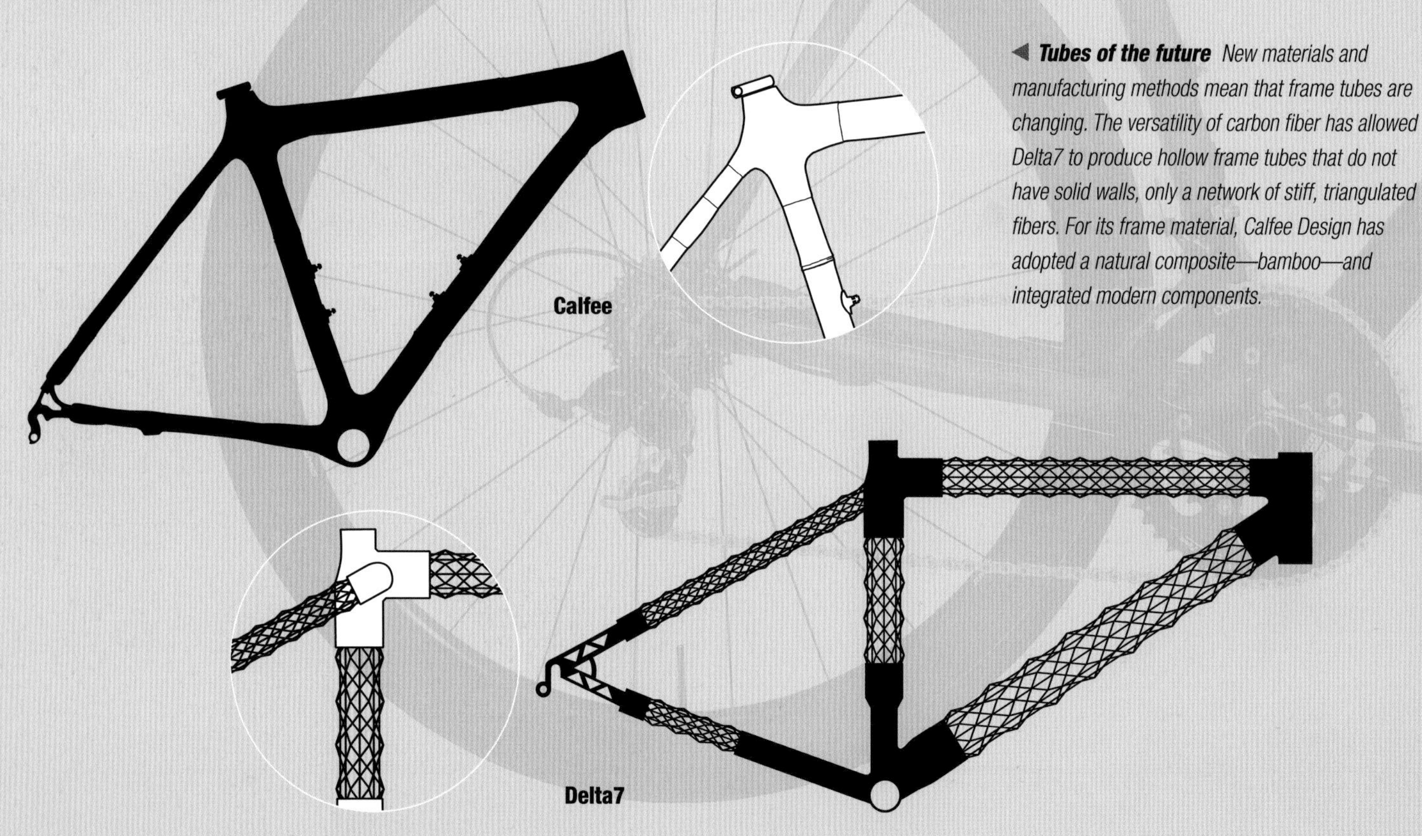

◀ ***Tubes of the future*** *New materials and manufacturing methods mean that frame tubes are changing. The versatility of carbon fiber has allowed Delta7 to produce hollow frame tubes that do not have solid walls, only a network of stiff, triangulated fibers. For its frame material, Calfee Design has adopted a natural composite—bamboo—and integrated modern components.*

Frame tubing, however, needs to be strongest at its ends, so it is often made thinner along its central section to save weight. This is done by "butting." In effect, the shaped mandrel, which defines the internal diameter and profile of the tubing, gets trapped inside because the walls at each end are too thick for it to be removed. The tube is then spun between offset reels. This is the clever part: While it has no significant effect on wall thickness or overall profile, it increases the external and internal diameters of the tube so the mandrel can slip out. The result is a thin-walled central section and thicker-walled ends.

Aluminum alloy tubing can be shaped more easily than steel. Profiles are hydroformed, with the sheet material placed inside a resilient mold and pressurized from inside by hydraulic fluid. This can also leave the aluminum alloy with a cosmetically better surface than other shaping methods.

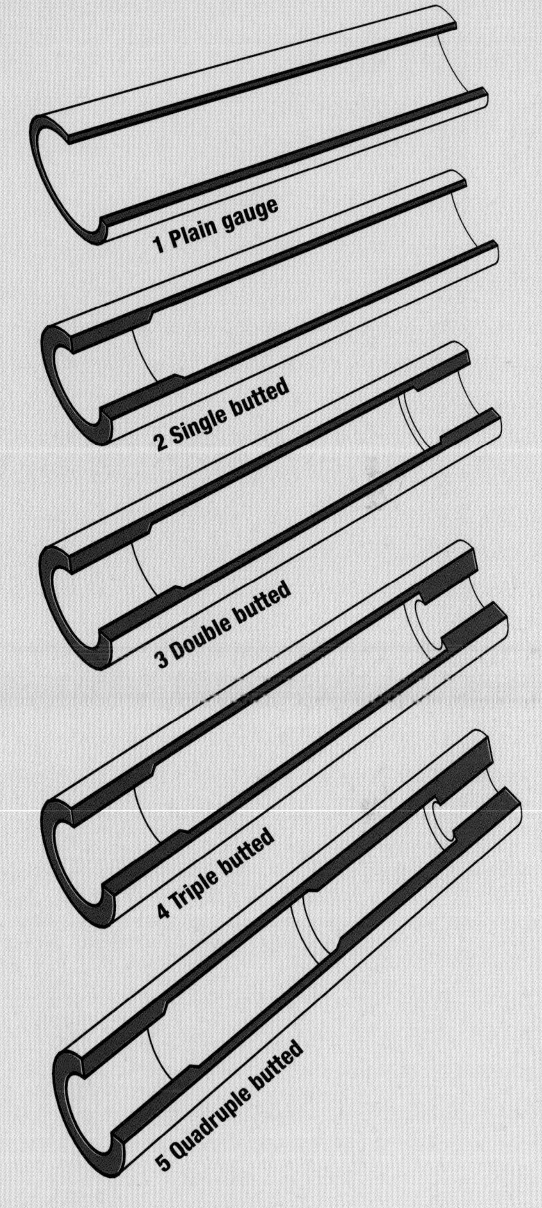

▲ ***Tube profiles*** *Conventional steel frame tubes have various profiles to save weight and retain strength and stiffness where they are most needed:*

1 *A tube without a butt has constant wall thickness, and is called a plain gauge tube.*
2 *A single-butted tube is thicker at one end.*
3 *Double-butted tubes are thin in the middle and thicker at both ends, with each end the same thickness.*
4 *Triple-butted tubes have ends of unequal thickness.*
5 *Quadruple-butted tubes have ends and midsections of varying thickness.*

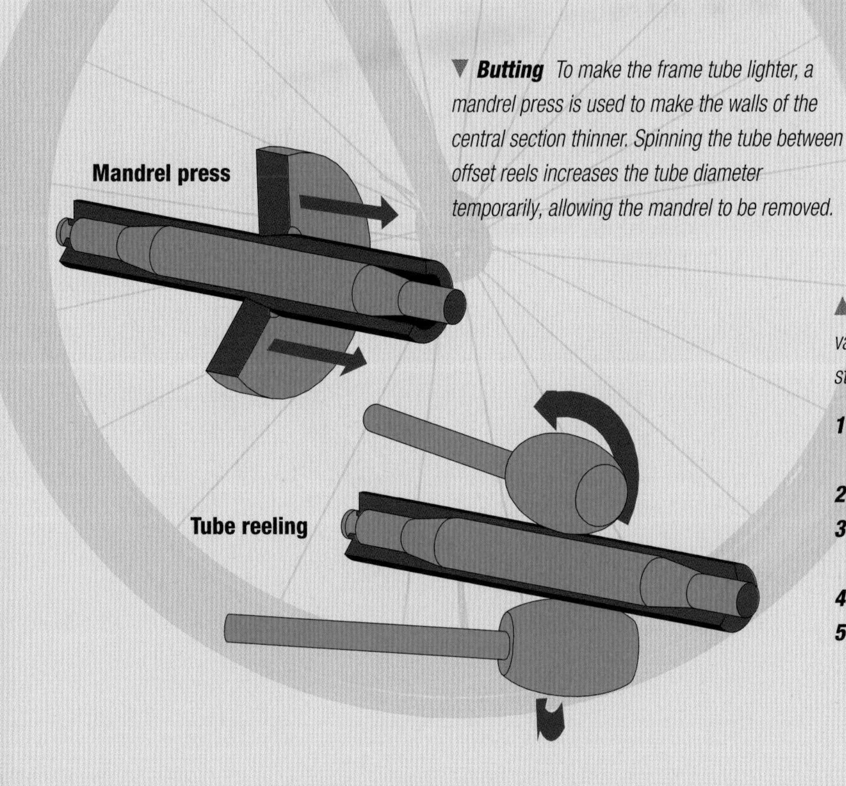

▼ ***Butting*** *To make the frame tube lighter, a mandrel press is used to make the walls of the central section thinner. Spinning the tube between offset reels increases the tube diameter temporarily, allowing the mandrel to be removed.*

What is a polymer? → What makes tires bendy?

Metals are not the only solid matter in bikes. Every single bicycle made also relies on polymers. If it's on your bike and not made of metal, there's a 99.5 percent chance it's a polymer. Bikes were among the earliest inventions to utilize them; the natural rubber used for the first tires is a polymer. Other varieties such as plastics and foams are used for mudguards, saddles, cable sleeves, lever hoods, grips, bar tape, seals, valves, valve caps, jockey wheels, reflectors, and pedal platforms. Polymers are also crucial for some carbon fiber frames, disc wheels, and rims.

A polymer is a very large molecule (or "macromolecule") built up by chemical combination of many small repeating units. The small molecules from which they are made are called monomers. Each monomer typically contains a "backbone" of up to around a dozen atoms in a line, with various side groups attached; the most common "backbone" atoms are carbon and oxygen. When thousands of monomers are linked together, the chain is referred to as a polymer. There are many kinds of polymers, because the chains can be built from different monomers, in different ways, and with different sequences, connections, and branches.

Polymer parts

▶ ***Polymer parts*** *On conventional bikes, parts that aren't metal are almost certainly made of polymers. Two of the rider's contact points—the saddle and handlebars—have polymer covers for comfort. The tires and tubes are made of polymers because of their elasticity. Other parts are made of polymers because they are cheap, impermeable, or lightweight. High-end machines with carbon fiber frames are more polymer than metal, as are carbon fiber disc wheels.*

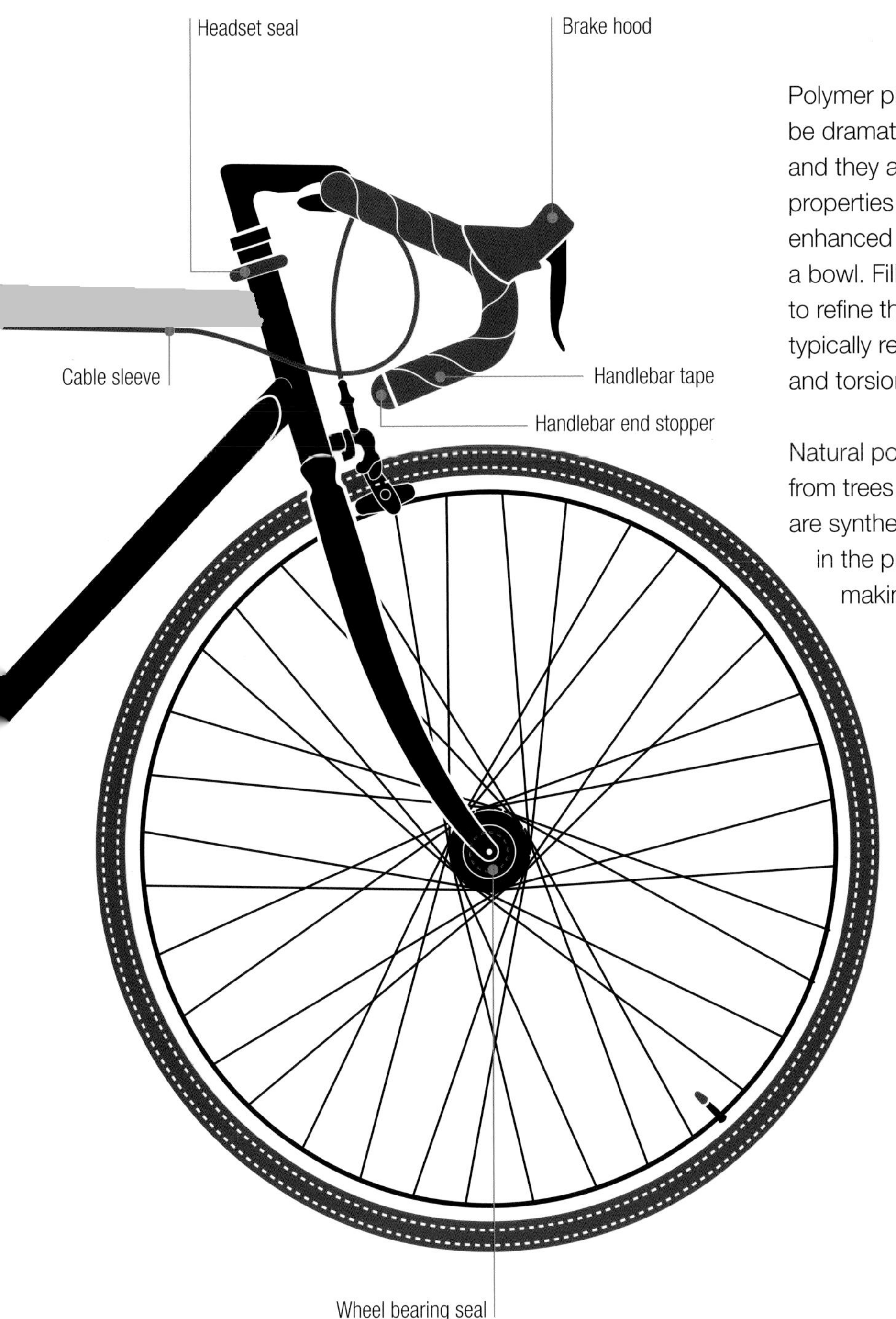

Polymer properties such as strength and melting point can be dramatically different from the properties of their monomers, and they also depend on the length of the chains. Certain key properties such as elasticity and resistance to breakage are enhanced when these chains get entangled, like spaghetti in a bowl. Fillers, reinforcements, and additives are introduced to refine these properties, but the secret of polymer behavior—typically recovering undamaged from compressive, tensile, and torsional loads—is the length of their chains.

Natural polymers have shortcomings. For example, latex rubber from trees is too sticky and rots quickly, so all polymers on bikes are synthesized from petroleum. Synthetic rubber is heated in the presence of sulfur to cross-link the polymer chains, making it more resilient and durable—yet still elastic.

Polymer properties

Polybutadiene (C_4H_6)

▲ ***Polymer shapes*** *Polymers can be linear or branched and their shapes affect their properties. Chemical diagrams such as this one communicate how they are arranged; the n indicates that this basic arrangement repeats.*

How are composites constructed?

What's so special about carbon bikes?

Composites—two dissimilar solid materials that are bound together yet remain physically distinct—are increasingly used in bike frames, forks, stays, stems, bars, rims, disc wheels, and small components. One material is embedded in the other to combine the strength and low density of the former with the toughness of the latter and get something stronger, lighter, and stiffer than any single material. The most common pairing is a polymer matrix reinforced with carbon fibers. The fibers provide tensile strength (the fibers are typically more than a hundred times stronger in tension than the matrix alone) and the matrix is strong in compression.

The carbon fibers actually start out as a solid polymer, called polyacrylonitrile, which is melted so that strands can be drawn from it and heated to about 2,200°F (1,200°C). This drives off hydrogen and nitrogen atoms and leaves carbon filaments with diameters of around 200 to 300 millionths of an inch (5 to 8 micrometers, or millionths of a meter), a fraction of the diameter of a human hair. If heated higher, up to around 4,500°F (2,500°C), they become stiffer, "high modulus" filaments. The filaments are wound into threads that can be woven into fabric or spun around a mold. Fabric layers are then fashioned into the shape of the bike part. The matrix is usually a polymer known as an epoxy, a plastic that is melted to coat the fibers and then solidifies irreversibly. The quantity and location of the carbon fibers and epoxy matrix are adjusted to maximize strength and minimize weight. The length and alignment of the fibers can also be adjusted, in advance, to fine-tune the strength, weight, stiffness, and other properties of the finished product.

The number of possible combinations of fiber, matrix, alignment, processing, and shape means "carbon" frames can be phenomenally diverse in their appearance, properties, and performance.

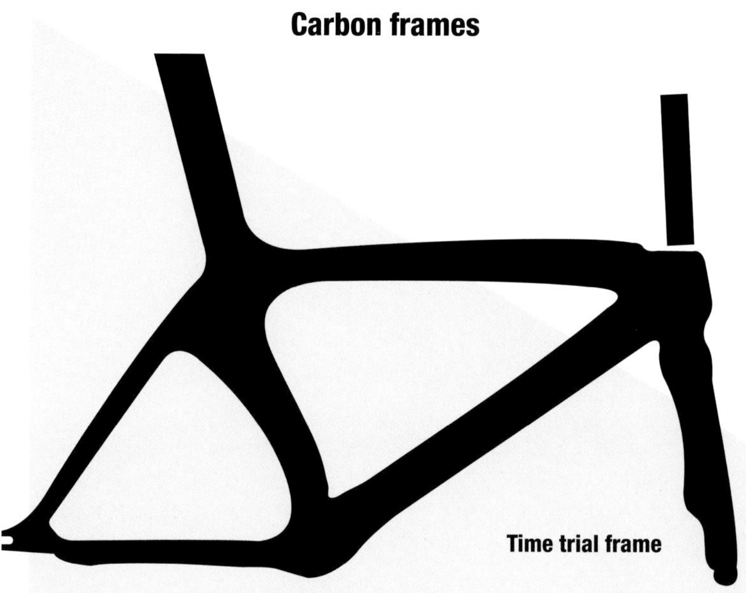

▲▼▶ ***Carbon rules*** *Once a black art, carbon fiber frames and components are now mass-produced, and their prices can rival those made of metal. Young's modulus (stiffness) ranges from 20 to 90 Mpsi (135–600 GPa). Ultimate tensile strengths can be between 300 and 500 kpsi (2,200–3,500 MPa). Compare this with 4130 chromoly steel, which has a modulus of 28 Mpsi (190 GPa) and ultimate tensile strength of 109 kpsi (750 MPa). A basic plain-weave carbon fiber sheet has a stiffness-to-weight ratio 14 percent greater than chromoly steel and 18 percent greater than aluminum.*

29er mountain bike frame

Carbon fiber structure

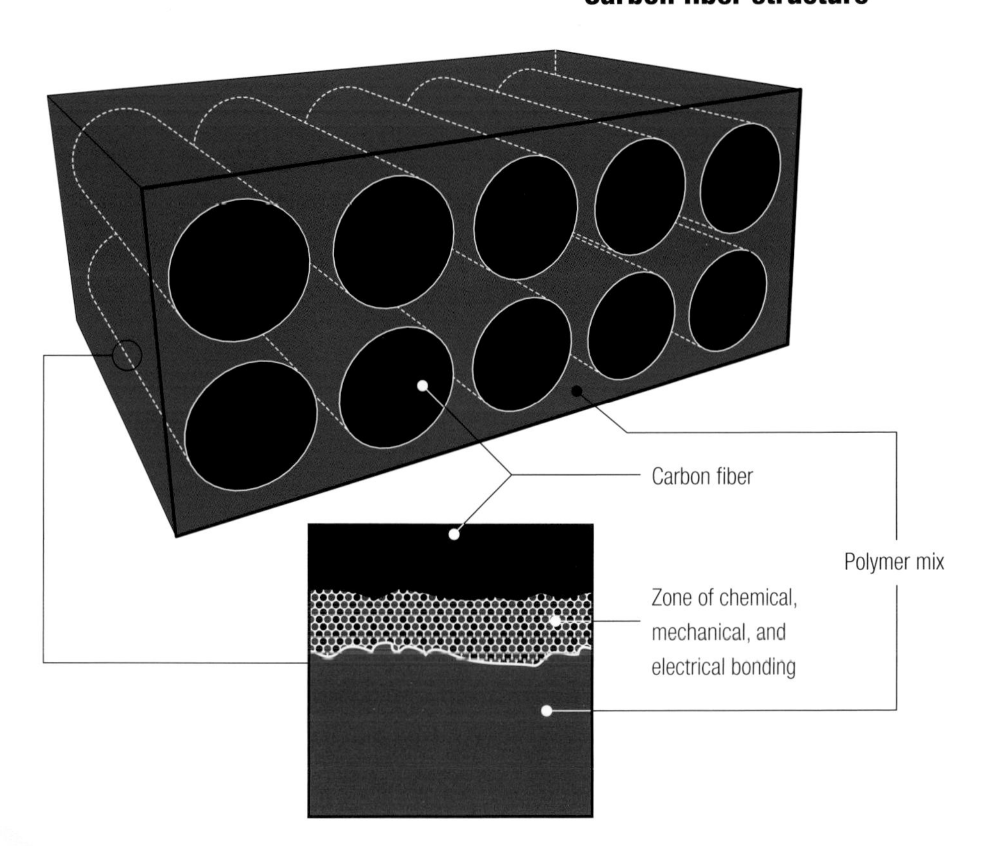

Twill weave

Plain weave

▲ ***Carbon fiber magic*** *The properties of carbon fiber-reinforced polymer varies according to weave pattern, fiber density, and matrix polymer. Carbon fibers can be aligned and woven to optimize the magnitude and direction of the tensile strength of a component. Compressive strength is provided by the plastic matrix, and it is important that this bonds well with the fibers. Many methods have been developed, including wet layup, vacuum bagging, resin transfer, matched tooling, insert molding, and pultrusion. Unlike metals, carbon fiber reinforced plastic doesn't yield under excessive loads, so it bends and then snaps without the intermediate stage of permanent deformation. A broken piece of carbon fiber is not easily repaired, and a repaired component cannot be used with confidence if it is exposed to large dynamic loads, so it is usually discarded. Some bike manufacturers now recycle discarded fragments into smaller items.*

Need to know

Carbon fiber-reinforced plastic contrasts with the limitations of the very first composite used to build frames—wood. This predecessor is a combination of cellulose fibers in a matrix of lignin but its formation and structure are determined by nature, not by bike makers.

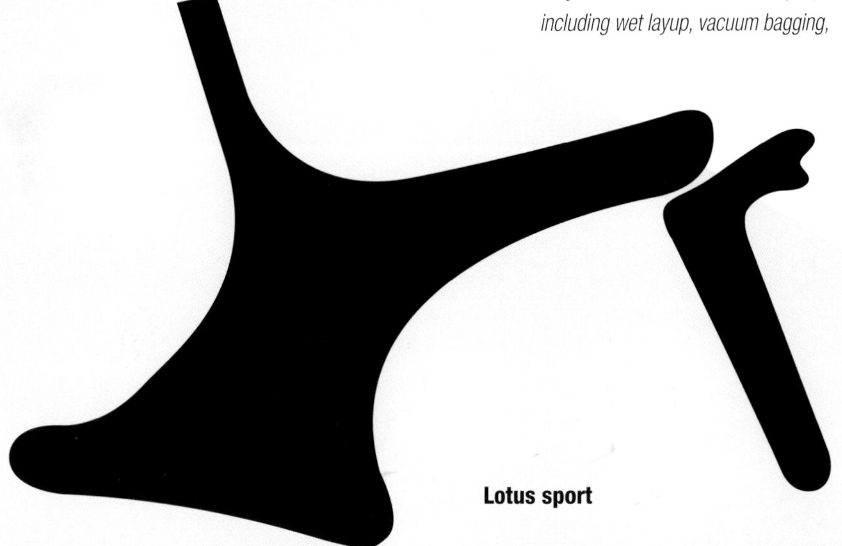

Lotus sport

What role do liquids play in cycling?

Can I just go with the flow?

Almost every material on a bicycle is liquid at some point during its manufacture. Metals and polymers go through liquid states as they are produced, and hydraulic fluids are pumped into some aluminum frames to expand them to fit their molds. Rubber and alloys are usually worked when they are molten. However, the only liquid found on most finished bikes is the lubricating fluid.

Under normal conditions, a liquid maintains its volume, but it can change its shape because its molecules are bound together more weakly than those in a solid, allowing it to move or flow. The rate of flow is affected by the forces between its molecules; these forces determine its viscosity, or "thickness." Lubricating oil has a viscosity that is just low enough for it to flow easily into the gaps of a chain or bearing; it remains there because of forces of adhesion between the oil and the metal parts. The oil provides a thin film that reduces friction between metal surfaces which would otherwise rub and wear each other. It also provides a barrier between the metal and oxygen in the air, which would otherwise react in the presence of water to form a complex mixture of iron oxides (including FeO, Fe_2O_3, and Fe_3O_4) collectively referred to as rust. Another important function of oil was revealed in tests in a clean lab. Brand-new unlubricated chains were shown to be as mechanically efficient as lubricated ones, suggesting that, on the road, a lubricant's main function is to fill microscopic gaps in the metal surfaces and stop contaminating particles getting in.[2]

A growing number of bikes also use hydraulic fluid in sealed hoses to transfer pressure from the brake levers to the pads or from the gear levers to the derailleurs. The fluid is usually a mineral oil of a viscosity chosen to transfer pressure consistently and repeatedly. The fluid must have a high boiling point for riding long descents—the constant pressure when braking continuously can raise the fluid temperature significantly.

Hydraulic fluids

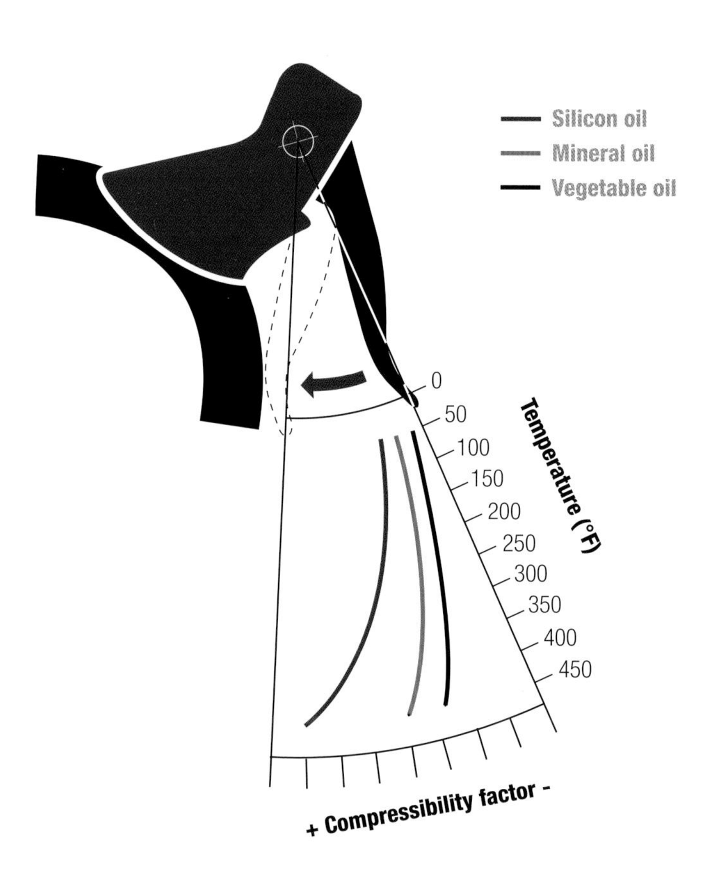

▲ ***Hot and bothered*** *Fluid compression varies with temperature and the type of fluid used. The warmer it gets, the more the hydraulic fluid can be compressed. This reduces its response under load so brakes can "fade" at high temperatures.*[3]

Need to know

At 68°F (20°C), water, a "thin" liquid, has a viscosity of about 0.000021 lb·s/ft^2 (0.001 pascal second, Pa·s or N·s/m^2) while that of "thicker" light machine oil is around 0.002 lb·s/ft^2 (0.1 Pa·s).

Power remaining after loss through friction

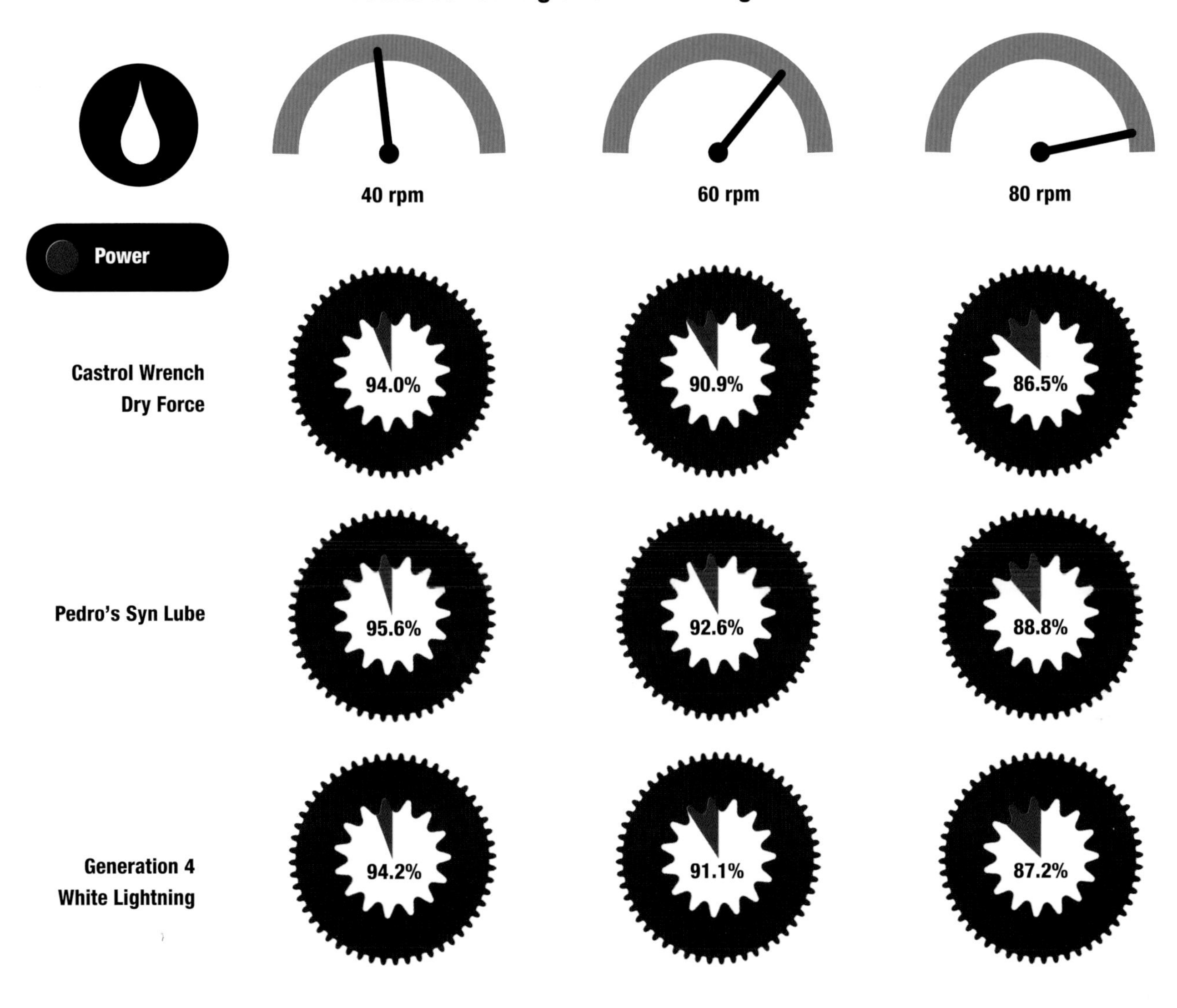

Metals in close-up

▲ ***One lube as good as another?*** *In the lab and with a brand-new chain, it's been shown that lubricants are not significantly different from one another when it comes to reducing the friction of a chain. When 74 ft·lb/s (100 W) of power was applied at the front chainring, the loss through friction at the rear cluster was very similar whatever the lubricant (the percentages indicate the power remaining after the loss through friction). However, real life is dirtier and harsher than in a laboratory and, as a lubricant ages and becomes contaminated, it can affect the longevity of the moving components.*[4]

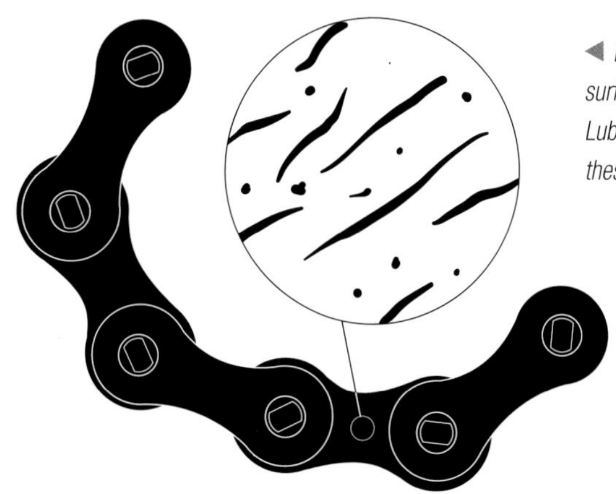

◀ ***Roughed up*** *A smooth-looking metal surface has microscopic ridges and pits. Lubricant serves to reduce the friction of these elements.*

science in action defeating the atmosphere

The biggest factor retarding fast cycling is air. The denser the air, the more the rider is slowed. Good aerodynamics reduces this effect, but there is little a rider can do to change this because air density is governed by the most basic natural systems so large and complex they thwart scientists—temperature and humidity.

With its combination of nitrogen, oxygen, and other trace gases, dry air at sea-level pressure (about 14.7 psi) and at 70°F (21°F) has a density of about 0.075 lb/ft^3 (1.202 kg/m^3). This is an internationally agreed value, which, unfortunately, is rarely respected by the world's chaotic weather systems and their variations in temperature, humidity, and pressure. At higher temperatures, the air molecules get more energetic and move around more, spreading themselves about and, thus, reducing their density (and, therefore, their retarding ability). If water vapor is added, these extra molecules are lighter than those of the components of air so, again, overall density is reduced. And if a low-pressure weather pattern develops, the mass of air pressing down is reduced and the air density again drops. All of these make riding easier.

Inside a building, steps can be taken to influence air density. The London 2012 Olympic velodrome was engineered to raise track temperatures to 82°F (28°C) and make the air thinner, boosting the likelihood of world record times. Outdoors, however, the only strategy is to go up and race at higher elevations, where the air is less dense. That's one reason Francesco Moser chose Mexico City in 1984 for his world hour record bids, as did Jennie Longo in 1989 and 1996. At 7,342 ft (2,238 m) above sea level, the density of dry air at 60°F falls below 0.062 lb/ft^3 (1 kg/m^3). This lower density gave Moser an extra 1.04 mph and Longo an extra 0.93 mph compared to recent rides they'd made at sea level.[5] The downside of going much higher, however, is that there is less oxygen available to breathe.

▶ ***On track*** *Aerodynamic wheels, helmets, and skin suits are crucial equipment choices for track riders, but the choice of velodrome can play a significant part in helping them to slice through the air. Outdoor tracks at high altitude offer thinner air, although they do remain exposed to variations in wind direction, temperature, and humidity. Fully enclosed tracks can provide conditions that are not only more predictable but can be controlled in such a way as to make the air thinner.*

sky

How do gases affect riding?

Why does my pump get hot?

Gas both impedes and facilitates cycling. Felt as a headwind, air is a gas that gets in the way, but without it cyclists would suffocate. Apart from all of the bike's solids, and their own body, the average rider is also transporting about 0.1 ft^3 (3 liters) of air. Roughly half of this is trapped inside the hollow frame tubes, forks, stays, and handlebars, with the other half inside the tires.

A gas has mass, although, luckily, very little compared to most solids and liquids, but it does count. A material becomes a gas when the kinetic energy of its molecules (which depends on temperature) is sufficient to overcome the attractive forces between them.

There is a specific relationship between pressure, volume, and temperature for any gas, and this is demonstrated every time you inflate a tire. Molecules are in constant motion; in a solid, they may just jiggle, but in a gas they hurtle about, colliding and transferring kinetic energy (energy of motion) to one another and to the walls of their container. Pushing the piston of the pump reduces the space in the barrel and, therefore, raises the density of molecules, so they suffer more frequent collisions. The classical theory of gases associates this frequency with pressure, so pushing gently on the piston makes the pressure in the pump (and the tire) go up.

What's more, when you push the piston in the pump, you give some of your own kinetic energy—through the moving piston—to the gas molecules in the pump. Using a simple model of a gas as tiny elastic spheres in random motion, one can show that its temperature is proportional to the average kinetic energy of those molecules: the faster they're moving, the warmer the gas. Some of this energy is imparted to the walls of the pump, so the pump warms up, too. The harder and faster you pump, the more energy you provide and the more noticeable the rise in temperature.

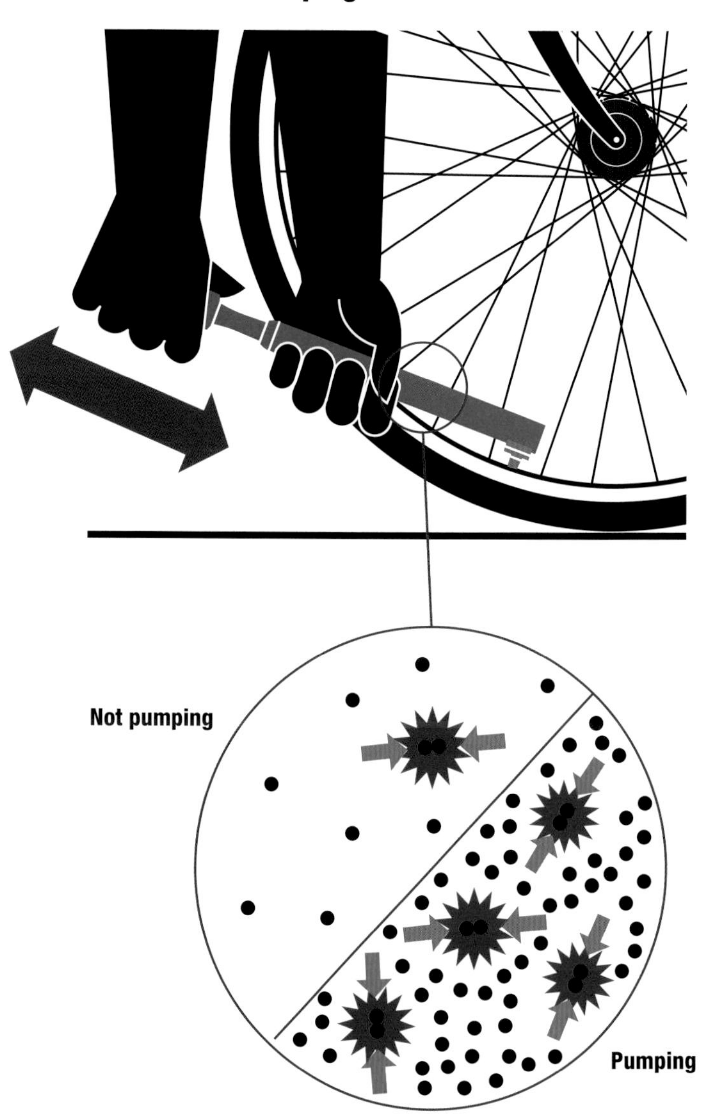

▲ ***Pressure up, temperature up*** *Put gas suddenly under pressure with a hand pump and it heats up as molecules acquire more kinetic energy. That's what happens when you pump up your tires.*

Does helium help?

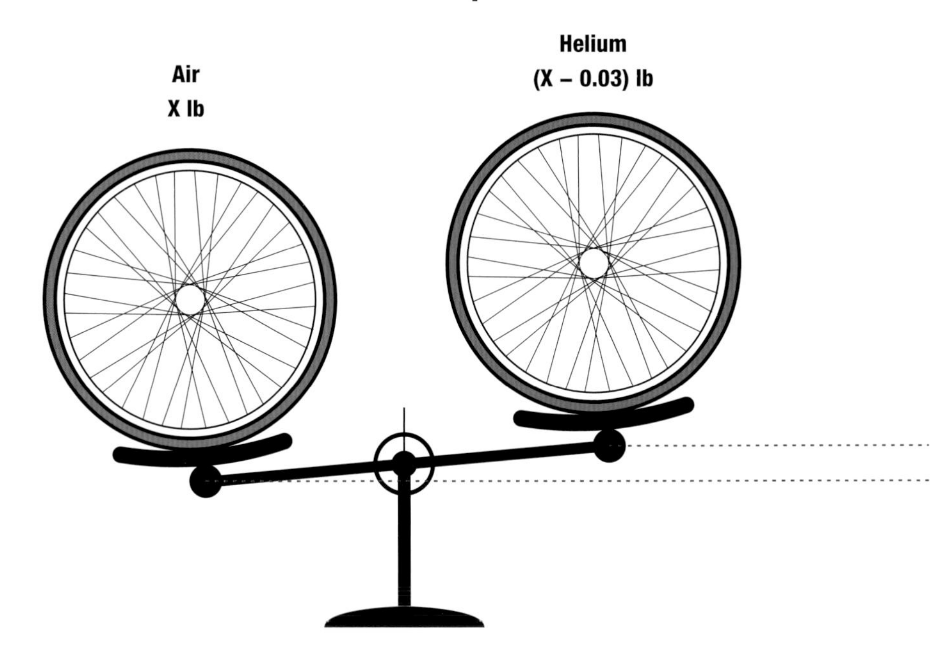

◀ ***Up, up, and away*** *As he was preparing for a world hour record bid in Mexico in 1972, Eddy Merckx, the great racing cyclist, decided he wanted helium in his tires because it would cut weight by about 0.03 lb (14 g). But according to his bike builder, Ernesto Colnago, he couldn't get any.*[6] *Nevertheless, Merckx set a new record and it stood for 12 years.*

The difference in mass is less than that of the air contained in the frame

▼ ***Unbearable lightness*** *Frame weights vary over time, due to advances in technologies and regulatory restrictions. They do not correlate directly with record times.*[7]

Men's Hour Records

	Bike weight (lb)	Distance (miles)	Rider	Location	Date
	15.4	30.232	Ole Ritter (DEN)	Mexico City	Oct 10, 1968
Lightest	12.1	30.716	Eddy Merckx (BEL)	Mexico City	Oct 25, 1972
	17.3	31.571	Francesco Moser (ITA)	Mexico City	Jan 19, 1984
	16.5	31.784	Francesco Moser (ITA)	Mexico City	Jan 23, 1984
	18.7	32.060	Graeme Obree (GB)	Hamar	Jul 17, 1993
	15.7	32.479	Chris Boardman (GB)	Bordeaux	Jul 23, 1993
	18.7	32.758	Graeme Obree (GB)	Bordeaux	Apr 27, 1994
	15.9	32.958	Miguel Indurain (ESP)	Bordeaux	Sep 2, 1994
	18.3	33.450	Tony Rominger (CH)	Bordeaux	Oct 22, 1994
	18.3	34.356	Tony Rominger (CH)	Bordeaux	Nov 5, 1994
Fastest	15.8	35.030	Chris Boardman (GB)	Manchester	Sep 6,1996

Why might plasma be the future of bike materials?

Will the stuff of stars boost bike speed?

Plasma is commonly known as the fourth state of matter. It's a little like a gas in that the particles move around relatively freely, but it differs because thermal or other energy is intense enough to strip electrons off some of its atoms, leaving a swarm of ionized (charged) and nonionized particles. In fact, this mix makes up some 99 percent of the matter in the observable universe—it's what stars are made of—and could, one day, help you ride faster.

Scientists have successfully used plasma to influence the airflow over aircraft wings and control the flight path without using wing flaps. The plasma is created using electrodes that pass electrical energy through the air, similar to the way a charge passes through the gas of a neon tube. By precisely positioning the electrodes and controlling the energy they discharge, plasmas of the desired volume, location, and duration are formed—and modify the airflow over the plane. The technology has been trialed with small unmanned drones and the power requirements are not large, so it is plausible that the technique could be applied to cyclists and their bikes.[8]

A mesh of plasma generators could be embedded in a composite frame, fork, handlebars, and disc wheels, even in the cyclist's clothing and helmet. These could help create dynamic virtual streamlining, which could change shape and location as the bike changes orientation in relation to the air flow, potentially reducing turbulence and optimizing the cyclist's aerodynamics in any gusts and breezes. Although still hypothetical, the suggestion is that it could reduce the power needed to overcome aerodynamic drag by some 40 ft·lb/s (50 W). Aerospace brought advances in alloy and composite technology to cycling; plasma could follow in their slipstream.

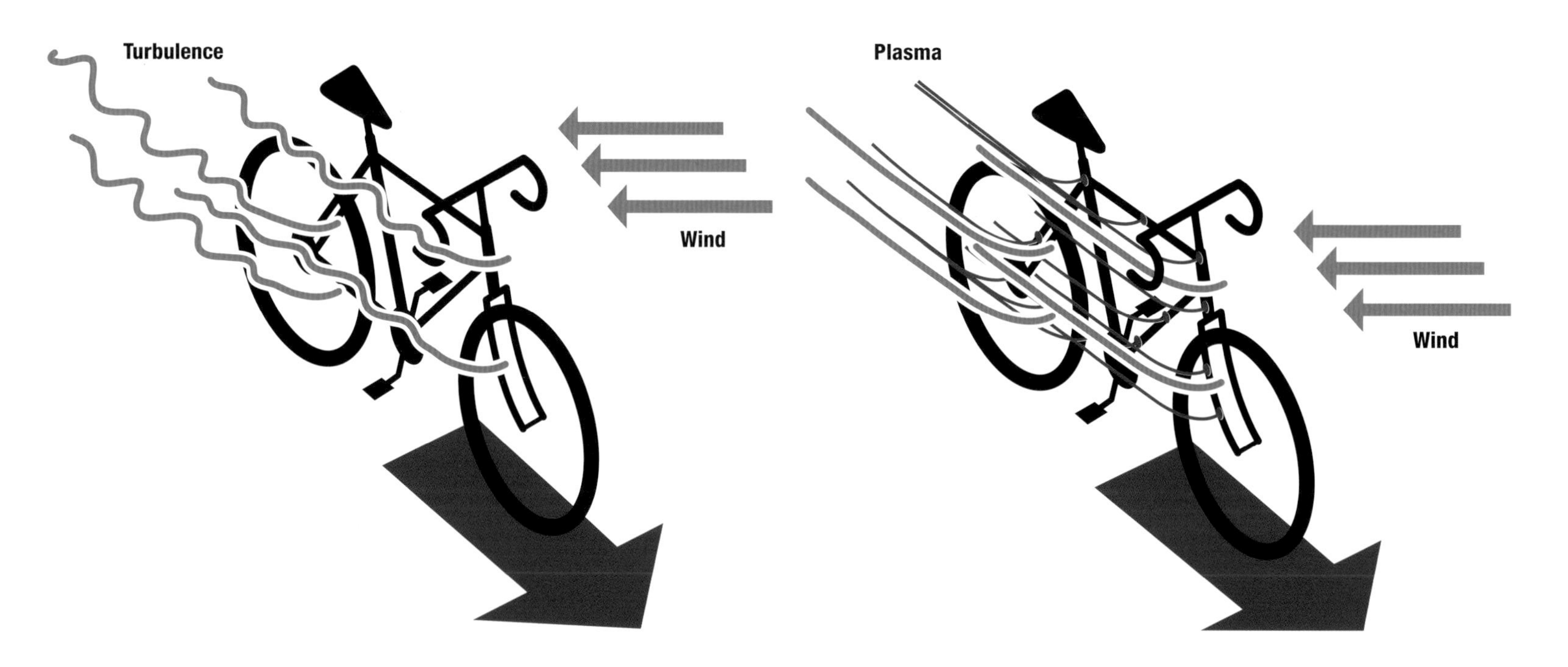

Plasma creation

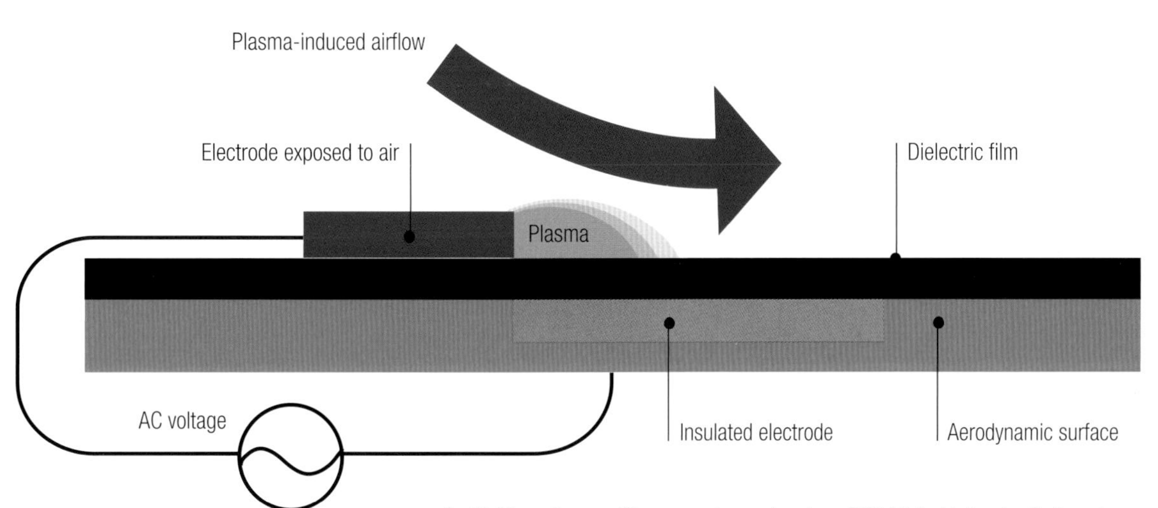

▲ ***Making plasma*** *Plasma can be made using a DBD (dielectric barrier discharge) plasma actuator. When an AC voltage of sufficiently large amplitude is applied across two electrodes, one exposed to the surrounding air and the other encapsulated in a dielectric material, the air ionizes in the region of the strongest electric field. This typically occurs at the edge of the electrode that is exposed to the air. In the presence of the electric field gradient, the ionized air (plasma) exerts a force on the ambient air, which has the effect of altering the virtual aerodynamic shape of the surface around the actuator.*

Plasma effects

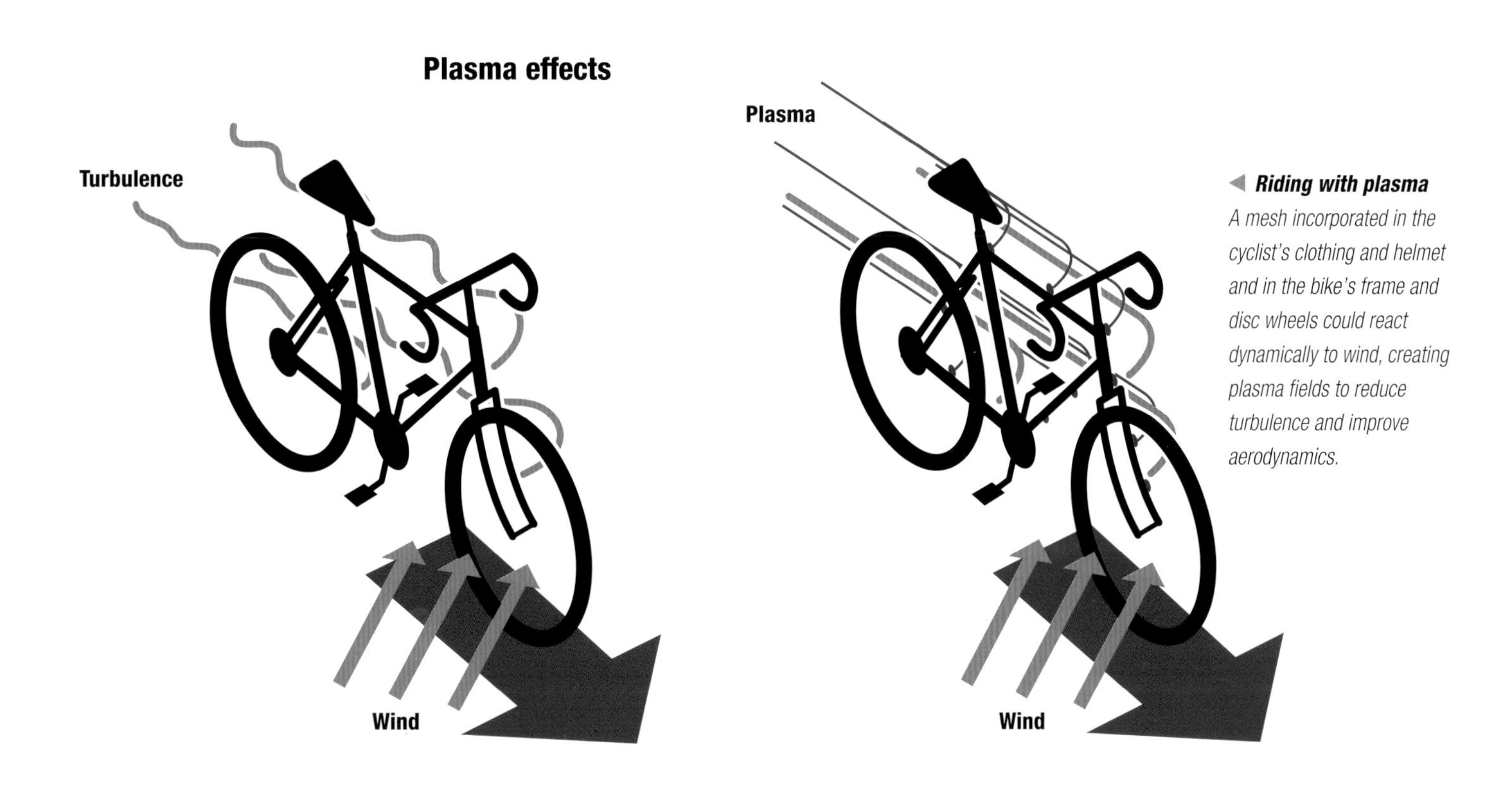

◀ ***Riding with plasma*** *A mesh incorporated in the cyclist's clothing and helmet and in the bike's frame and disc wheels could react dynamically to wind, creating plasma fields to reduce turbulence and improve aerodynamics.*

Legs evolved to carry our upright bodies; but by shifting that burden to a saddle and handlebars, their strength can be used to turn the pedals with a force and duration unmatched by any other muscles. Over time, the power train components of a bicycle have evolved to harness as much of that pedaling energy as possible and convert it into forward motion. Piece by piece, engineering and design improvements have been made so we can ride in a controlled way without wasting our effort. We push down and make progress because there has been incremental progress in each item. From the pedals right through to the tires, excess weight has been eliminated. The tension-spoked wheel supports large, dynamic loads while weighing very little itself. Friction in the drive train and wheel bearings has been reduced to such a low level that for most riders it is negligible. Gears accommodate our preference to work at a constant rate whatever the gradient. Everybody can see how it works, but to know why it works so well the devil is in the detail of this chapter.

chapter four

power on

How does a bike turn effort into speed?

Why do I need all those components?

The mechanical composition of a bicycle is visible and compelling. There is elegance in the way the machine's function demands the components be stark and hardworking. Together, they transfer kinetic energy from the rider into forward movement; they do this by converting a circular pedaling motion into linear travel. This is achieved by using many specialized parts with different functions and interactions.

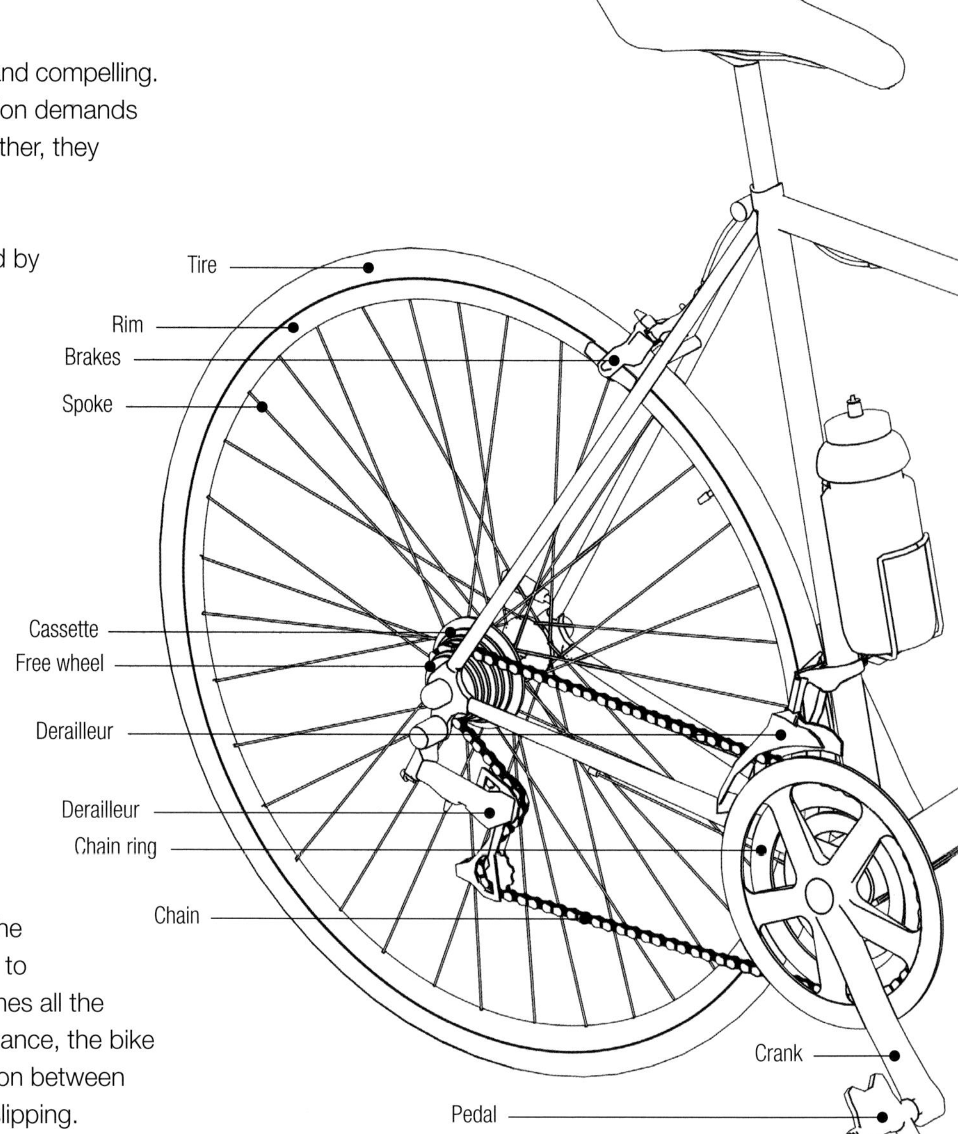

The pedals are the contact points between the rider and the machine's primary rotary levers, the cranks. Two largely hidden parts are important here: pedal spindles allow the pedals to revolve so they remain in the same position relative to the feet—mainly horizontal—and the bearings within the bottom bracket allow the cranks to rotate freely. Chainrings bolted to the cranks convert their rotary motion to linear motion of the chain. At the rear wheel, the chain engages with the teeth of the cassette sprockets, converting linear motion back to rotary. The sprockets are attached to the hub and rotate it on bearings around the fixed axle. The hub's rotation is transferred via the wheel spokes to the rim and the tires. If the cyclist's effort overcomes all the energy losses in the power train and the air resistance, the bike will move forward, assuming there's enough friction between the rubber and the road to prevent the tire from slipping.

This process can be modified by two key operations. The first is to move a gear shifter or derailleur, altering the relationship between the rotational speeds of the cranks and the rear wheel. The second is to squeeze a brake lever. This action is transmitted via a cable to apply brake pads that convert kinetic energy to heat via friction, and slow the wheel—and bike.

The major mechanical components

▼ ***Working together*** *The entire power train relies on upward of a thousand components. Most are in view, apart from bearings and lengths of cables, usually covered for protection. The visibility makes troubleshooting easy, as do the obvious mechanical functions of most components. Most of the main movements are rotary, with only the levers, cables, derailleurs, brake arms, and chain being linear.*

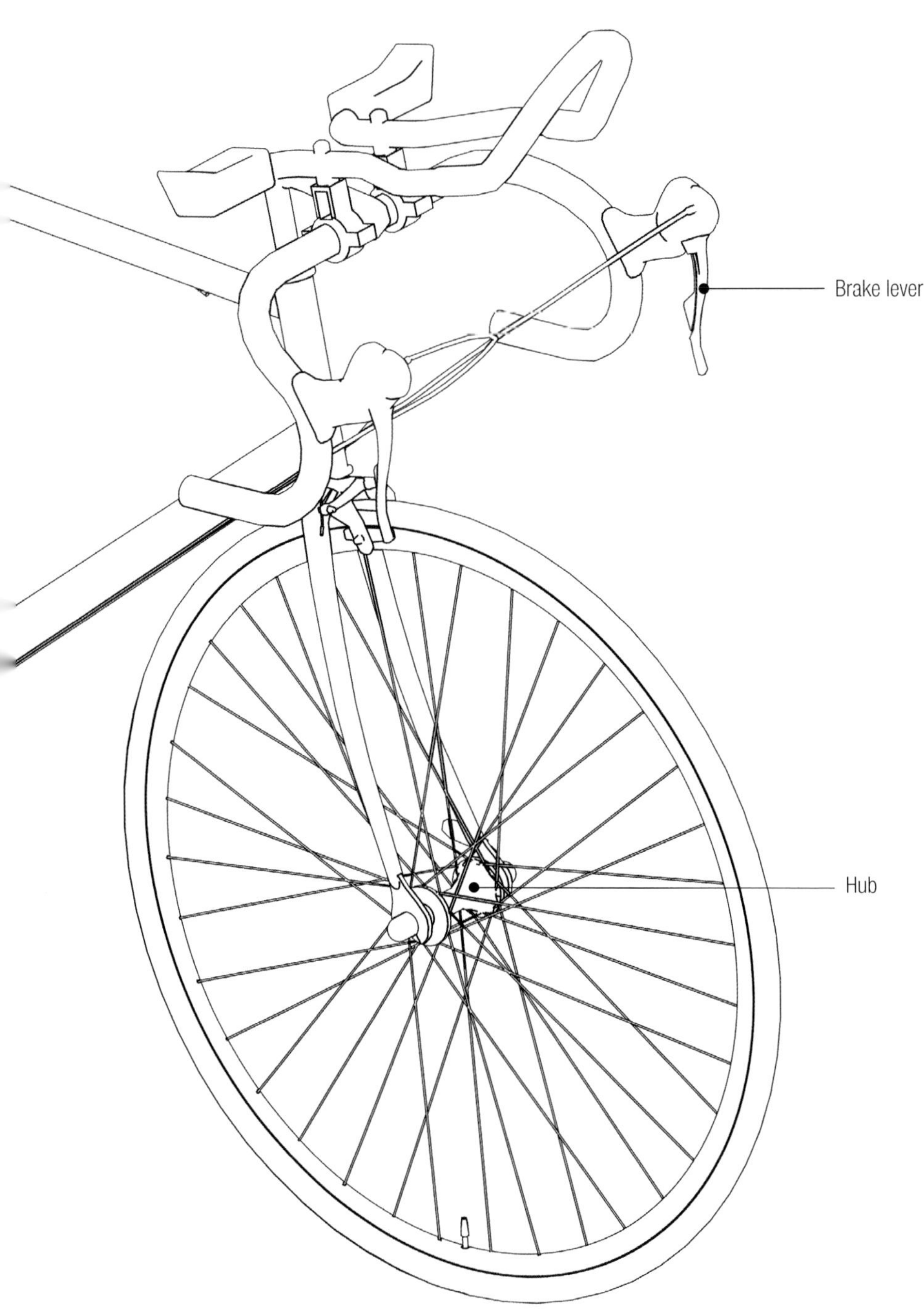

Alternative technologies

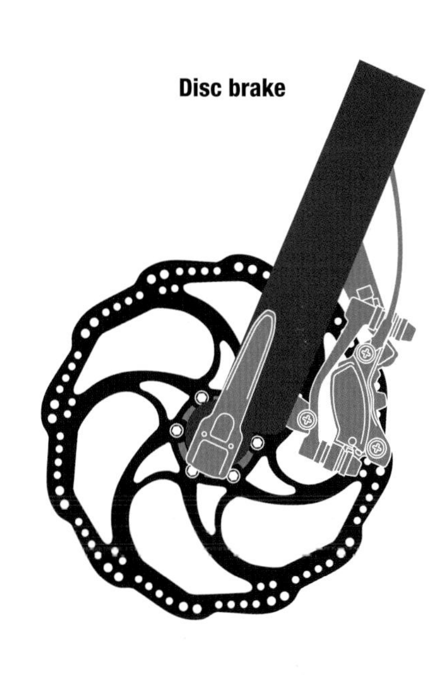

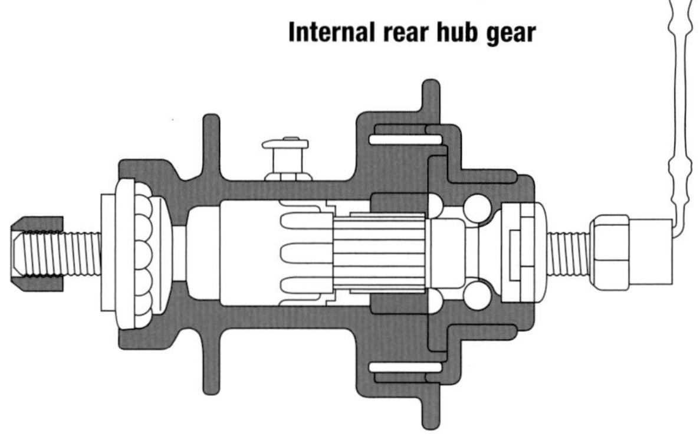

▲ ***One thing leads to another*** *Not every bike is the same, and variations in components have emerged during the 150 years of their existence. Changing one component often requires changing another. For example, nobody has yet perfected a derailleur to move a toothed belt from one cog to another, so internal rear hub gears are required. Other changes occur when better solutions are invented. Water and dirt on rims reduce the effectiveness of brake pads, so disc brakes—with a braking surface farther from the ground—are becoming increasingly common.*

What is the power interface between human and machine?

These shoes are made for cycling?

Efficient cycling begins at the interface between the foot and the pedal. If the cyclist's feet don't engage properly with the pedals, then energy is wasted or, even worse, there can be injuries. This means it's important to wear shoes that fit, have stiff soles, and that stay securely on the pedals.

The difference in stiffness can be significant. Researchers have discovered that the heel of a rider with nylon shoe soles was displaced 40 times farther than with soles made entirely of stiffer carbon fiber reinforced plastic, which eliminated bending almost completely.[1] The energy expended in bending a sole may be converted into heat, or it may be returned to the rider when the sole straightens out. Either way, it isn't turning the pedals, and another research team found that a specialist cycling shoe produced an improvement of 1.9 percent in the force transferred to the pedal compared to a running shoe.[2] However, a stiff sole might also exacerbate problems. A third team tested shoes with nylon and carbon soles to measure plantar (foot) pressures and shoe stiffness. "Large stiffness discrepancies, especially between shoes that are otherwise identical, lead one to believe that stiffness differences are responsible for the increased peak plantar pressures," the team concluded, and warned that the wrong insole contours could cause pain.[3] This is because the pressure applied by the foot is considerable: a fit cyclist generating 300 ft·lb/s (about 400 W) of power puts a massive 67 psi of pressure on their hallux (big toe) as they push down on the pedal on each stroke.[4]

To complete the link between rider and bike, a rider whose shoes are locked to the pedal doesn't tire as quickly. In particular, the muscles in the quadriceps group are less fatigued, because the rider can use the flexor muscles of one leg to pull one pedal up, alleviating peak demand on the quadriceps of the other leg pushing down.[5] Inexperienced cyclists who adopt this technique have found their mechanical effectiveness increased by 57 percent, and elite cyclists have fared even better, enjoying an 86 percent gain.[6] There is, however, a cost to giving some of the work to the leg that would otherwise be resting during the pedal upstroke. While the bicycle receives more propulsive power from the rider, it consumes disproportionately more of the rider's resources.[7]

Foot pressure distribution

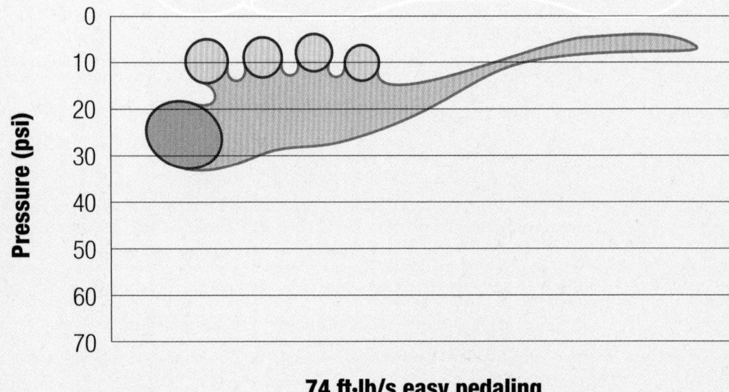

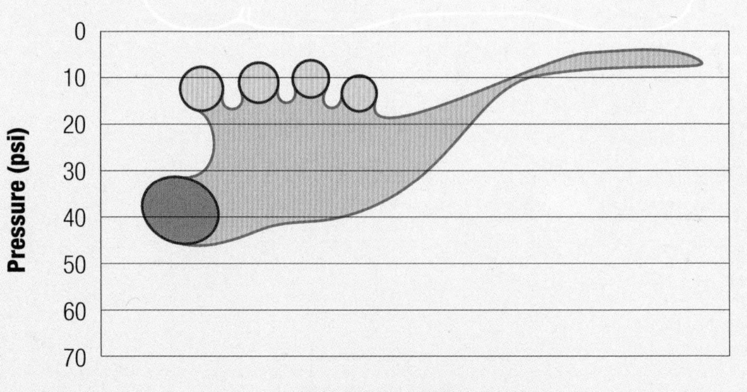

Effect of shoe material on flexibility

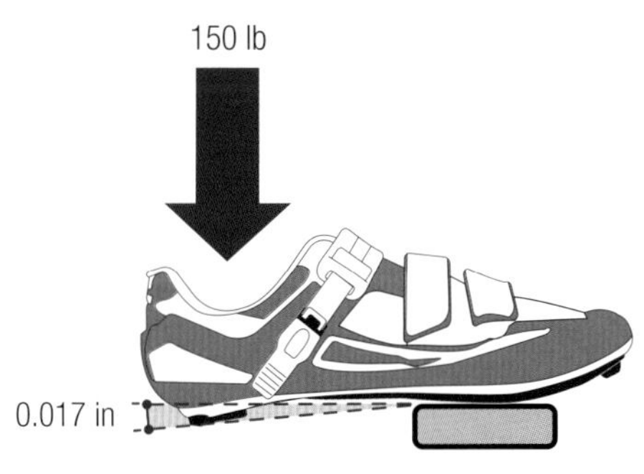

100% carbon outsole

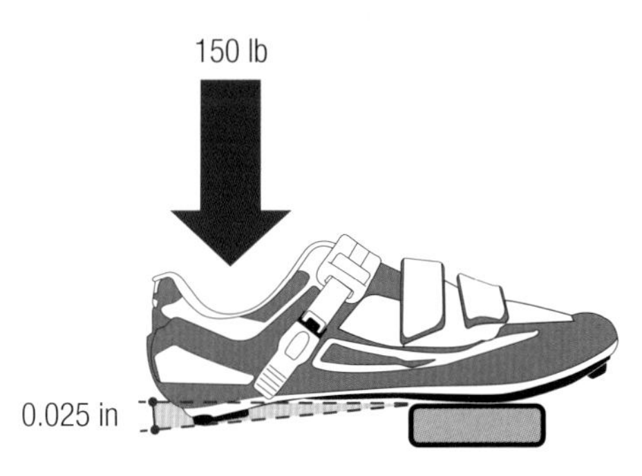

50% nylon outsole

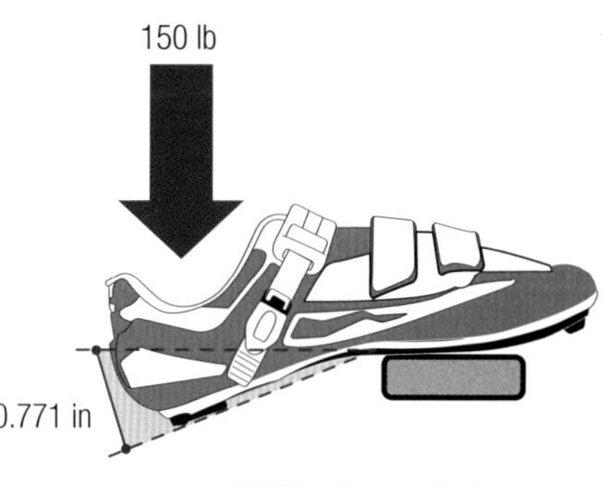

100% nylon outsole

▲ ***How flexible are your shoes?*** *Replacing just half of the nylon of a sole with carbon fiber-reinforced plastic can increase its stiffness dramatically and it is only marginally less flexible than a 100 percent carbon sole, according to calculations made using finite element analysis. Less bending increases the chance of power being transmitted from the rider to the pedals (the deflections are not to scale).*[9]

▼ ***Hard-working big toe*** *The pressure exerted by different parts of the foot during the downstroke varies greatly. The greatest load is applied through the hallux (big toe), several times greater than the load through the heel. As power increases, the proportion of the total force being applied by the hallux and its pad increases, while the share exerted by the little toe and pad decreases. The diagrams show measured pressure (in psi) at various points for four different power levels.*[8]

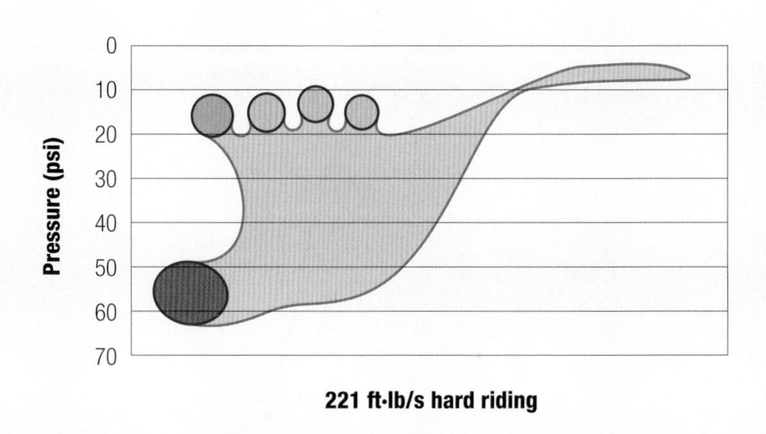

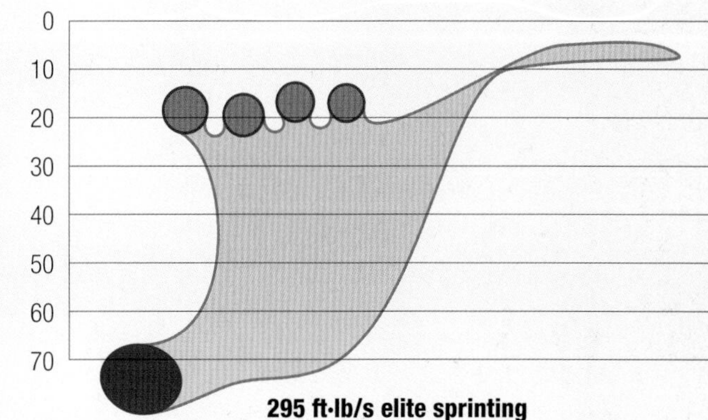

How does gearing help efficiency? → Why pedal faster going up hills?

The toothed cogs that drive the chain on a bike are usually called gears by cyclists, but engineers call this a pulley system, with the chain transferring rotational motion from the front cogs (chainrings) to the rear cogs (sprockets). Gears accommodate the natural limits of human pedaling power. They make cycling easier because the chain can be moved to rings of different sizes when the rider wants to change speed or pedal more easily. They allow the rider to make best use of their optimum pedaling rate (cadence). Leisure cyclists average 60 rpm and peak at 90 rpm, while an elite racer can achieve 175 rpm.[10]

If the chainring and the rear sprocket had the same radius, they would spin at the same rate, completing a full rotation in the same amount of time. However, the rear sprocket is generally smaller and, therefore, has fewer teeth in its full circumference. Consequently, the passage of a particular length of chain corresponds to a greater rotation of the rear sprocket than of the chainring. In other words, the rear sprocket—and, therefore, the hub and wheel to which it is attached—spins faster than the chainring.

To enable a variation in road speed without changing the pedaling rate, the rear derailleur moves the chain onto a rear sprocket of a different size. If the cyclist pedals at a fixed rate, when the chain drives a smaller sprocket (or internal hub gear) the wheel spins faster and so the bicycle moves faster. This may sound like getting something for nothing, but to achieve this the cyclist has to do more work—the smaller sprocket with its shorter radius is effectively a shorter lever, so the cyclist must push harder on the pedals. It's like trying to open a door by pushing closer to the hinges—you must push harder than if you push near the handle. To reduce the force needed to achieve the same speed, the cyclist could shift the chain to a larger rear sprocket (or change down with a hub gear), or use the front derailleur to shift the chain to a smaller chainring.

Understanding gears

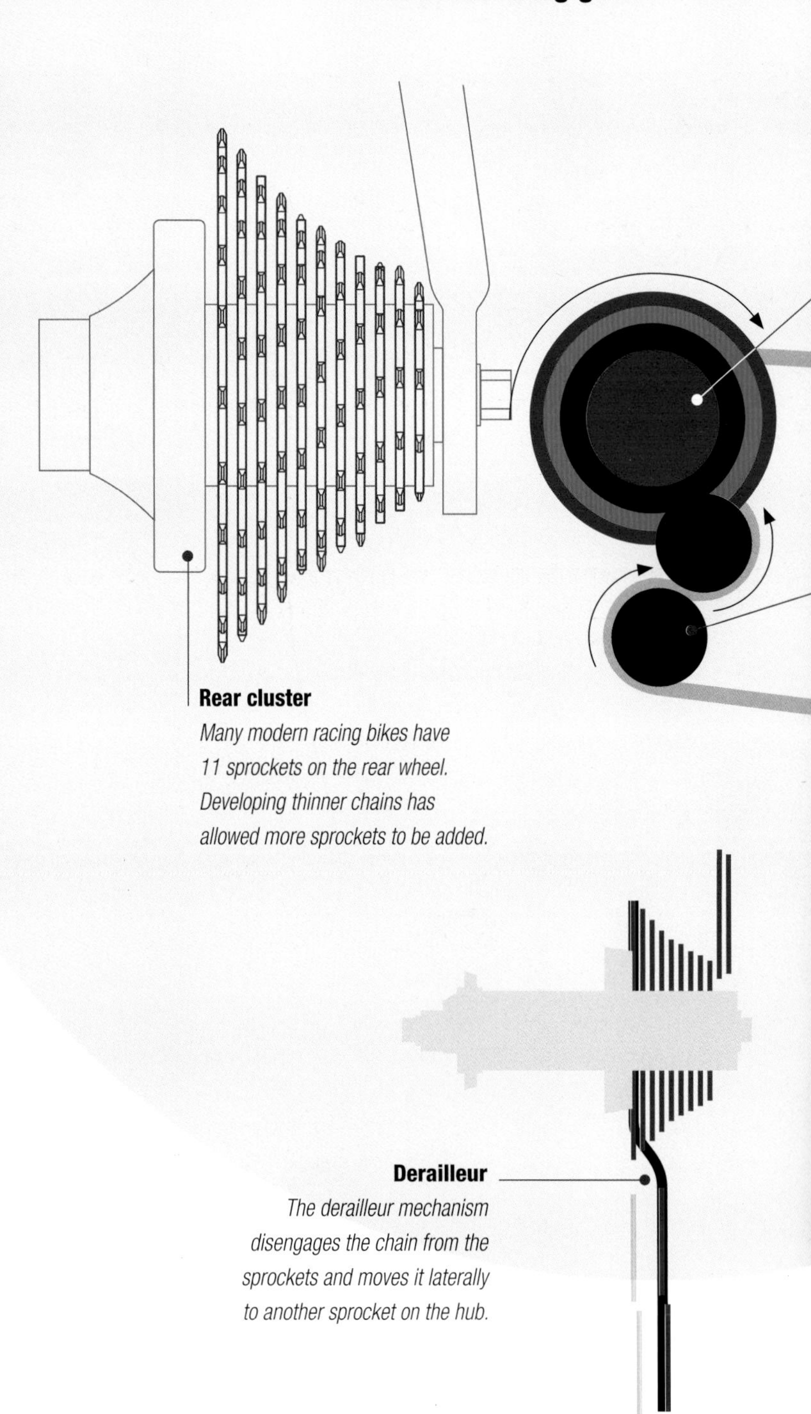

Rear cluster
Many modern racing bikes have 11 sprockets on the rear wheel. Developing thinner chains has allowed more sprockets to be added.

Derailleur
The derailleur mechanism disengages the chain from the sprockets and moves it laterally to another sprocket on the hub.

▶ ***Gear inches*** *A gear chart shows the relationship between the forward motion of a bike and one revolution of the pedals. The gear size is expressed in terms of the diameter of a notional front wheel driven directly, like the wheels of the old-fashioned Ordinary bicycles. The values in the table are derived by multiplying the wheel diameter by the number of chainring teeth and dividing by the number of sprocket teeth. The figures are for a 700C wheel.*

Sprocket teeth \ Chainring teeth	46	47	48	49	50	51	52	53
13	93.6	95.7	97.7	99.7	101.8	103.8	105.8	107.9
14	86.9	88.8	90.7	92.6	94.5	96.4	98.3	100.2
15	81.1	82.9	84.7	86.4	88.2	90.0	91.7	93.5
16	76.1	77.7	79.4	81.0	82.7	84.3	86.0	87.6

Rear hub
The rear hub can be a freewheel block with fixed sprockets or a splined cylinder that allows individual sprockets to be fitted.

Cage
The cage takes up the slack in the chain after it has moved from one sprocket to another.

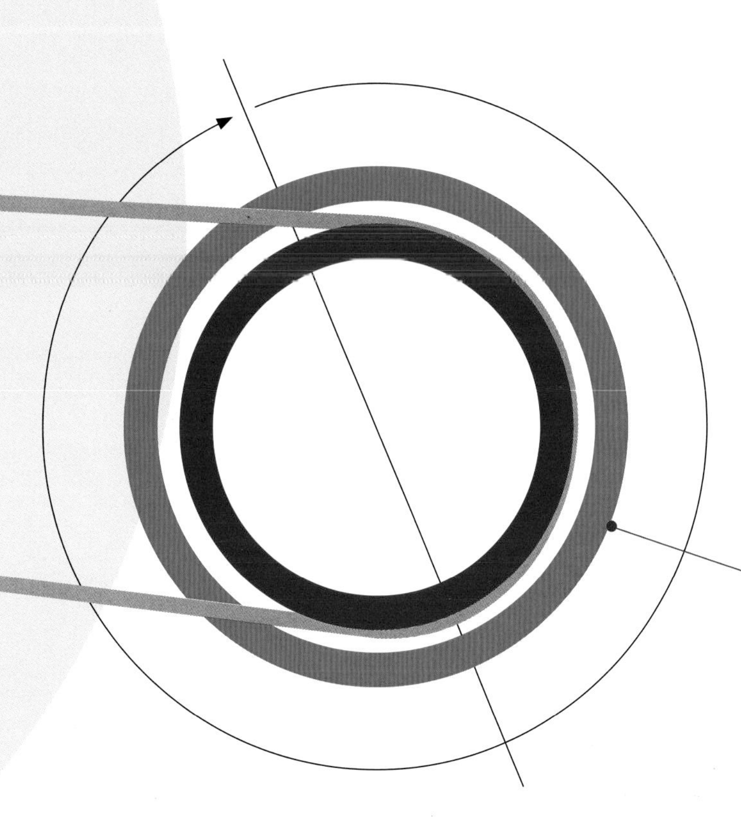

Chainrings
Modern chainsets can have one, two, or three chainrings. The number of chainrings multiplied by the number of rear sprockets gives the maximum possible number of gear ratios that are available.

▶ ***Gear rollout*** *For a given chainring and rear sprocket combination, this table shows the distance in feet a bike with 700C wheels will travel in one revolution of the cranks.*

Sprocket teeth \ Chainring teeth	46	47	48	49	50	51	52	53
13	7.4	7.6	7.7	7.9	8.1	8.2	8.4	8.6
14	6.9	7.1	7.2	7.4	7.5	7.6	7.8	8.0
15	6.4	6.6	6.7	6.9	7.0	7.1	7.3	7.4
16	6.0	6.2	6.3	6.4	6.6	6.7	6.8	7.0

equipment: gear changing

▲ ***Kangaroo high wheeler*** *(1886)*
With chain drive gearing, it became possible to reduce the diameter of the front wheel compared to ordinary high wheelers.

Gears let cyclists trade effort for speed and so make best use of the rider's range of power output. For many years in the first half of the twentieth century, gear mechanisms were banned by race organizers (riders would run different sprockets on each end of the hub and turn the wheel around to get a different gear), so it was mainly leisure cyclists who used early derailleurs and hub gears. Sense eventually prevailed and demands by professional racers drove innovation and improvements. This development continues and is fiercely competitive, with electronic shifting now taking the lead and narrower chains giving the option to contemplate rear cassettes with as many as 12 sprockets.

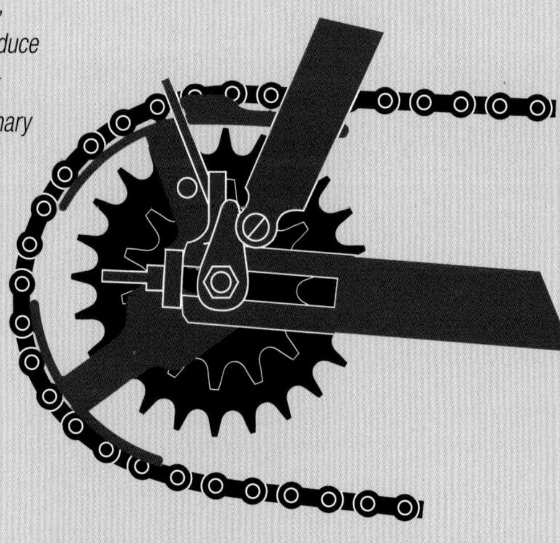

▲ ***Hodgkinson's Gradient derailleur*** *(1899)*
Sprockets of different sizes, attached to the rear axle, could slide in and out when the chain was lifted clear.

▼ ***Super Champion Osgear*** *(ca. 1936)*
A cable-operated fork below the chainstay shifted the chain between sprockets, while the tension arm eliminated any slack.

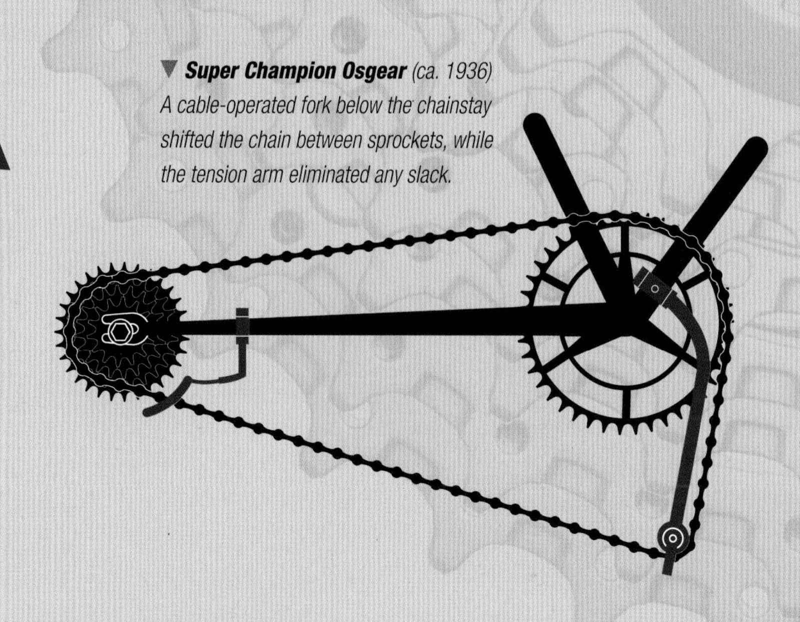

Shifting a moving chain sideways from one rotating sprocket to another is tricky. It has to be picked cleanly from the teeth of one spinning cog, pushed sideways a precise distance, and released onto the chosen sprocket without interrupting the load being applied to the pedals. The mechanism must be sufficiently robust to perform the operation repeatedly, in both directions, from large sprockets to smaller ones and back again, without failing. It must also take up any slack produced by smaller cog choices so the chain never droops onto the ground. And it has to be controlled by a remote lever that is within easy reach of the rider's hand. A derailleur does all this.

The first attempts emerged toward the end of the nineteenth century. They added complexity and cost but the benefits were huge—as shown by the leap in average race speeds when the governing body of cycle racing, the Union Cycliste Internationale, finally allowed competitors to use them in 1938. Early versions often used a separate jockey wheel to take up the chain slack, but companies such as Simplex and Campagnolo brought rapid improvements, along with the evolution of the parallelogram actuating mechanism for both rear and front ring changing. Today's derailleurs are so highly developed that they are fitted—and work reliably—on the cheapest of bikes. The most expensive now have electronically operated shifters.

The popularity of hub gears has waxed and waned since they were invented in the 1890s. Initially offering two speeds, they now have up to 14, and offer a range equivalent to a 27-speed derailleur system. What they lose in mechanical efficiency, they gain in reliability.

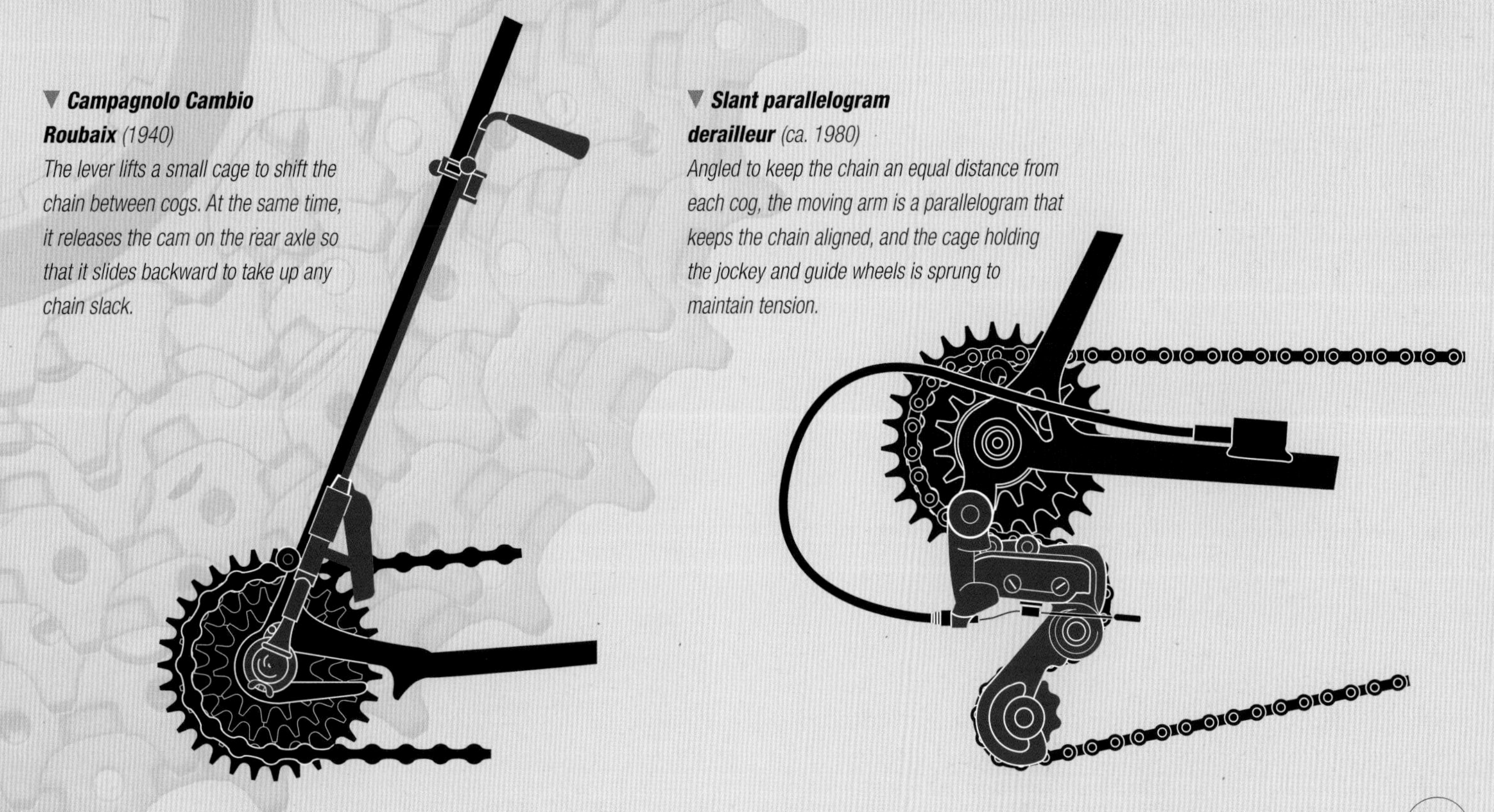

▼ ***Campagnolo Cambio Roubaix*** *(1940)*
The lever lifts a small cage to shift the chain between cogs. At the same time, it releases the cam on the rear axle so that it slides backward to take up any chain slack.

▼ ***Slant parallelogram derailleur*** *(ca. 1980)*
Angled to keep the chain an equal distance from each cog, the moving arm is a parallelogram that keeps the chain aligned, and the cage holding the jockey and guide wheels is sprung to maintain tension.

How efficient is a bike chain?

Why do we still use oily chains?

The significance of the bicycle chain is underestimated because it is oily, unglamorous, and sometimes dirty. Unfortunately, for such an important item, it's noticed only when there is a problem. Rarely does a well-maintained chain demand attention. In 1886, Hans Renold's newly invented chain with roller bushings was specified for the mold-breaking Rover Safety bicycle, and since then nobody has designed anything that transmits pedal power as efficiently.[11]

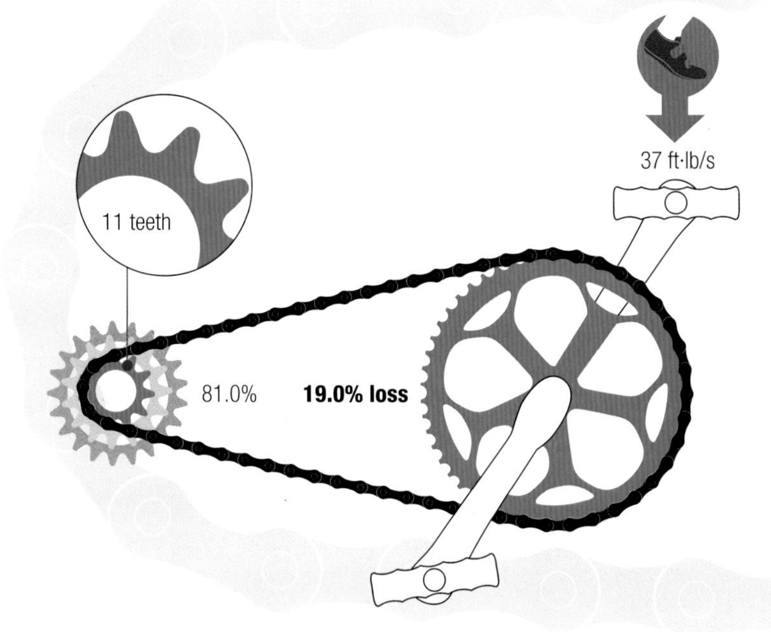

New 114-link chains (the standard length) range in weight from 9 oz to 18 oz (250 g to 500 g). They have eight parts per link, all of which must work smoothly when the pedals are turned. Some riders suspect that this continually flexing linkage converts a lot of their energy into heat through friction between moving parts. Engineers at Johns Hopkins University, in conjunction with Shimano, put brand-new chains to the test and found that power transmission from the front chainring to the rear sprocket can be as high as 98.6 percent. They found that the key to efficiency—apart from aligning the chain correctly—was to keep it under the greatest practical tension, which depends on the choice of gears. The highest efficiency measured in the study, 98.6 percent, was recorded at a chain tension of 68.6 lb (305 N), while the lowest, 80.9 percent, was seen at a slack 17.1 lb (76.2 N). A chain that engages and departs efficiently from the teeth during the high-tension part of the drive (the top run from rear sprocket to front chainring) makes best use of the rider's energy. Of the energy that's lost, infrared measurements showed that friction does indeed heat the chain.

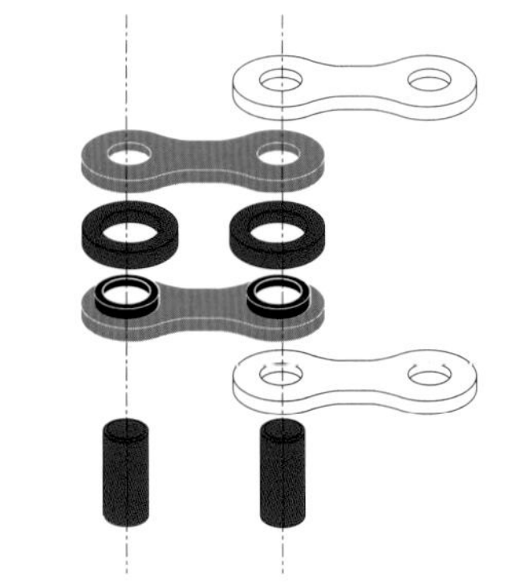

▼ ▶ ***Chain*** *A chain is a sequence of outer plates held to inner plates via a pin, firmly attached to the outer plate and loosely attached to the inner plate, and a roller that holds the two inner plates apart.*

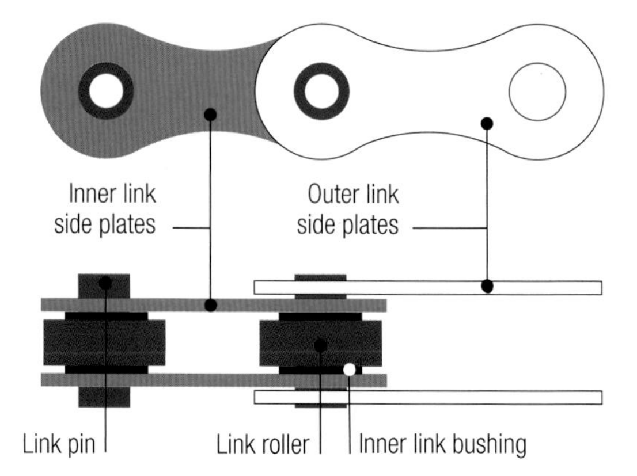

There are alternatives to the chain. The most successful are toothed belts, finding favor on commuter bikes. They use hub gears, and their tension is constant. Shaft drives on some public rental bikes are heavier and less efficient but can reduce maintenance requirements. Engineers in South Korea have recently designed a sophisticated electromechanical hydraulic transmission, and efficiency values are eagerly awaited.[12]

How much energy is lost?

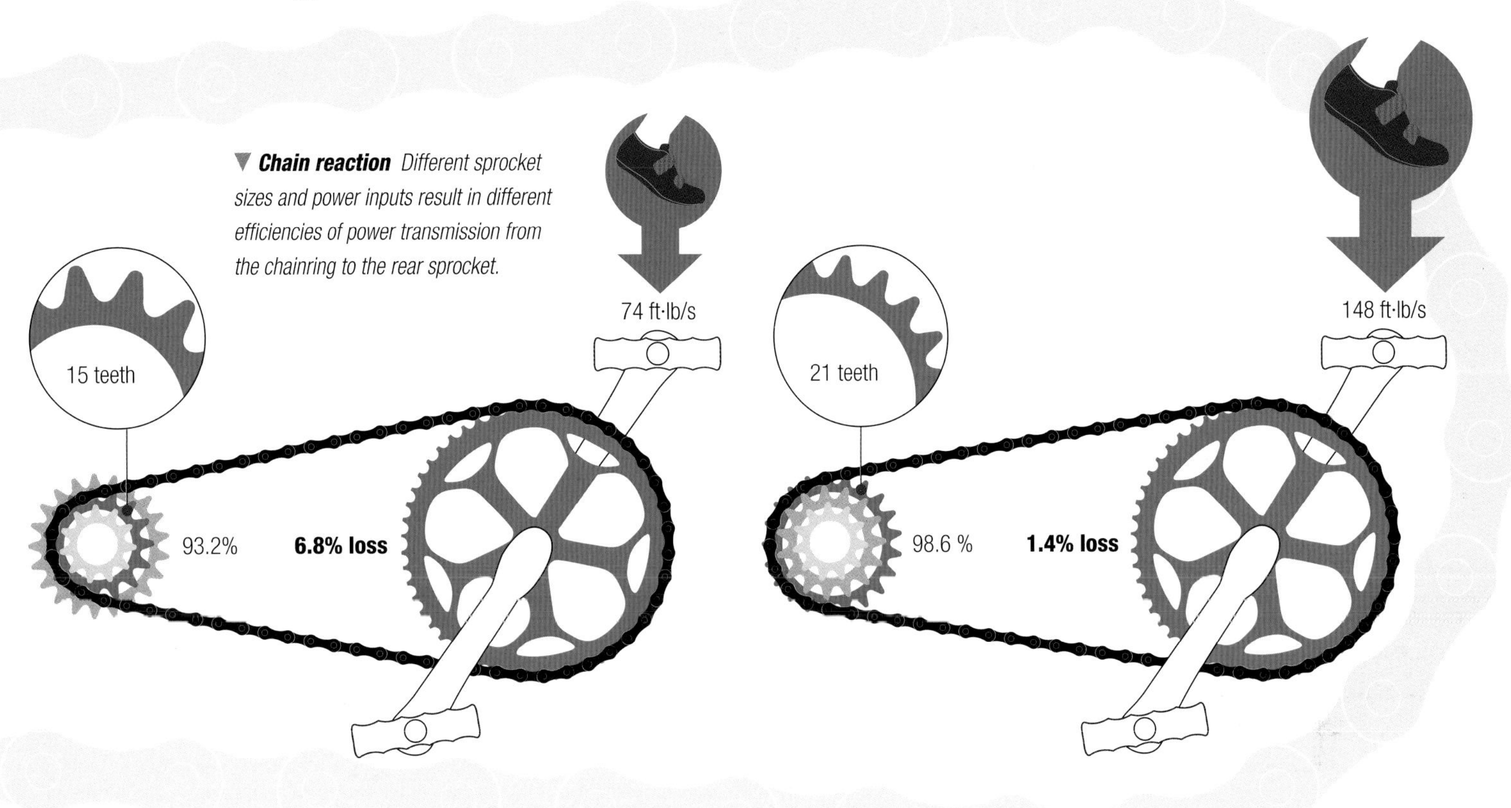

▼ ***Chain reaction*** *Different sprocket sizes and power inputs result in different efficiencies of power transmission from the chainring to the rear sprocket.*

▼ ***Shaft drive*** *This loses about 10 percent of a rider's energy because it involves two right-angle direction changes, requiring bevel gears or the like, meshing at very low speed with consequentially high tooth pressures. Gears (particularly bevel gears) exhibit high frictional losses under such conditions.*[13]

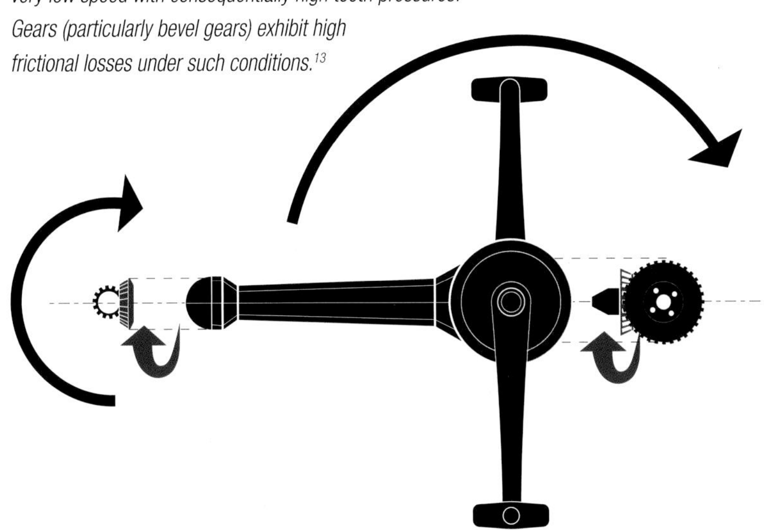

▼ ***Toothed belt*** *This promises a quieter transmission that does not need lubrication. Because it cannot work with a derailleur, tension is constant, so there is no variation in energy loss as gears inside the wheel hub are changed.*

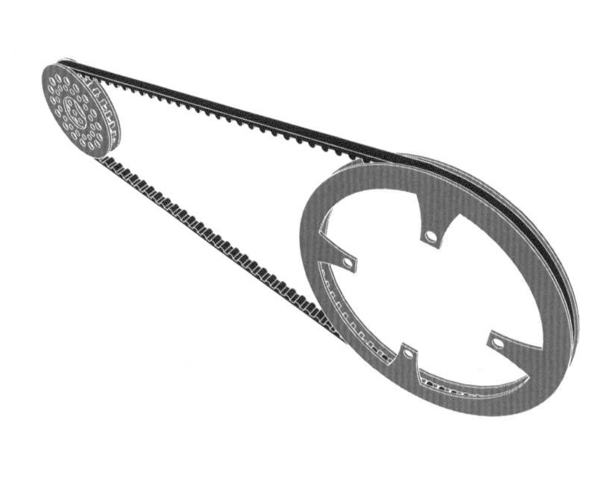

How important is wheel weight? → Should I buy lighter wheels?

Weight is important because a rider has to work harder to move a heavy bike. While heavier wheels do demand more effort, there's a popular myth that wheel weight is especially significant; hence the saying that "a pound off the wheels is equivalent to two off the frame."

Newton's second law of motion says that if an object experiences a force, it will accelerate, and that force equals mass times acceleration. This can be rephrased for rotating objects by saying that an object experiencing a torque—commonly referred to as a "twisting force"—undergoes angular acceleration; that is, it spins faster (or slower) around an axis. The rotational equivalent of mass here is the object's moment of inertia.

Just as mass indicates how hard it is to accelerate something in a line, moment of inertia tells you how hard it is to get it rotating (or slow it down). This depends not only on total mass but on how it is distributed: the farther it is from the axis of rotation, the harder it is to change its rotation. With two wheels of the same weight, the smaller one is easier to set spinning because its mass is closer to the axis. A wheel with a heavy hub and light rim is easier to set spinning than one with a light hub and heavy rim.

Clearly the effort to spin a wheel is sensitive to how its mass is distributed, and this is probably the origin of that old saying. However, its actual mass has the same significance as that of any other component—the lighter it is, the easier it is to accelerate. In fact, the rewards for saving weight in the wheels can be insignificant compared to the weight of the frame and rider. Also, lighter wheels produce most of their benefits only during periods of acceleration, which happens infrequently during most cycling. Other factors, such as aerodynamics and rolling resistance, are influential throughout a ride, so a more streamlined riding style on a well-maintained bicycle with properly inflated tires usually offers better return for your money.

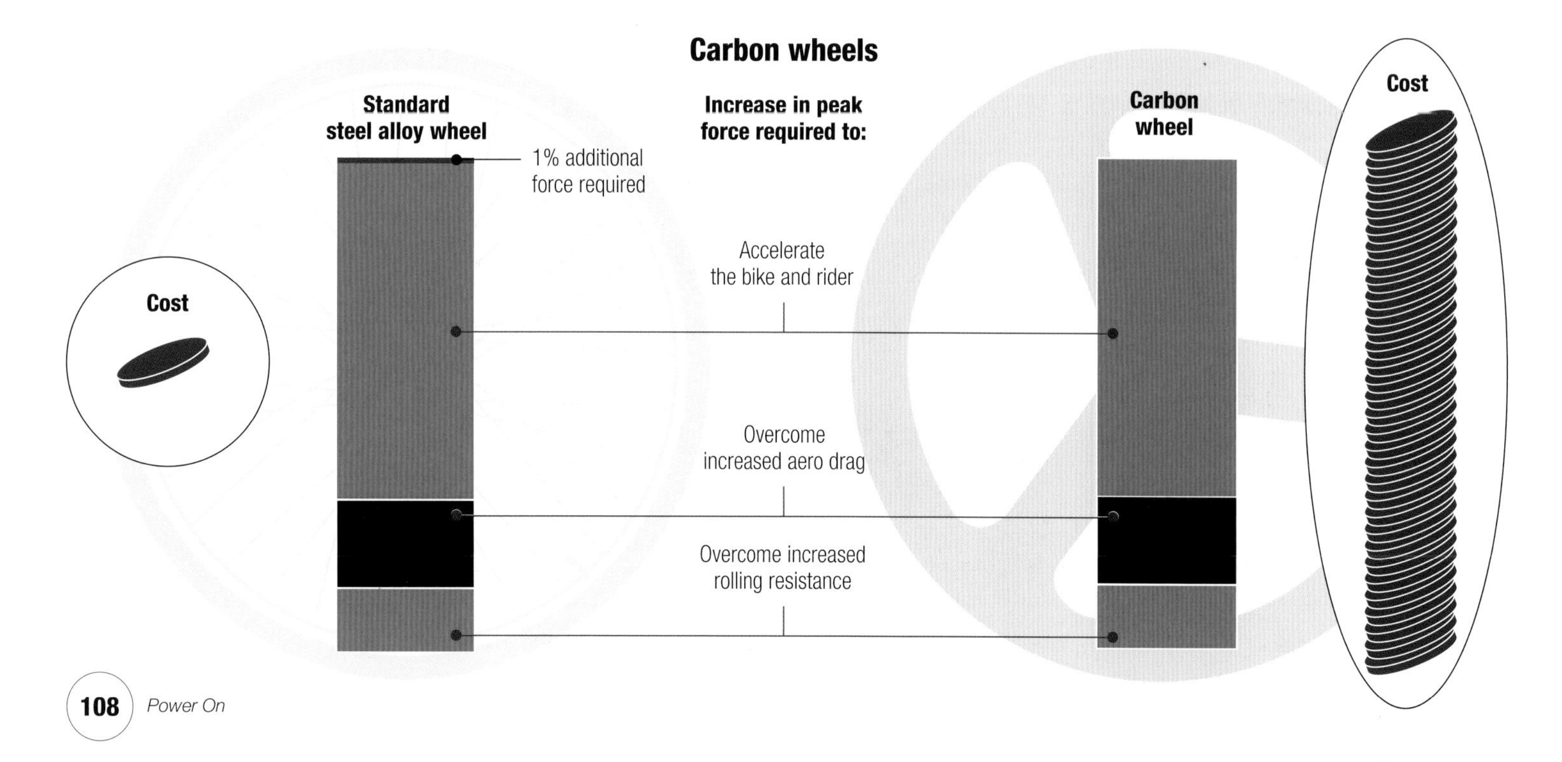

▶ ***How about a lighter wheel?*** *The benefits offered by reducing wheel weight may be smaller than those gained by improving aerodynamics, and it can cost more. With data derived from trials by Kraig Willett at* Bike Tech Review, *it can be seen that a rider training for two hours on a bike with a lightweight rear wheel uses less energy than when spinning a standard 32-spoke wheel. However, this isn't as efficient as if the bike were fitted with a deep-section front wheel. Of course, the ideal of reducing wheel weight and improving aerodynamics at the same time is promised by several wheel makers.*[14]

Need to know

Newton's second law of motion

$$F = ma$$

Newton's second law in rotational form

$$\tau = I\alpha$$

Moment of inertia

$$I = mr^2$$

where:

F = force (lb)

m = mass (slug)

a = acceleration (ft/s^2)

τ = torque (ft·lb)

I = moment of inertia ($slug \cdot ft^2$)

α = angular acceleration ($radians/s^2$)

r = radius (ft)

◀ ***Wheel of fortune*** *It has been calculated that a cyclist who switches from a carbon front wheel to a standard steel alloy wheel would sacrifice 1 percent of their energy output over a five-second burst of acceleration in order to overcome the additional weight. However, for many riders, the performance boost of a carbon wheel needs to be compared to the financial cost. (The cost ratio shown here is an estimate—actual costs fluctuate over time.)*[15]

Wheel weight versus wheel shape

How do tensioned spokes make the wheel work? → Can I make my spokes work in harmony?

A bike wheel is built by connecting the rim to the hub with tensioned wire spokes. They pull the hub and the rim toward each other, and each is under a predetermined tension so that they pull equally and keep the wheel in a true circle. This means the rim is kept a constant distance from the hub and can support the total load of rider and frame on any part of its circumference.

The loads the whole wheel can carry are impressive for a structure which can weigh as little as 18 oz (500 g). A wheel with wire spokes tensioned to 225 lb (1000 N) can support about 900 lb (400 kg), far greater than the weight of any cyclist and frame. The rim of a standard bicycle wheel experiences a compression force of about half a ton from its 36 tensioned spokes. If the spokes are kept at the right tension, a wheel will fail catastrophically only if it hits a massive bump at high speed while carrying a heavy load. For front wheels, the spokes on each side of the hub leave the hub flange at identical angles on their route to the rim. It's different on the rear wheel of a bike equipped with derailleur gears, because the cassette of sprockets must be accommodated; the flange on the gear side of the wheel is closer to the midpoint of the hub and the spokes must follow a path to the rim that is steeper than on the other side of the wheel. This means they have to be under greater tension to keep the rim in equilibrium.

Although developed initially by trial and error, virtual models of bike wheels have been put through finite element analysis (FEA) since the development of computers. These help pinpoint the strains and stresses on spokes as the wheel rolls. Despite the results of FEA, a vigorous debate continues about whether the hub hangs from the top spokes or is supported by those below. On one hand, FEA shows that while the stress is reduced in the five lowest spokes when the wheel is on the ground, this is balanced by a slight stress increase in all of the remaining 31 spokes, taking up the slack in varying amounts. Although this might imply that the hub is held in suspension by the spokes, the counterargument is that the lowest spokes provide 95 percent of the lift experienced by the hub. The debate continues.

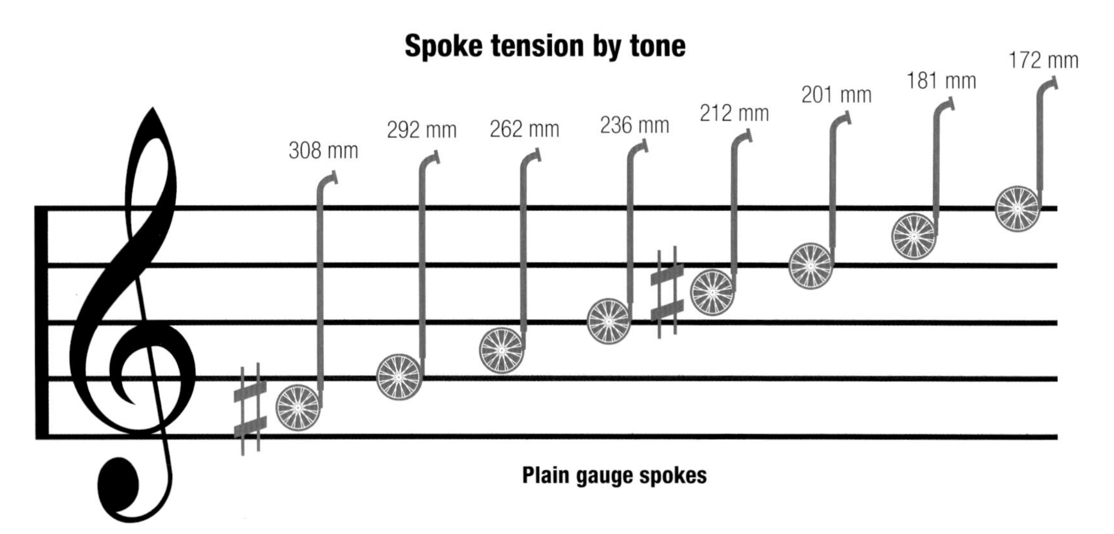

◀ ***Ringing spokes*** *Wheel builders check that spokes are under the same tension by plucking them so that they ring like a guitar string. Those with the same pitch are equally tensioned. With a set of pitch pipes or a tuning fork, it is possible to check by ear if a wheel has spokes tensioned to the recommended optimum. The notes shown here are the optimum for a selection of plain gauge spokes.*[16]

Spoke tension

▶ ***In suspense*** *A wire-spoked bicycle wheel is a structure under tension throughout its life, even when not carrying any load. When a load of 225 lb (1000 N) is applied through its axle and the wheel is on the ground, the lowest spokes undergo some compression, although not enough to overwhelm the original tension that was introduced during wheel building. Nevertheless, the rim deforms, increasing the tension in spokes on both sides of the contact area. When the wheels are rotating, the tension in the spokes is constantly increasing and decreasing. (The distortions are exaggerated here for clarity. The values are an average of the tension in two spokes of a 36-spoke wheel superimposed on a reproduction of a finite element analysis [FEA] of an 18-spoke wheel.)*[17]

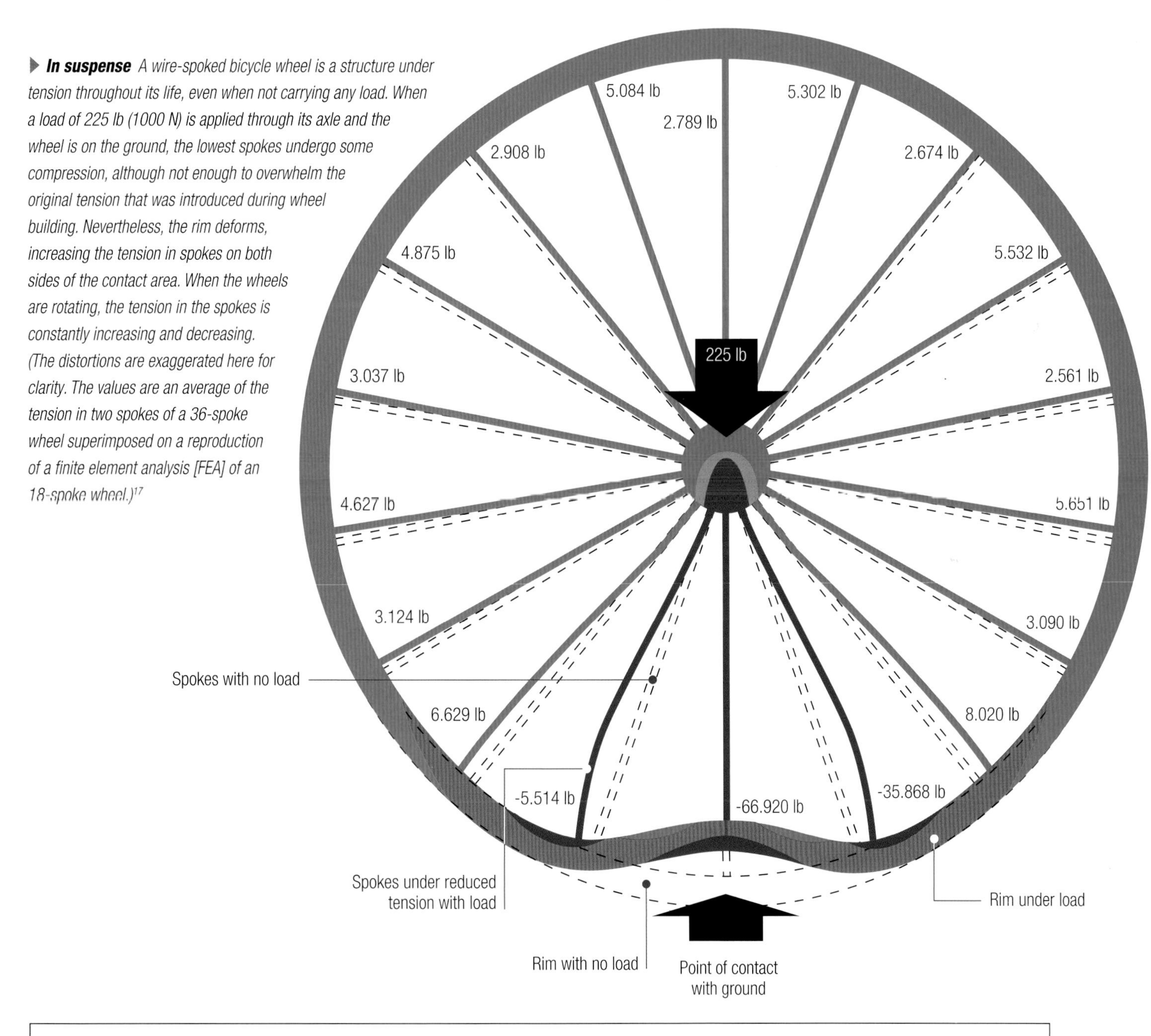

Need to know

The tension in spokes can be measured with a tensiometer, or assessed by plucking them so that they ring like a guitar string—when they all play the same note, they are in equal tension. The expression for the fundamental vibration of a tensioned spoke is:

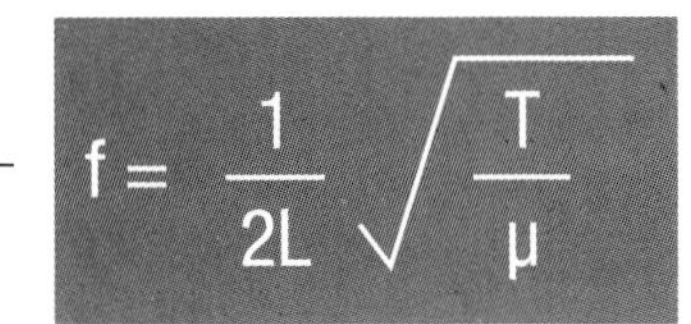

$$f = \frac{1}{2L}\sqrt{\frac{T}{\mu}}$$

where:

f = fundamental frequency (cycles per second)
L = length of the spoke (ft)
T = tension of the spoke (lb)
μ = mass per unit of length (slug/ft)

equipment: the wheel

The bicycle wheel is such an obvious and essential component of the machine that it is easy to overlook what a remarkable feat of mechanical engineering it is. The modern bicycle wheel is one of the strongest of all man-made constructions, able to support approximately 400 times its own weight. It is also capable of absorbing stresses in several different planes—radially, from the weight of the rider and the consequent upward pressure of the road surface; laterally, from sideways pressure as the rider leans into a bend or stands up on the pedals for extra power; and torsionally, from the torque exerted by the chain as it drives the wheel forward. The history of bicycle wheel development reveals a process of finding ways to make lighter and lighter wheels without compromising the necessary strength.

▼ ***Draisine*** *(ca. 1817)*
With wooden wheels derived directly from carriages, the most advanced components were brass bushings on the axles.[18]

▲ ***Boneshaker*** *(ca. 1860)*
This style of bicycle had pedals attached to the front axle, but it still had wooden wheels, although they were lighter than those of the Hobby Horse.

▼ ***High Wheeler*** *(ca. 1870)*
The metal spokes continue to replicate the radial pattern of its forebears and are so numerous.

The first wheels—hubs, spokes, and rims—were made entirely of wood, like cart wheels. Rims continued to be made of wood well into the twentieth century, particularly for racing bikes, which would have rims made from strips of laminated hardwood. The development in the 1870s of High Wheeler bicycles, sometimes called "Penny-Farthings" or "Ordinaries," led to the use of steel spokes to reduce the weight of the wheels.

At first, steel spokes were connected radially from the hub to the rim, but later wheels connected them tangentially, meaning that the spokes run out at an angle from the hub and cross over each other. This configuration is more efficient in transmitting the rotational motion of the hub to the wheel rim. Tangential spokes also resist deformation of the wheel under stress. The immense strength of the bicycle wheel comes from the tension under which the rim is held by the spokes.

The spokes are not acting as columns to support the rim, which becomes obvious if you remove a spoke and note how easily it can be bent. Spokes have little compressive strength, but they do have immense tensile strength. It is virtually impossible to stretch a spoke, and this is the key to the spoked wheel's rigidity. The spokes are pulling the rim toward the hub; the whole structure is in tension, so it resists deformation.

Lighter materials such as aluminum, carbon fiber, and Kevlar are now commonly used by designers working for competitive teams, and racing wheels often have deeper rims, with a V section instead of the more familiar shallow U. There are wheels with spokes that span the diameter of the wheel and bend around the axle so as to avoid the stress fracture point where the spoke attaches to the hub. Everything is geared toward reducing weight while retaining rigidity.

▼ ***Light wheel*** *(ca. 1980)*
Concave metal rim to accept inflatable tire. Spokes are tangential, interlaced, and tensioned steel.

▼ ***Disc wheel*** *(1984)*
Francesco Moser broke the hour world record and made discs popular.

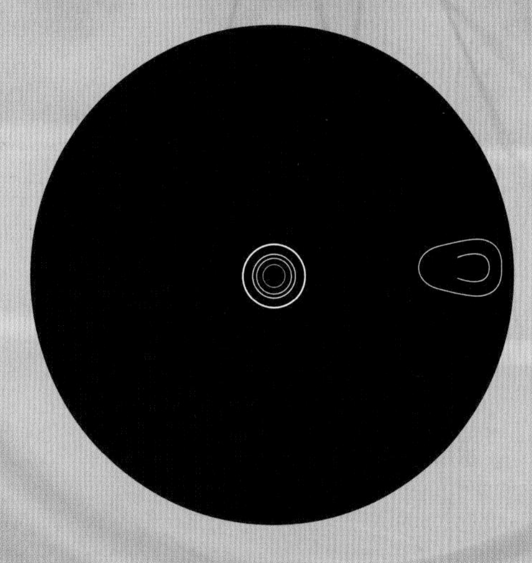

▼ ***Tri-spoke wheel*** *(1989)*
The latest wheels feature carbon fiber, deep-section rims and few spokes to improve aerodynamics and reduce weight. Note the return of radial spokes.

What part does the tire play in moving the bike forward?

Why do mountain bikes have fat tires?

The grip of a tire on the road surface (friction) is crucial for creating controlled motion. If there isn't enough friction between rubber and road when you start to pedal, the rear wheel will spin. If there's not enough when you brake, you'll slide. And if there isn't any friction when you steer, at best the bike won't turn and at worst you'll fall.

Friction is the force that resists the relative motion of two objects that are in contact with each other. It has two main components—adhesion and surface roughness. Adhesive friction happens at the molecular level. Inside a solid object, molecules bond with neighboring molecules. On the surface, molecules don't have neighbors on all sides, so they have some spare bonding capacity, the potential to adhere to other surfaces they come into contact with—such as the road, in the case of a tire. The other main component of friction, surface roughness, is as familiar as sandpaper. It's the millions of physical barriers created by all the tiny pits and projections, which are sometimes large enough to see and feel.

Both adhesion and surface friction work to make the tire "grip" as a bike moves. The strength of friction between two given materials is described by their coefficient of friction, usually represented by the Greek letter μ. This is defined as the ratio of the frictional force preventing two surfaces from sliding over each other to the force pressing them together—the ratio of the tire's grip on the road to the weight of the bike—and is determined from direct measurement. The larger the coefficient of friction, the better the grip.

One way to increase friction is to put more rubber on the road, either by increasing the load or by reducing tire inflation pressure. This creates a larger contact patch, which makes for more friction. Apart from comfort, that's why mountain bikes have wider tires.

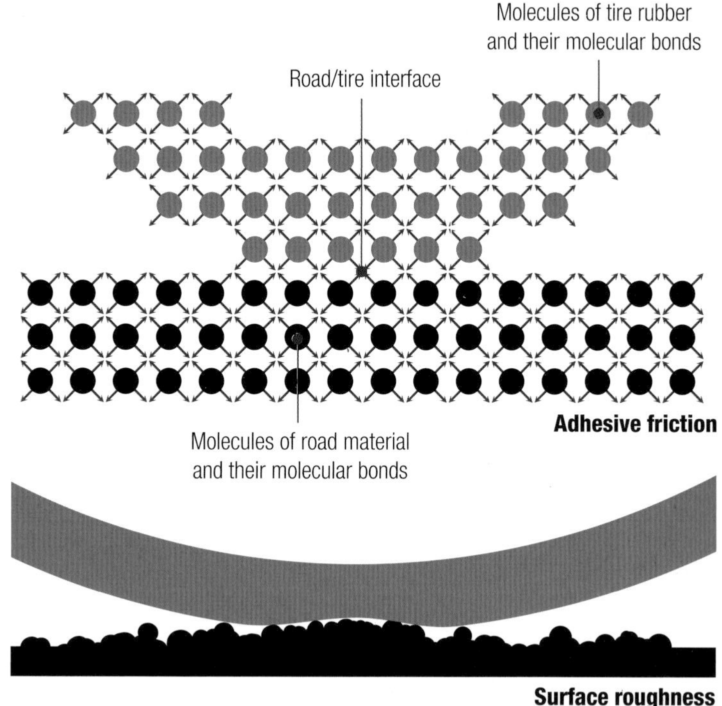

▲ ***Sticky and rough***
Adhesive friction occurs as a result of the tendency for molecules at the surface of a material (such as asphalt or rubber) to bond with the molecules at the surface of another material that it comes into contact with. Friction also arises from microscopic and macroscopic surface roughness, giving flexible rubber tires the opportunity to gain a grip on protrusions in the road.

Need to know

The strength of friction is different for each pair of materials. Different compounds of rubber and different road materials result in different coefficients of friction. The coefficient of friction also depends on environmental factors such as temperature, humidity, and surface wetness, so even a given tire and a given road surface will have a range of possible coefficient values.[19]

Coefficient of friction for:

Rubber and dry asphalt	0.5–0.8
Rubber and wet asphalt	0.25–0.75
Rubber and dry concrete	0.6–0.85
Rubber and wet concrete	0.45–0.75

Front view of a cross section through rear tire and road

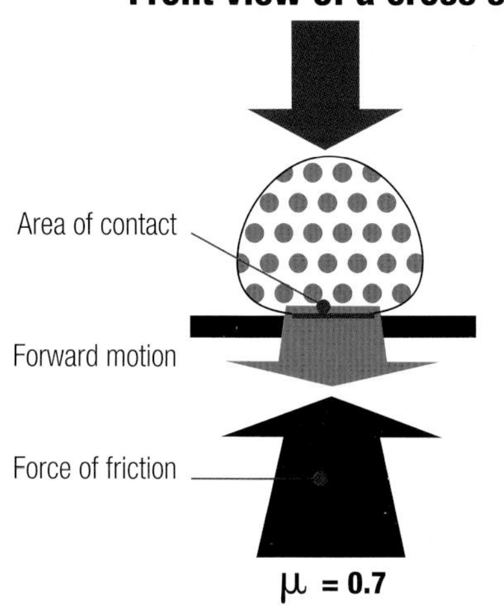

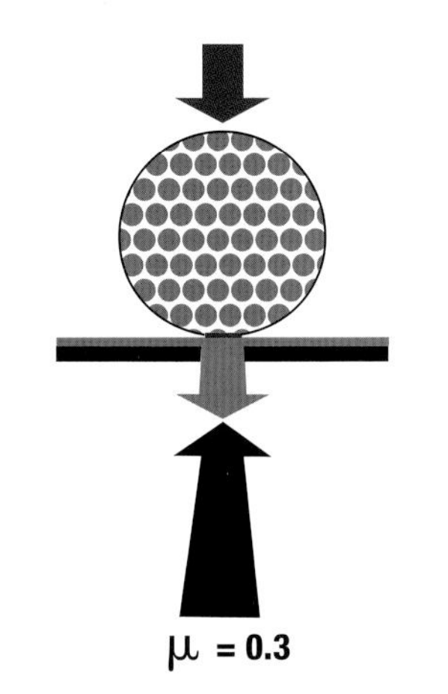

◀ ***Factors affecting grip***
Far left: Thin tire, low pressure, heavy load, large contact patch, dry road—plenty of friction between rubber and road, so tire grips when accelerating and wheel torque gets converted to forward motion.

Left: Thin tire, high tire pressure, light load, small contact patch, wet, oily or icy road—low friction between rubber and road, so wheel torque simply spins the wheel on the spot.

Contact patch for various loads and tire pressures

▶ ***Tire footprints*** *The area of a rubber bicycle tire that is laid on the road is remarkably small. The size of this contact patch depends on three things—its structure, how hard it is inflated, and the load it is carrying, which is mostly the rider's weight. On a serious road bike with high-pressure tires, the contact area for the rear tire is only about 1 in² (7 cm²), not much more than a thumbprint. For a leisure cyclist who keeps their tires soft to absorb more bumps, it's still only twice or maybe three times that. An approximation of contact area can be calculated by dividing the tire pressure by the load, not forgetting that the rear wheel usually carries about twice the load of the front (which is why it is common practice to put 10 percent less pressure in the front tire).*

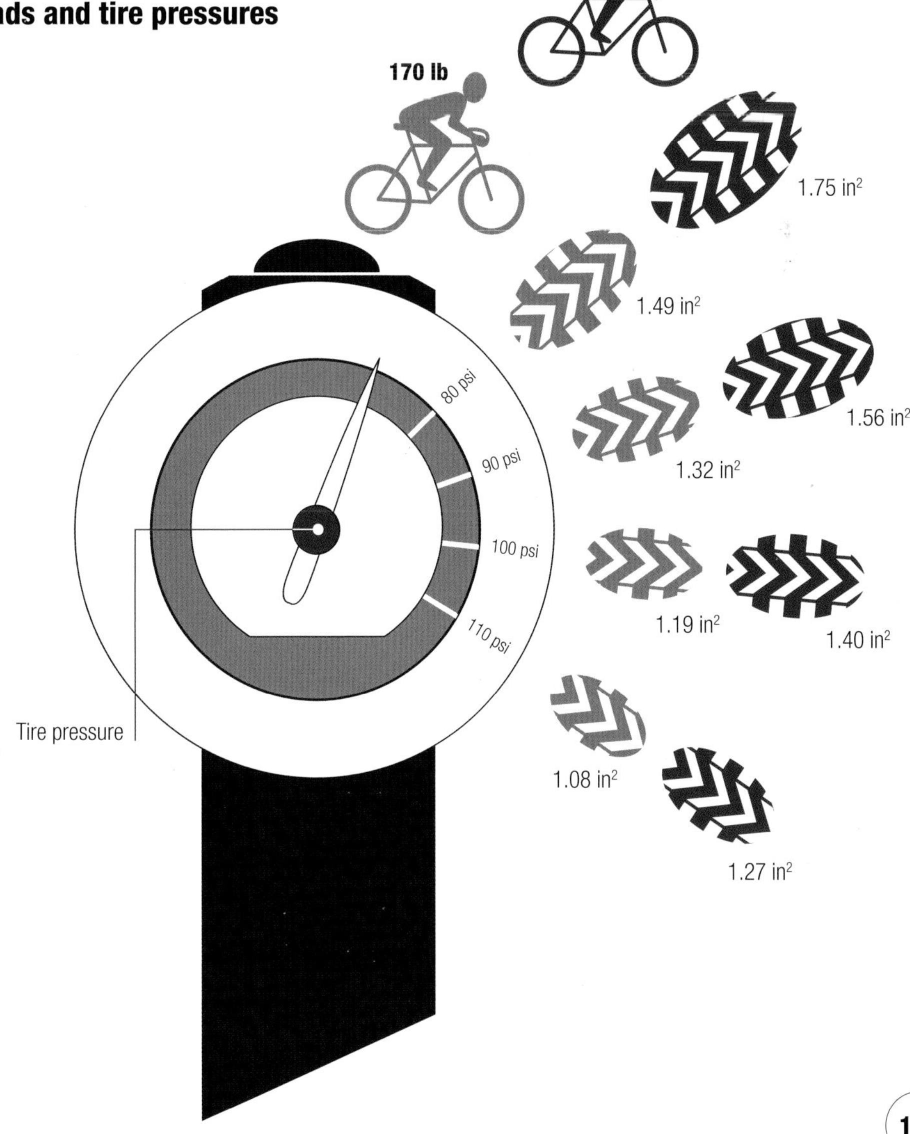

Need to know
The frictional force can be calculated using the following equation:

$$F = \mu N$$

where:
F = frictional force (lb)
μ = coefficient of friction
N = normal force (lb)

What happens to energy during braking?

Are safe brakes a hot item?

Cyclists don't take slowing down for granted. For virtually all riders, safe, predictable braking is more important than accelerating. Friction, so important in accelerating the bike, is also essential for stopping it.

A rider uses friction when braking by pulling a lever to squeeze a brake pad against a braking surface, either the wheel rim or a disc. The brake pad then rubs against the rim or disc, creating a frictional force that slows down the rotating wheel. This contact excites the surface molecules of the pad and braking surface so they become hot: kinetic energy is being converted to thermal energy. Having lost some of its kinetic energy, the bike is slower and the rider can release the brake. If the rider hangs onto the lever, the wheel will lock and the bike will skid (generating more friction in a different place and even possibly melting rubber onto the road). The downside to all this hot action is that some kinetic energy is always diverted to the physical erosion of material from the pad and braking surface—so sooner or later they will need to be replaced.

The efficiency of bike brakes is dependent on how clean and dry the pad and braking surfaces are, and on their coefficient of friction. Actual stopping distances and times also depend on the mass of the rider and bike, their initial speed, and the gradient. High values for any of these variables will require a greater braking force to achieve the same rate of deceleration. This can lead to more heat buildup, with too-hot rims leading to tire failure. Riders on long, switchback descents must judge how to keep to a safe speed without risking a blowout.

▶ ***Stop sooner*** *The stopping ability of only the rear brake is approximately 76 percent of the front brake, and, more significantly, a mere 65 percent of the front and rear brakes applied at the same time. The graph clearly illustrates how quickly the brakes reduce speed and is derived from timed experiments on 33 yd (30 m) of mountain track descending at an average gradient of 10.96 percent. Brakes were applied at around 25 mph (40 km/h).*[20] *While it is possible to reduce speed by performing a controlled skid, this not only increases tire wear but also places greater dependence on the consistency of the road or track surface.*

Need to know

The maximum amount of energy that can be converted to heat in braking is the kinetic energy of the bike plus rider:

$$E_K = \frac{1}{2}mv^2$$

where:

E_K = kinetic energy (ft·lb)

m = mass (slug)

v = velocity (ft/s)

▶ ***Face plant*** *When braking only the front wheel, there is a danger that the rider will pitch over the handlebars. When the brake is applied, inertia means the rider continues to move forward. The center of mass of the combined rider and bike also moves forward. If the center of mass moves ahead of the point where the front tire touches the ground, the rider and bike will rotate around this point and perform a potentially painful "face plant." The chance of this happening is greater if there is a down slope, if the wheelbase is short, or if the rider is perched high up and well forward.*

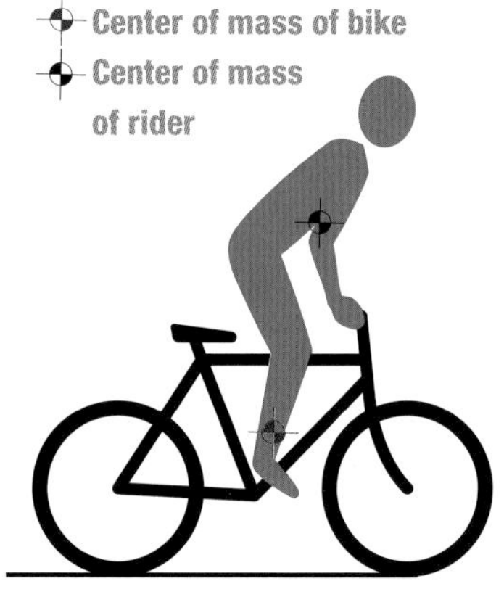

Braking deceleration times

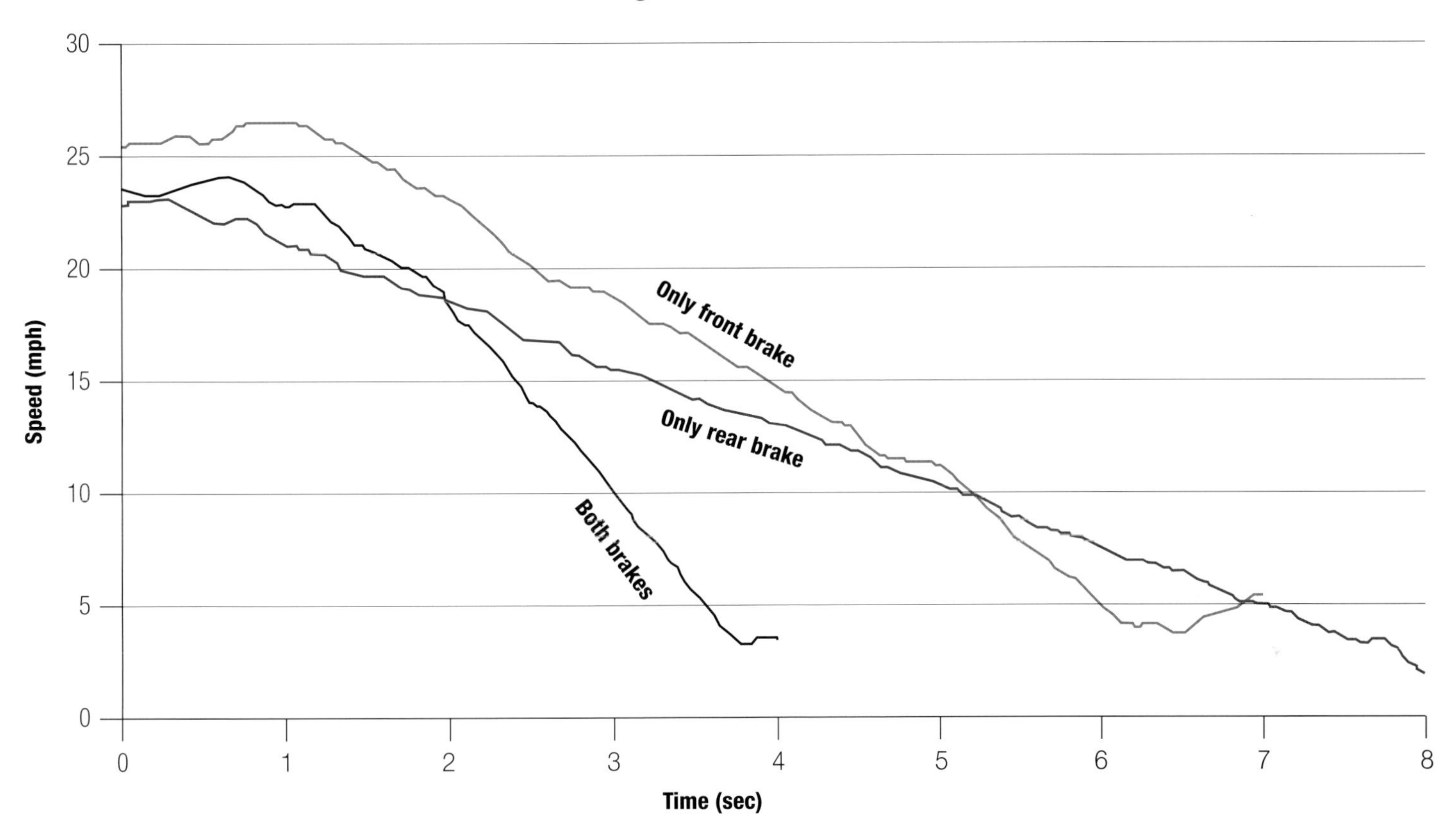

Inertia disaster

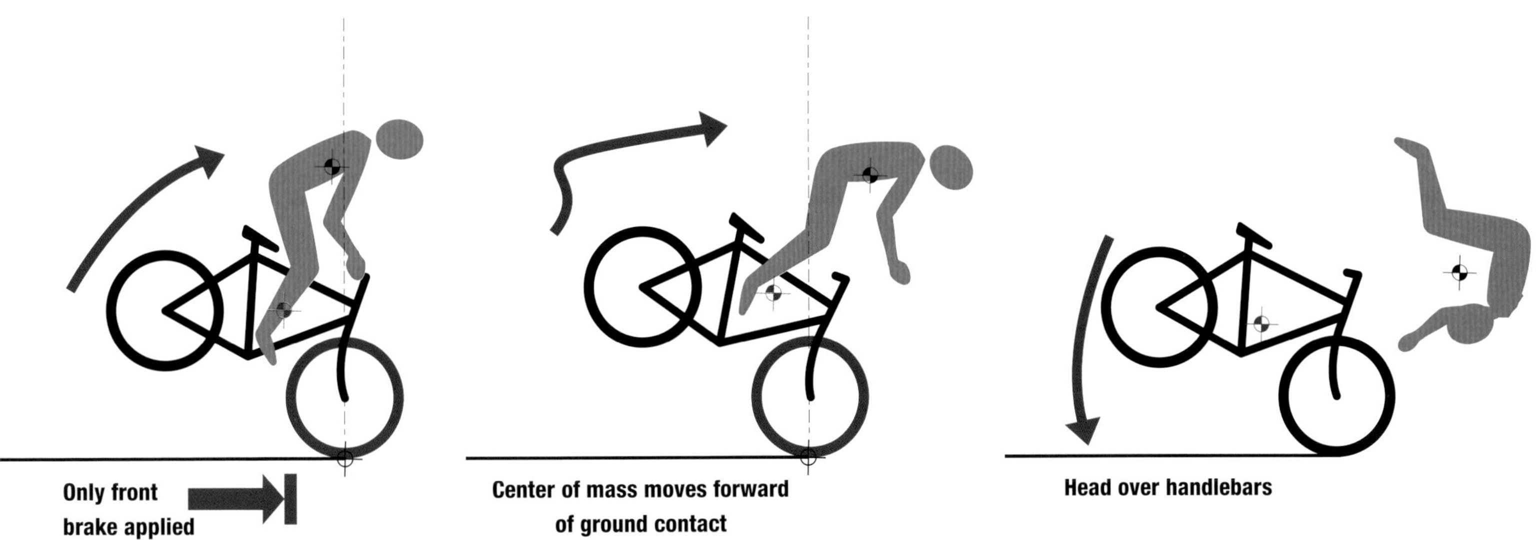

How does mechanical advantage aid braking?

How can I brake with only two fingers?

A bike stops because the wheels are slowed down by the brakes. A lever on the handlebars is squeezed and this moves a cable (or hydraulic fluid), which pushes a pad against the spinning surface of a disc or wheel rim. Brake systems use the principle of mechanical advantage, making sure that the squeeze of your hand is converted into enough frictional force to slow the wheel.

Mechanical advantage refers to the use of a lever to multiply a force. The simplest lever is a long, rigid arm that pivots at a point, called the fulcrum. Conservation of energy means that work done at one end causes the same amount of work to be done at the other end. Physicists define work as the product of force times the distance over which it acts, so a small force pushing over a large distance can be converted to a large force pushing over a small distance. Bike brakes are more complex but the principle is the same. The force of the brake pad on the disc or rim is many times greater than the force you apply, because levers at each end of the system move through different distances. Look at your hand levers: The part you squeeze is set up to travel several times farther than the little arm that pulls the cable or shifts the fluid. At the other end another lever, the brake arm, steps this motion down even farther to the very small distance that the brake pad moves.

Riders who cycle only slowly don't need much stopping power, so the length of the hand lever can be similar to that of the pad actuator. Riders who must make sudden stops from speed need their squeezing force to be multiplied by a much higher factor, so the shorter the braking pad movement relative to the lever movement, the better.

In reality, some of the mechanical advantage of the lever system is lost between the hand and the pad. Internal friction in the connecting cable reduces efficiency. Even hydraulic disc brakes steal some of the force, although they do promise more control while braking because fluids move more consistently than cables.

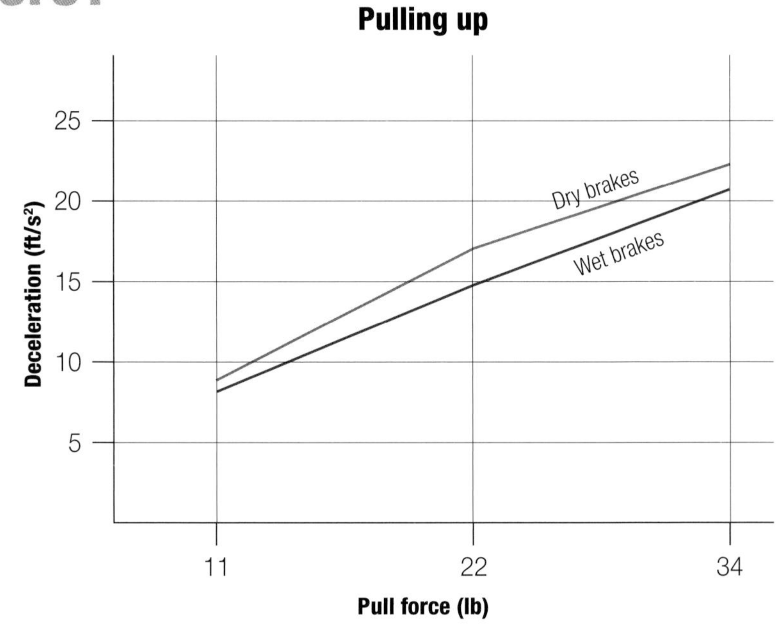

▲ ***Slippery when wet*** *Lab tests of brakes on mountain bikes consistently prove that greater pull force on the levers is needed to slow a bike when the brake pads are wet. Although each combination of brake pad material and braking surface has its own coefficient of friction, any water present lubricates the braking surface, reducing the friction that reduces the wheel torque and slows the bike. The relationship between pulling force and deceleration with dry brakes is linear, but with wet brakes there can be a step change when braking improves as friction dries the pad and braking surface.*[21]

Need to know

The mechanical advantage of bicycle brakes is usually between 3:1 and 6:1, because this is an acceptable compromise between the reach of a hand, its strength, and the clearance between brake pad and braking surface. With a mechanical advantage of 4:1 and negligible losses to internal friction, a cyclist squeezing with a force of 10 lb (44 N) applies 40 lb (178 N) to the braking surface, gently slowing down. For firm braking, about 20 lb is (89 N) applied to the lever. An emergency stop, complete with risks of skidding or pitching over the handlebars, is pretty much guaranteed with a lever squeeze of 30 lb (133 N).

Multiplying forces

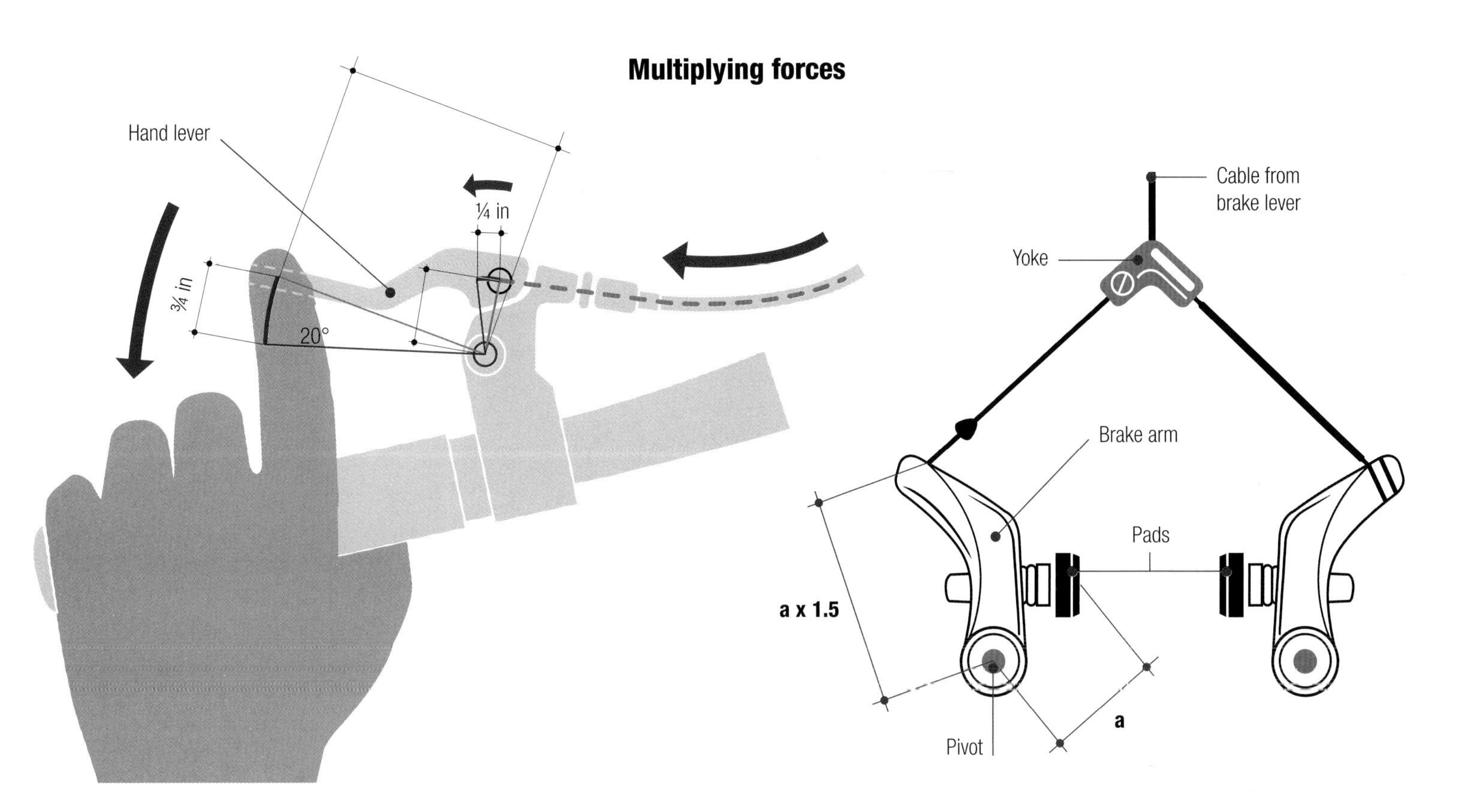

▲ ***Cable brakes*** *A cable brake system has levers at both ends and their design affects the stopping power they offer. On the left is a hand lever attached to a brake cable. The rider's finger in this case is about twice as far from the pivot as the cable end is, so if the finger moves ¾ in, the cable is pulled by about one-third of that, or ¼ in. However, mechanical advantage means that the force exerted at the cable is tripled.*

At the other end of the cable, the force is multiplied again by the mechanical advantage of the brake arm. What counts here is the ratio of the distances from the pivot to the yoke cable end and from the pivot to the face of the brake pad. In this example, the ratio is about 1.5:1, so that, for example, for a 3⁄32 in movement by the cable, the brake pad travels just 1⁄16 in. However, mechanical advantage means the force has been multiplied again, by 1.5.

The final result is that a light force on the brake is doubled by the hand lever and then multiplied by 1.5 by the brake arm, so the braking force on the rim is three times as strong.

▶ ***Principle of the lever*** *If a steady force F acts on an object, causing it to move a distance d, the work W done on it is defined as F × d, and its energy increases by an amount equal to that work. Without friction or other losses, conservation of energy requires that the work done on one end of the lever equals the work done by the other end. Imagine a lever with one arm three times as long as the other. If you push down on the long end with a force of, say, 10 lb (44 N) and it moves a distance d, the short end moves a distance of just one-third d, but exerts a force of 30 lb (133 N). The energy in is the same as the energy out, but the force is increased. A lever allows you to move something very heavy, but only by a short distance.*

Mechanical advantage

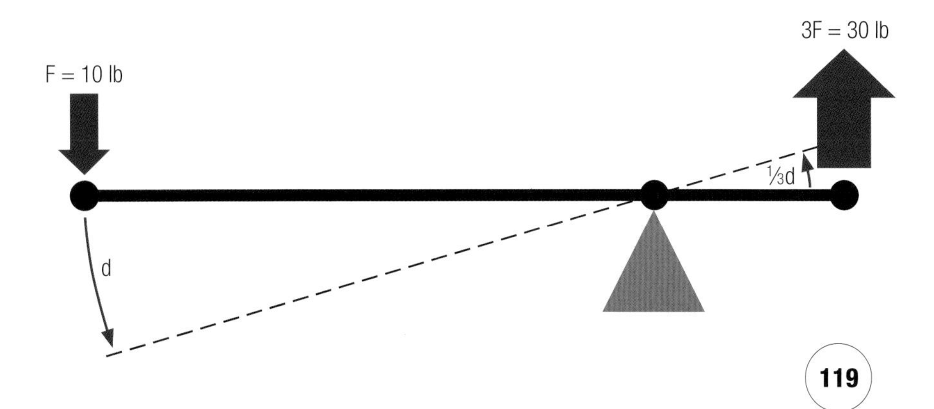

science in action mixing and matching

Bicycles have always attracted innovators, all intent on creating new ways to make bikes better and faster. Unfortunately, this process of constant evolution undermines mass production's goal of standardization. As a result, it can be very difficult to swap components between two bikes that might appear to be similar.

Innovations in bottom brackets, chains, and shifting mechanisms just in the first few years of this millennium means they are not compatible with older frames. Disc brakes continue to oust rim brakes, suspension forks are spreading, and hydraulics are replacing cables. Hollow axles, oversized off-road rims, tubeless tires—the list will continue to grow as long as there is a demand for improved designs. Not all find favor. Automatic gearing, complex tire tread patterns, and rollercam actuators have all come and gone.

For most cyclists this rarely matters. Over time, it may become harder to find a replacement part or it may cause a small annoyance, such as when you discover that each tire has a different valve. For professional squads, however, compatibility can make the difference between success and failure. During a race, a fault might arise when the support car carrying spare bikes isn't nearby. The quickest way to rejoin the action is to take a good component from a teammate's working bike. This happens frequently in major races when the team leader has a problem such as a puncture. A domestique in the squad pulls up and donates a good wheel to the leader, sacrificing his own chances of glory. Consistency and standards are essential in this case.

▶ ***Wheel problems*** *The ambitions of professional racers can be thwarted by a hazard that all cyclists face every day they ride—punctures. Team cars or neutral support cars may be on hand to help by providing a replacement wheel, but when the incident happens to a team leader and far from such assistance, teammates are expected to sacrifice their own chances by donating components from their own bikes. It can only happen if the wheels are compatible, which is one reason why every member of the team uses the same type of bike.*

CAMPAGNOLO
LCL
CAISSE D'EPARGNE
CAISSE D'EPARGNE
CAISSE D'EPARGNE
Nalini

How do bearings reduce friction?

How come my wheels are full of marbles?

Friction allows the tires to grip the road and the brakes to slow your bike—but is the enemy of all other moving bike parts. It saps kinetic energy that would otherwise help you go faster. One of the best ways to cut the main source—rotational friction—is to use bearings. Bearings are small, round components that both bear the load of a moving part and reduce friction compared to direct rubbing. The key is that, instead of rubbing, they rotate. Instead of one surface having to slide over another, it can roll—and rolling parts can be made much more precisely to minimize imperfections and ongoing wear.

To achieve this, bearings are spheres, cylinders, or tapered cylinders, designed to spin when in contact with a turning part, but also able to stay in the place where the support is needed. This is done by containing the individual bearings in a restraining support, or race, and sometimes kept apart from each other by a spacer, or cage. Bearings typically have to deal with one of three kinds of loading—radial, thrust, or a combination of these. Radial loads are perpendicular to the shaft or axle; thrust or axial loads are parallel to it. The headset of a bicycle allows the steering to turn while carrying the loads from the handlebars, frame, and front wheel. Because the head tube is not vertical, the bearing must deal with high thrust loads plus some radial loads. Hubs and bottom brackets deal largely with radial loadings. The metal spheres in ball bearings can experience greater pressure at the point of contact than cylindrical elements in roller or taper bearings—but can tolerate more movement, which makes them good in wheel hubs. There are also bearings in pedals and freewheels.

A good bearing assembly has little or no play, and needs to be lubricated so is often sealed to retain the lubricant.

Forces on bearings

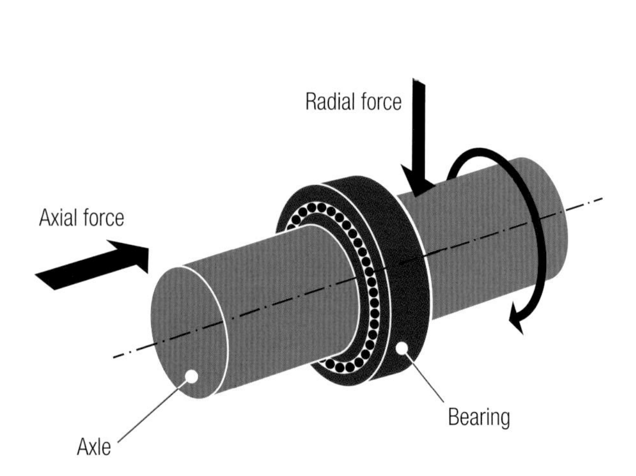

▲ ***Radial and axial forces*** *Bearings must withstand forces that run parallel to the axle, perpendicular to the axle, or both.*

Retarding forces

Bearing friction
0.003 lb
Bearing friction
0.003 lb
0.2 to 0.7 lb rolling resistance
0.2 to 0.7 lb rolling resistance

▲ ***What's holding you back?*** *Efficient, well-lubricated, and well-maintained bearings sap very little of a rider's energy. The retarding force caused by the bearing friction of one wheel is about 0.003 lb (0.014 N), which is negligible compared to typical tire rolling resistance of about 0.2 to 0.7 lb (1 to 3 N).*[22]

Bearings inside a freehub

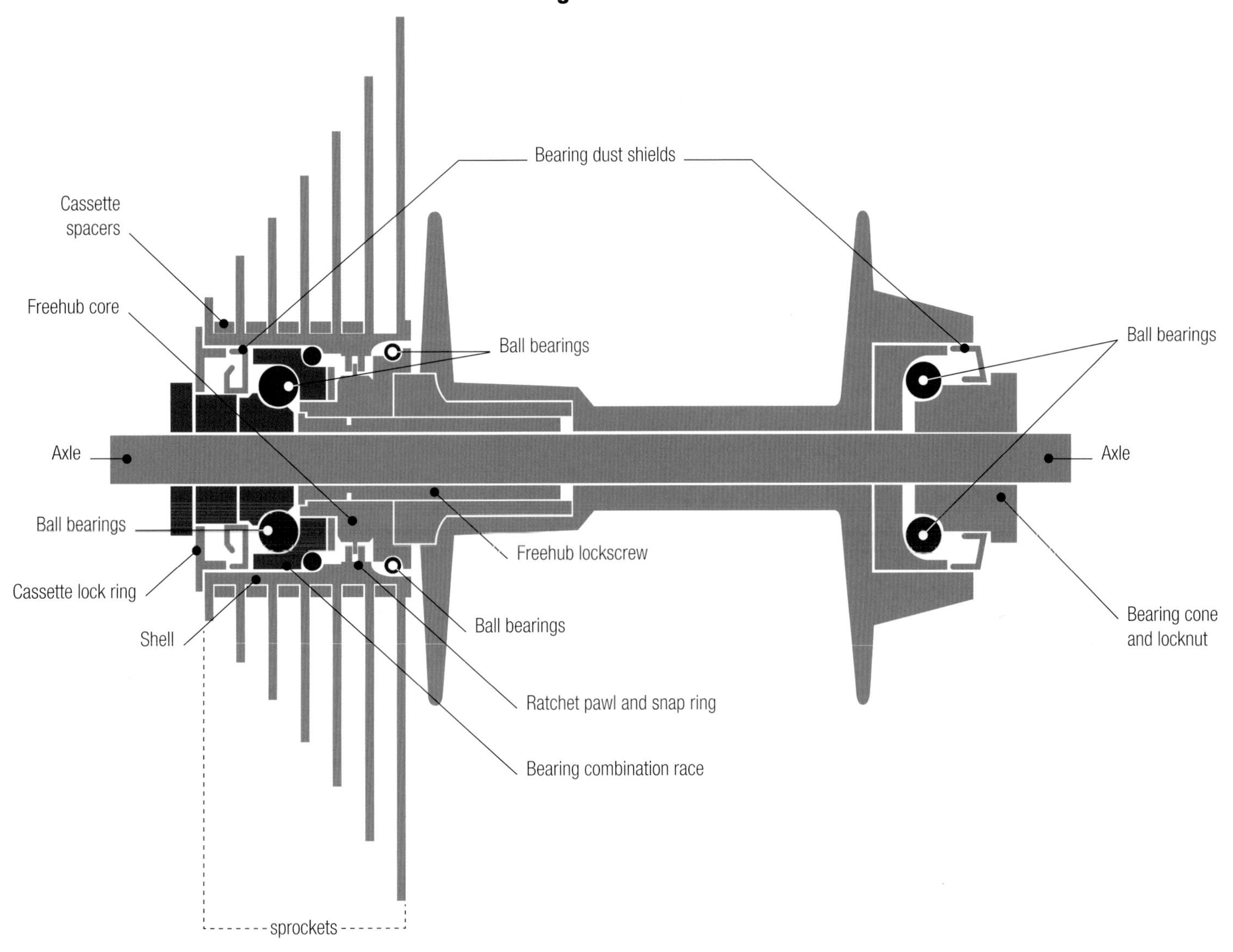

▲ ***What's going on inside?*** *The first steerable two-wheeler, the Draisine of 1817, probably used brass bushings to allow the axles to spin in the forks. But since the first ball bearings were applied to bicycles in 1869, they have become increasingly sophisticated and reliable. This is a typical freehub from a rear wheel of the mid-1990s, with cassette, sprockets, and integrated bearings.*

Need to know

During Cadel Evans' victory in the Tour de France in 2011, his hub bearings facilitated an average wheel rotation speed of 301 rpm. If he had kept the same wheels throughout the three-week race, the bearings would have rotated more than 1.5 million times. Yet the wheels were changed frequently, only because other components, including the rim and spokes, were more vulnerable to failure. The bearings could have been expected to function without problems for four more Tours.[23]

Everything that is aerodynamic immediately gives the impression of speed. We see it in the swept wings of an eagle diving to seize its prey. It's apparent in the sleek lines of a sports car and of a jet fighter, announcing their intention to cut through the air as swiftly as possible. So it's surprising how long it's taken for this rarified branch of science to be recognized as of great significance for cycling. Yet it is now so important that even amateur racers are renting wind tunnels to learn how best to leave rivals in their wake. Equipment has been redesigned to make it more streamlined. Computers have been programmed to generate virtual wind flows around nonexistent components, so bicycles' shapes can be perfected before a single solid sample is made. Riders are finding out how to hold their heads, bodies, arms, hands, and even their fingers without triggering the chaos of turbulence. Aerodynamics in cycling is a classic tale of how the diligent application of science can disrupt the stately progress of a traditional sport and turn it into a battle that only the knowledgeable will win. This chapter cuts to the chase.

chapter five

aerodynamics

Why bother with aerodynamics? → Why is the wind always against me?

Air is the cyclist's strongest opponent. It takes more energy to overcome this invisible foe than all of the other resisting forces—even on a flat road and at a modest speed, on a day when there is no hint of a wind. It is air that prevents cyclists from riding at highway speeds. Elite riders at their peak may look as if they're flying, but they are struggling continually against the atmosphere. Humans on bikes are simply the wrong shape when it comes to aerodynamics.

The problem is that air isn't weightless, although it's just thin gas. At sea level and a temperature of 68°F (20°C), a cubic foot of dry air weighs about 1¼ oz (35 g). It may not be solid and is less viscous (sticky) than water, but every rider has to work to push it out of his or her way. The science of aerodynamics examines how air moves around objects (rider, bike) and can help produce better bike designs that make bikes more slippery and reduce the work a rider has to do. Aerodynamics has had a profound impact on cycle sport in the last three decades. Riders now know which rims, wheels, frames, helmets, and riding positions will slice through the air to the finish line most easily. Aerodynamics has affected the results in classic races, world records, and championship titles and has resulted in cycling's governing body, the Union Cycliste Internationale (UCI), changing its rules several times.

Aerodynamics is a complex subject because it involves dynamic bodies, changing shape as the rider pedals, moving through a gas (air) whose own speed, direction, temperature, and density are not necessarily constant. Like all decisions in cycling, compromises have to be made in the pursuit of the optimum efficiency. This chapter explains why, how, and when those choices are made.

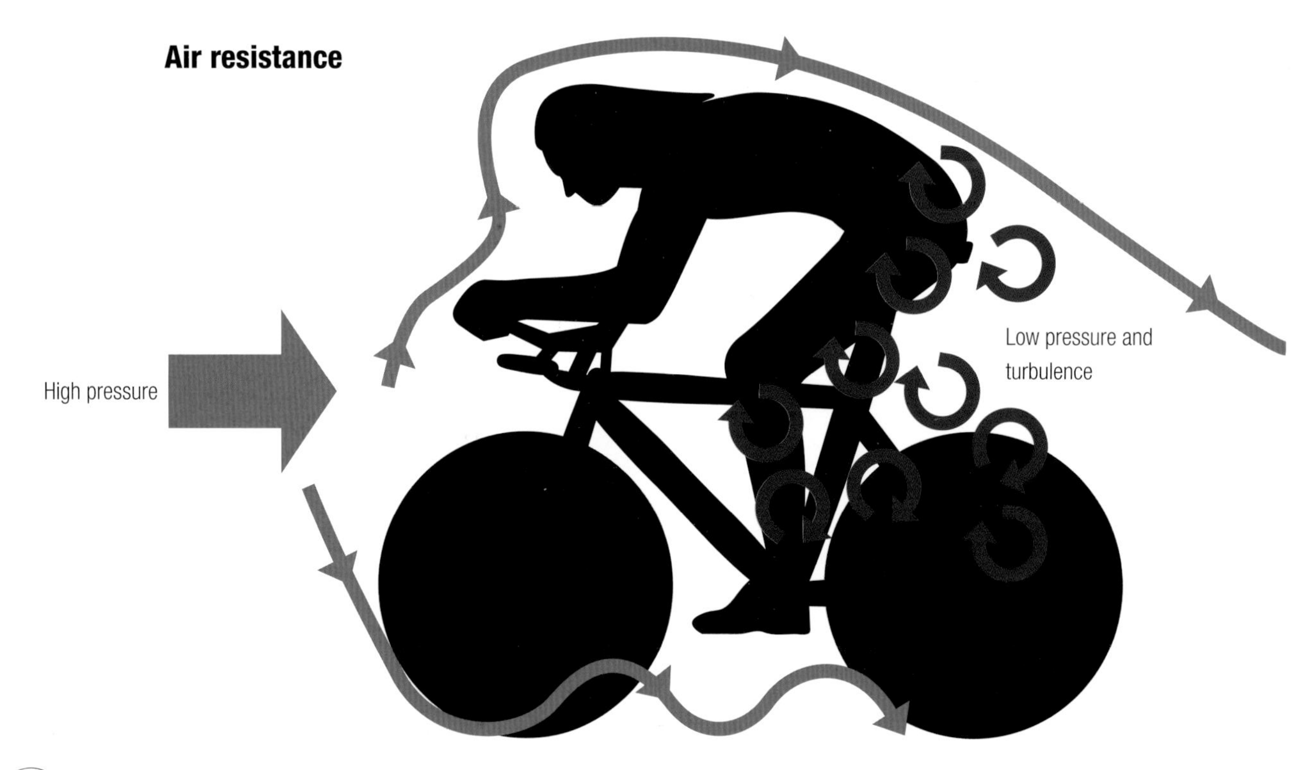

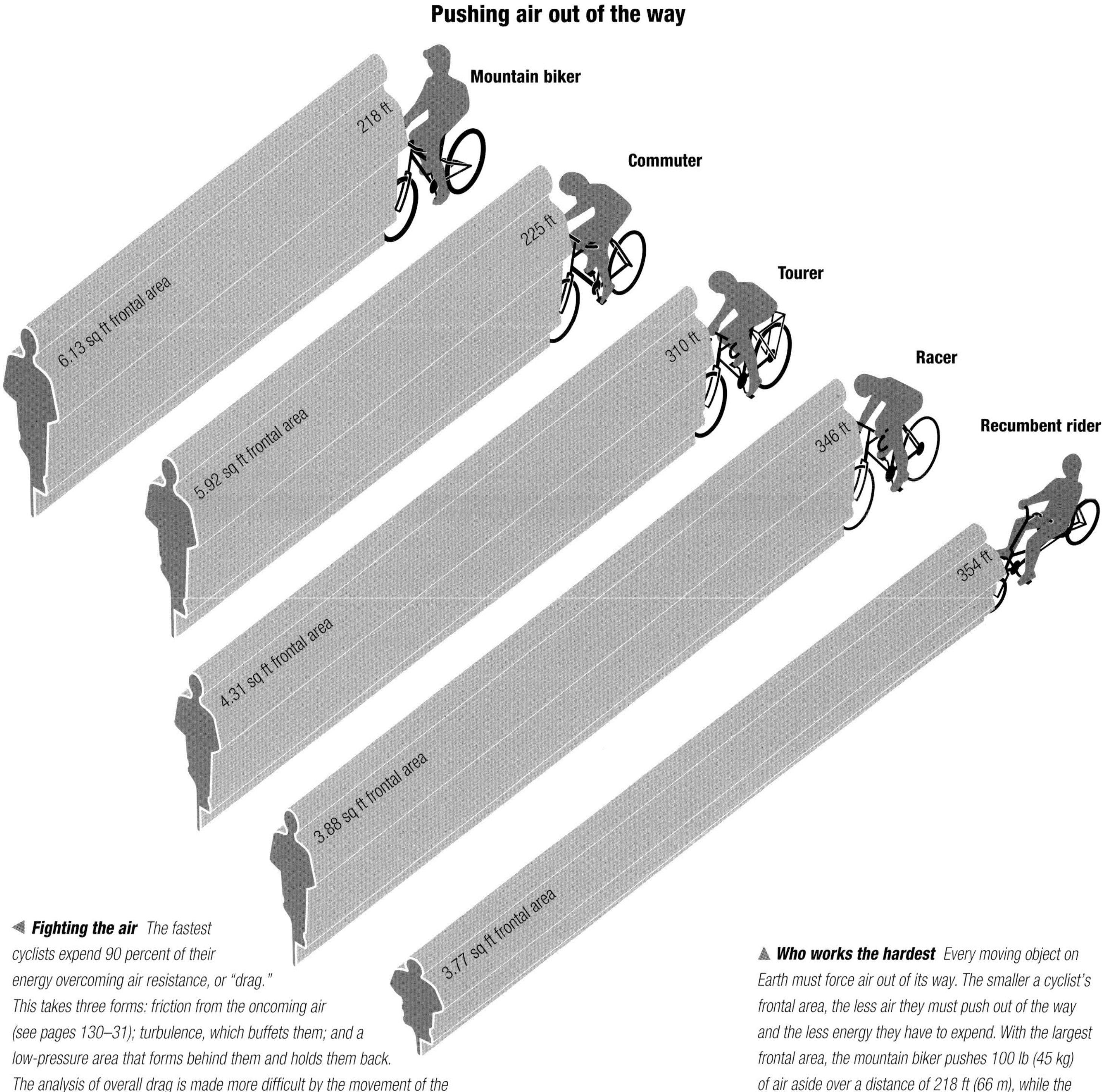

◀ ***Fighting the air*** *The fastest cyclists expend 90 percent of their energy overcoming air resistance, or "drag." This takes three forms: friction from the oncoming air (see pages 130–31); turbulence, which buffets them; and a low-pressure area that forms behind them and holds them back. The analysis of overall drag is made more difficult by the movement of the wheels, pedals, legs, and feet, which disturb the air and create additional complex flows. Resistance can be reduced by changes to the design of the bike (including an aerodynamic frame and wheels) and the rider's position (aero handlebars and a crouched pose), and by using a more streamlined helmet and clothing. All can contribute to a faster ride without requiring additional energy from the cyclist.*

▲ ***Who works the hardest*** *Every moving object on Earth must force air out of its way. The smaller a cyclist's frontal area, the less air they must push out of the way and the less energy they have to expend. With the largest frontal area, the mountain biker pushes 100 lb (45 kg) of air aside over a distance of 218 ft (66 m), while the sleekest recumbent rider can travel 63 percent farther before they have shifted the same amount of air. It may not seem such a large difference, but over a ride of an hour, the mountain biker will have to shift hundreds more pounds of air and will be more tired, sooner.*[1]

How are aerodynamics assessed?

How can invisible air currents be measured?

Measuring the aerodynamics of a complex dynamic object such as a bike rider is not straightforward. Ever since the significance of cycling aerodynamics was realized, scientists and engineers have been refining data collection and analysis techniques—but there are still important elements to be established.

The obvious way to compare the aerodynamics of different designs is to put them in a wind tunnel, where the strength and direction of the wind can be controlled. Using a variety of sensors and instruments, researchers can see how each design performs. Low-speed wind tunnels are used, together with devices to make the bike's wheels spin (if a rider isn't being included in the test). However, a lot of testing is done in "virtual" wind tunnels, using so-called computational fluid dynamics (CFD) software. Researchers create a digital model of the bike, component, or cyclist and run the program to reveal the pros and cons of various designs in different air streams. It's quicker and more convenient than real wind tunnels—but digital models may not be as accurate as they could be. For example, there are still debates about how best to measure a cyclist's frontal area.

As the commercial market for aerodynamic cycling equipment grows, the amount of research is increasing and being used to market aero products. Unfortunately, tests may differ and some information, such as error margins, may be omitted. And don't forget that the wind on the road is unlikely to behave as obediently as that inside either a wind tunnel or a computer. So real-world performance may not match the published results.

▼ ***Aerodynamic measurements using a wind tunnel*** *The flow of air in a wind tunnel has to be controllable and smooth, with all of the air mass moving at the same speed and direction—until it meets the object being tested. There are two ways this is done in low-speed wind tunnels. The first (below) is to use fans to force a steady flow of air through a large tube, shaped to smooth it before it is released into the wind tunnel. Alternatively, air can be sucked through the wind tunnel using fans placed behind the test subject; the incoming air flows over blades and foils to ensure the constant, laminar flow required. Aerodynamic effects are assessed in several ways. To quantify the forces, strain and other gauges are built into the bike and mounting platform. Once calibrated, they collect experimental data in real time so researchers can immediately see the effects of changes. To reveal the air flow, a smoke trail or laser system is used. When information is needed at a finer level of detail in specific areas around the cyclist and bike, a "tuft wand" can show how the air is flowing at specific points. This is little more than a short thread attached to a narrow rod, but it can reveal where laminar flow is separating and becoming turbulent.*

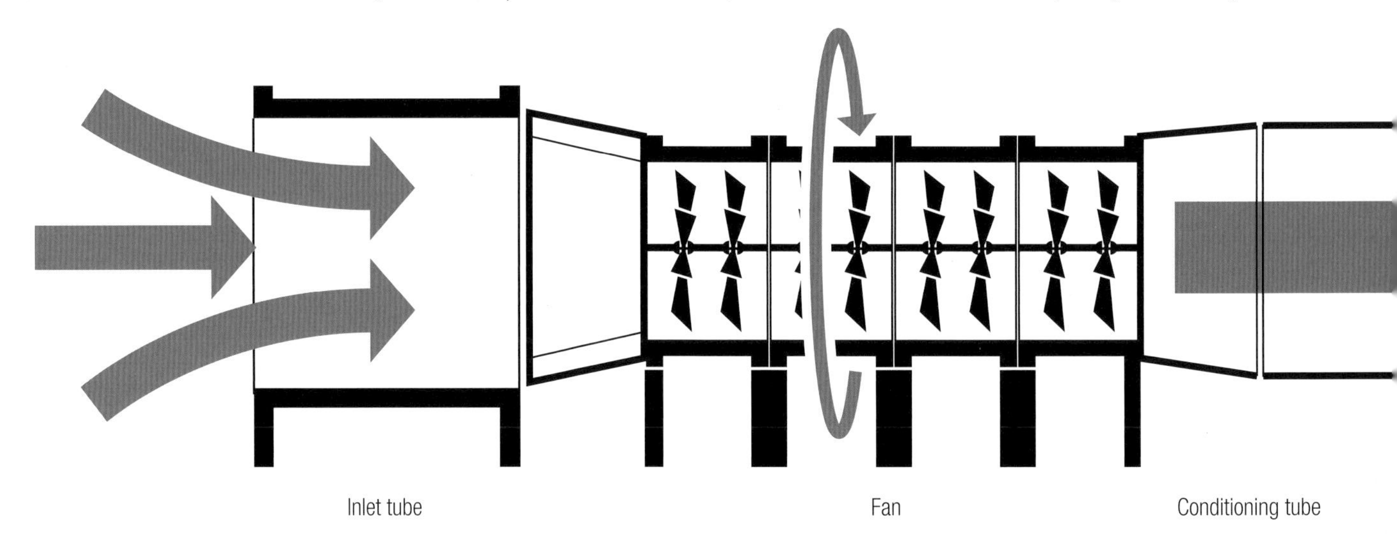

Tunnel testing

▲ ***Wind tunnel*** *Air flows over the rider, drawn through the tunnel by the fans behind. The turntable platform allows analysis of riding into winds offset by up to 25 degrees. The wheels can be powered without a rider, so the aerodynamics of the bike can be assessed seperately.*

▼ ***Computational fluid dynamics*** *CFD uses computers to analyze the flow of air around a virtual cyclist. The software uses the data to generate a visual model, so it is easy to see the results. CFD gives the researcher the ability to estimate air pressures and velocities at points where it would be impossible to measure in the real world. The accuracy of the virtual models is improving every year, and CFD is now used to study thc acrodynamics of thc dctails of componcnts, some of which are themselves moving relative to the bike frame and the wind.*

Flow simulation

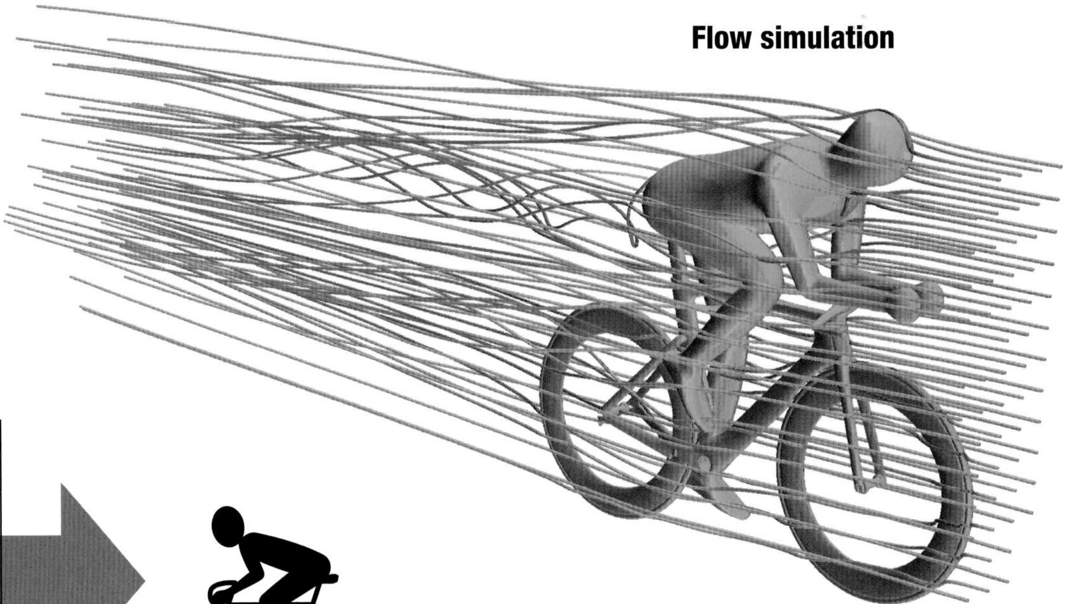

Wind tunnel

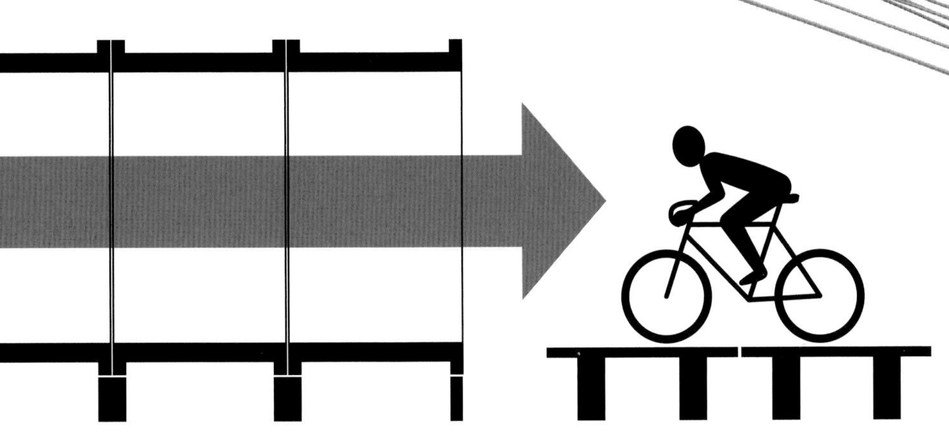

How does the air flow around a cyclist?

Why does air suck?

No matter how big or small a rider's frontal area, the airflow over the bike and body determines the level of frictional drag. There are two major components to this—pressure drag and skin friction drag. Pressure drag (also called form drag) is associated with the size and shape of the bike and rider. Skin friction drag is associated with the roughness or texture of the surfaces of bike and rider. Both also depend on the rider's velocity through the air.

To understand how these invisible drag forces affect cycling, it's necessary to focus on the thin layer of air in contact with the surfaces of the bike and rider. This is called the boundary layer, and it's crucial to aerodynamic efficiency. In a perfect world, the boundary layer would behave as if it consisted of thin layers of air sliding smoothly over each other with minimal friction—so-called laminar flow. There are a few areas that do experience such flow, such as along the rider's arms and shoulders, but this smoothness readily breaks down. It can fail as a result of surface irregularities, which may be as small as a crease in a sleeve or as big as a water bottle in the wrong position. These interruptions disrupt the laminar flow, replacing it with small, turbulent vortices that increase surface friction and make cycling harder. It gets harder still when the boundary layer separates completely from the surface, creating relatively large eddies and currents. If that happens, instead of the rider's energy propelling the bike forward, a fair amount of it is being diverted to churning the air, so the bicycle's speed is reduced.

Need to know

Depending on how well the bike is maintained, once a cyclist exceeds about 9 mph (15 km/h), the majority of their energy is used to overcome air resistance. And it doesn't get any easier the faster they ride. The power required to overcome drag is roughly proportional to the cube of their speed, so, for example, if you double your speed, you need eight times as much power.

Boundary layers

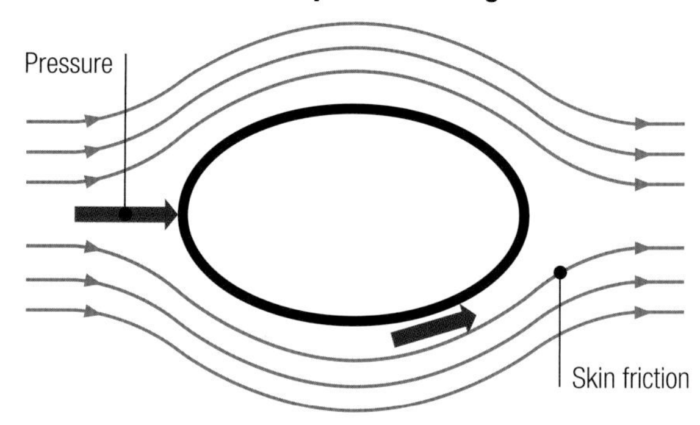

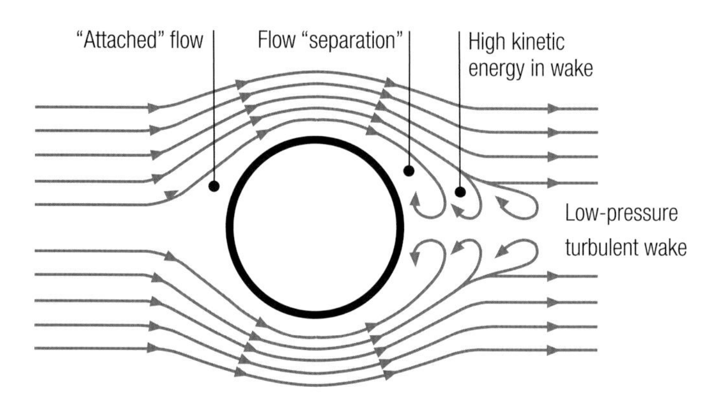

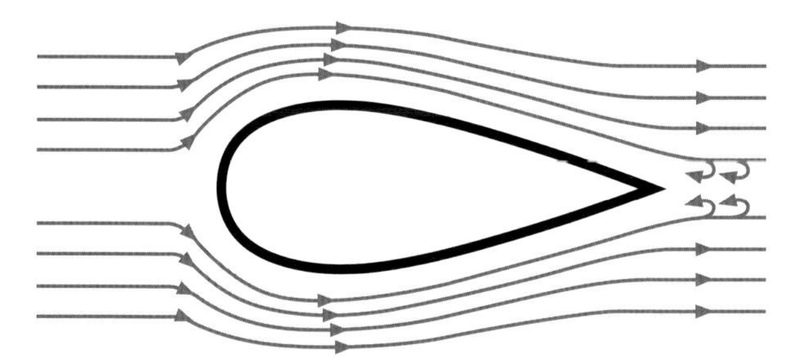

▲ ***Laminar and turbulent flow*** *Aerodynamicists know it's best to keep a laminar boundary layer "attached" to any moving surface for as long as possible. The first diagram shows the two components of drag against a fork blade that is oval in cross section, with pressure drag against the leading edge and skin friction drag along its surface. The second diagram shows a fork blade with a circular cross section. The boundary layer initially shows smooth laminar flow, but this separates into turbulence behind the blade, creating vortices with a low-pressure region behind the blade, adding to the retarding force. The third diagram shows a blade with an aerodynamic cross section. The airflow in the boundary layer stays laminar across most of the surface, with only a small amount of turbulence at the rear.*

Increase in drag with speed

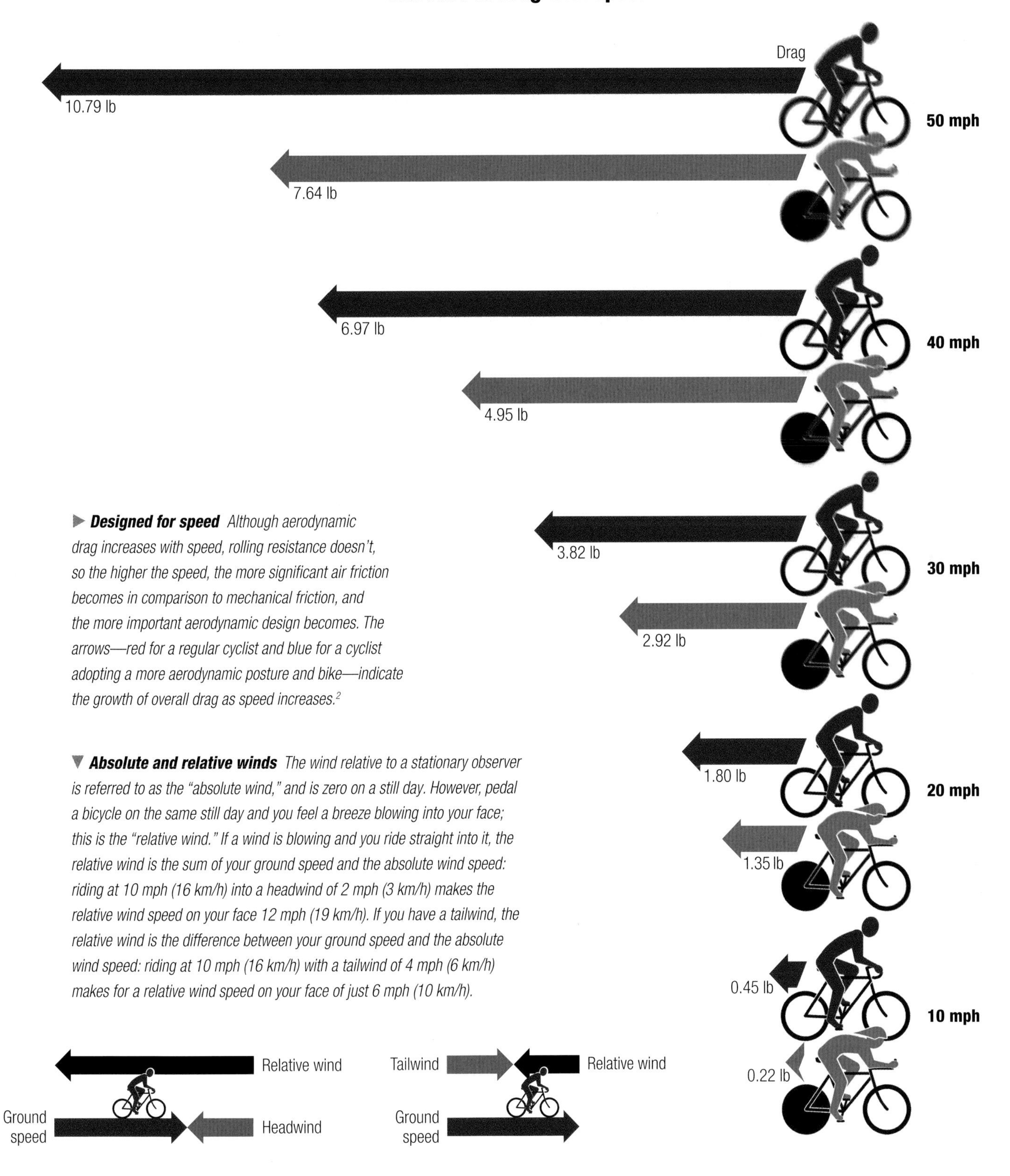

▶ ***Designed for speed*** *Although aerodynamic drag increases with speed, rolling resistance doesn't, so the higher the speed, the more significant air friction becomes in comparison to mechanical friction, and the more important aerodynamic design becomes. The arrows—red for a regular cyclist and blue for a cyclist adopting a more aerodynamic posture and bike—indicate the growth of overall drag as speed increases.*[2]

▼ ***Absolute and relative winds*** *The wind relative to a stationary observer is referred to as the "absolute wind," and is zero on a still day. However, pedal a bicycle on the same still day and you feel a breeze blowing into your face; this is the "relative wind." If a wind is blowing and you ride straight into it, the relative wind is the sum of your ground speed and the absolute wind speed: riding at 10 mph (16 km/h) into a headwind of 2 mph (3 km/h) makes the relative wind speed on your face 12 mph (19 km/h). If you have a tailwind, the relative wind is the difference between your ground speed and the absolute wind speed: riding at 10 mph (16 km/h) with a tailwind of 4 mph (6 km/h) makes for a relative wind speed on your face of just 6 mph (10 km/h).*

equipment: recumbents

The first recumbents appeared toward the end of the nineteenth century, and they rose to prominence in 1933 when one broke the world hour cycling record. However, within eight months the laid-back design had been banned from races by the Union Cycliste Internationale (UCI), thanks to pressure from conventional bike makers. The ban is still in place, although leisure, commuting, touring, and racing recumbents have been growing in popularity since the 1970s. Today, there are dozens of recumbent manufacturers around the world. Recumbent competitions are administered by the International Human Powered Vehicle Association (IHPVA) and fully faired recumbents continue to outdo bikes at time trials. Their aerodynamic advantage is so great that it would not be possible for a rider on a standard bike to achieve such speeds.

▼ ***Challand*** *(1896)*
Getting lower. In 1896, the Challand recumbent was exhibited in Geneva by M Challand but, allegedly, its weight stopped it from becoming popular.

▲ ***Mochet*** *(1933)*
Going faster. Francis Faure, a second-category rider, set a new cycling world hour record of 27.996 miles (45.055 km) on a recumbent built by Charles Mochet of France, smashing the time set by an elite cyclist. The following year, recumbents were banned from competitions regulated by the UCI.

▼ ***Street Machine GT*** *(2002)*
Going practical. HP Velotechnik's recumbent for touring, commuting, and leisure, with a small front wheel to shorten the wheelbase.

The advantage of a recumbent biycle compared to a conventional upright lies solely in the fact that it has a smaller frontal area—and, therefore, less drag. The rider's supine position doesn't give them any advantage in terms of power output, and the design is not significantly more mechanically efficient.[3] Instead of utilizing their body weight by standing on the pedals, they rely on different groups of leg muscles and the opportunity to use the seat's back support as an aid to pushing against the pedals.

There are many recumbent designs, with astonishing variations in wheelbase, wheel sizes, steering systems, aero fairings, and transmissions. In the mid-twentieth century, they almost vanished, but since Chester Kyle, a professor of engineering at California State University, Long Beach, started designing such machines with his students in the 1970s, there have been many improvements. The main reason now for their popularity is that the lower drag makes it easier to ride at a given speed. To travel at 30 mph (50 km/h), a cyclist has to expend about 310 ft·lb/s (420 W) of power, which is approaching the maximum power of an elite athlete for anything longer than five minutes. Yet superior aerodynamics means a recumbent rider can reach the same high speed with less than one-fifth of that effort, about 55 ft·lb/s (75 W). Similarly, the top speed for a conventional racing bicycle for a timed 200 m is about 45 mph (72 km/h), while for a recumbent it is close to twice as high, at just over 80 mph (130 km/h). For a cyclist to pedal at the recumbent's speed, they would have to expend some 4,300 ft·lb/s (5,800 W)—a feat no human could physically achieve. The streamlined recumbent rider, however, requires only 260 ft·lb/s (350 W).[4] This doesn't use the full potential power output of a human, but higher speeds haven't been achieved because even recumbents are constrained by aerodynamics: to go still faster would require an ever smaller frontal area and, so far, that has defeated the engineers.

Recumbents are made from the same materials and components as other bicycles, but the design challenges differ. For example, transmission chains are often far longer and need much more careful routing and tensioning. Stability is also an issue, because self-balancing qualities are usually significantly poorer. The UCI ban means recumbent devotees are not fettered by restrictions, so the diversity of shapes and styles is as wild today as it was for bicycles 130 years ago.

▼ ***Varna Tempest*** *(2009)*
The fully faired recumbent set a new world hour record of 56.3 miles (90.6 km) from a standing start. Inside the shell, designed by Georgi Georgiev, Sam Whittingham was the rider. At the same IHPVA meeting, he reached a speed of 82.819 mph (133.28 km/h) in a timed 200 m from a flying start.

▼ ***Velokraft No-com*** *(2012)*
Going sleeker. The frame is a carbon fiber monocoque (made in one piece). Splitter plates below the frame maintain a laminar boundary layer around the wheels. The rear brake is shielded inside the frame, along with all brake and gear cables. A narrow hub on the front wheel and forks minimizes frontal area. The frame and seat weigh 6½ lb (3 kg).

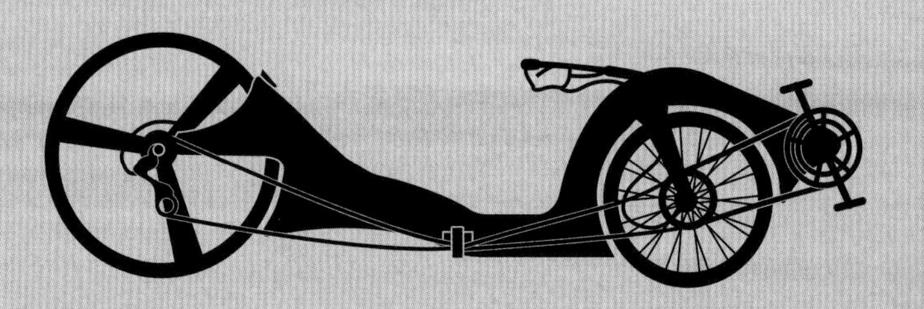

Which positions optimize aerodynamics and power? → How can arms and hands affect aerodynamics?

It's all very well being able to measure the invisible forces of aerodynamic drag that restrain a cyclist, but cycling scientists never forget that their ultimate goal is more speed. While aerodynamics is a large part of this, some of the most streamlined poses can prevent the cyclist from performing at their best. Following three decades of research into aerodynamic positions, the trade-off between the cyclist's aerodynamic body posture and their power output is beginning to be better understood.

A team at the University of Canterbury, New Zealand, has investigated and cataloged the relationship between a cyclist's shape in aerodynamic poses and their power output.[5] The team focused on torso and shoulder angles, because other angles are less variable due to restrictions on bike setup imposed by the Union Cycliste Internationale (UCI), the sport's governing body. They found that the most streamlined body shape is not necessarily the most powerful, and that the optimum position is with a low torso angle and a medium shoulder angle. While the lowest possible torso angle has sometimes been recommended as the most aerodynamic, it does reduce power output. They also discovered that there is no single optimum position for all riders. Interestingly, it appears that the shoulder angle has a greater influence on the rider's power output than the angle of the torso.

Other useful aerodynamic details were gathered by the New Zealand research team when a wind tunnel was used to work out the most aerodynamic hand grip for aero bars to reduce the creation of a turbulent wake.[6] They discovered that when a rider shapes their hands into an arrow, a region of smooth air recirculation develops behind the thumbs. This, they theorize, could be better than the region of turbulent separation produced when other hand positions were tested. While it was established that an arrow-style hand position does not compromise rider power output, the scientists noted that this hand position might not give adequate control for a bike on the track or open road. As ever in cycling, compromises may be needed.

▶ ***Winning hands down***

The normal hand position on aero bars, with thumbs on top and pointing forward, was compared with three others—thumbs inside, a fist grip, and fingers shaped to form an arrow. The tests were conducted in a wind tunnel and the results extrapolated to show how many seconds a rider could expect to save for every minute riding at 35 mph (56 km/h) for men and at 33 mph (53 km/h) for women. A positive time gain represents a time saving; a negative time gain represents a time loss.[7]

Effect of hand position

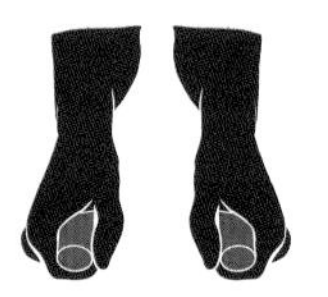

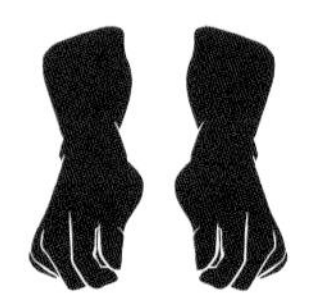

Thumbs inside				Fist grip				Arrow grip			
Athlete	Time gain (sec/min)	Drag (lb)	Power (ft·lb/s)	Athlete	Time gain (sec/min)	Drag (lb)	Power (ft·lb/s)	Athlete	Time gain (sec/min)	Drag (lb)	Power (ft·lb/s)
A	0.25	−0.049	−1.5	A	0.22	−0.042	−2.2	A	0.79	−0.154	−11.1
C	−0.01	0.002	−8.1	C	0.16	−0.309	−17.0	B	0.35	−0.077	−22.9
B	−0.11	0.025	2.2	B	0.03	−0.007	−13.3	C	0.35	−0.068	−22.9

Torso angle and power requirements

▶ ***Compromise*** *Torso angles affect not only a rider's aerodynamics but also their ability to generate power, so finding the best pose is difficult. Data for five positions shows that tucking right down isn't necessarily the right strategy when trying to minimize power requirements for completing a 4,374 yd (4,000 m) pursuit in 4 min 30 sec.*[8]

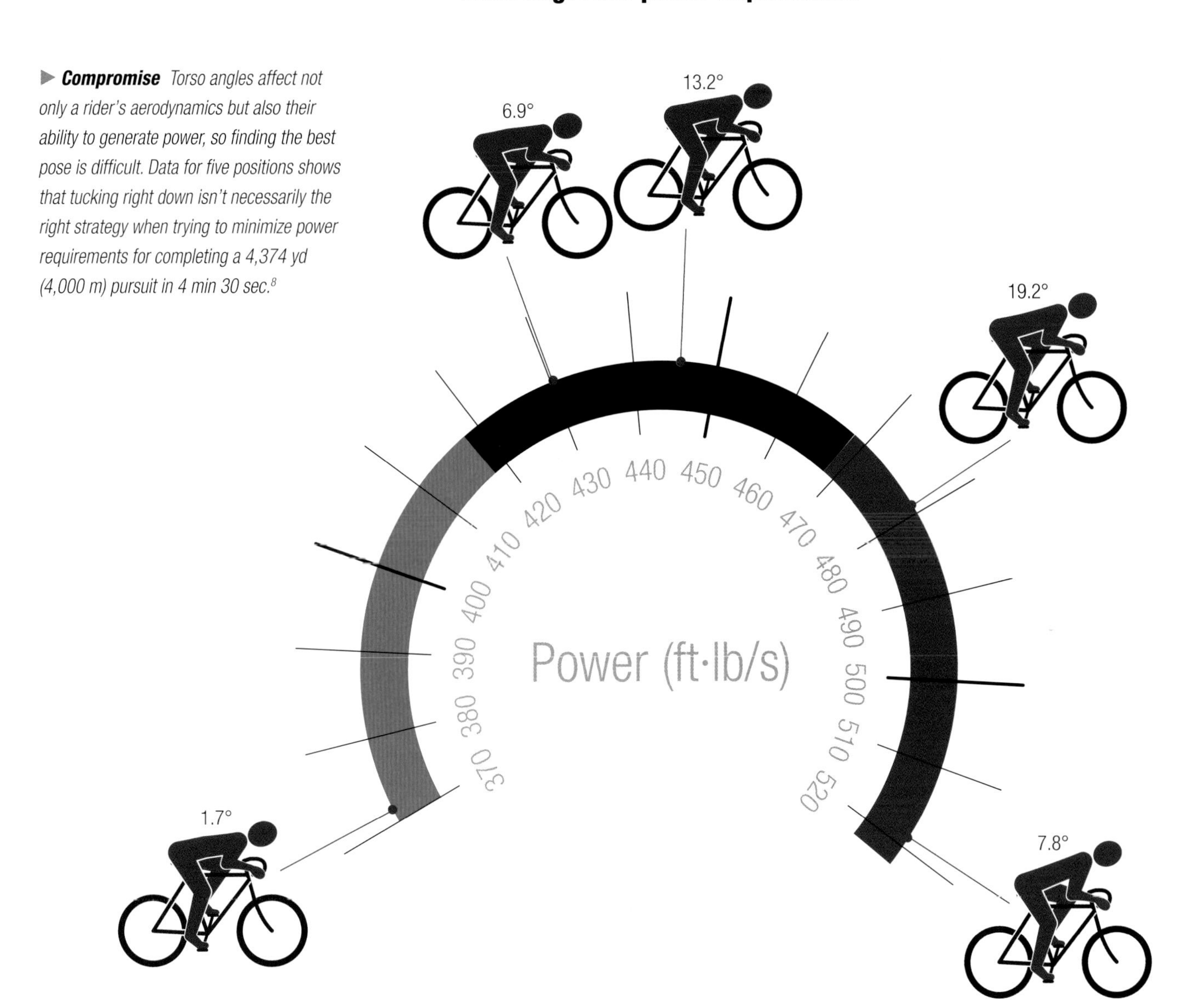

Position statement

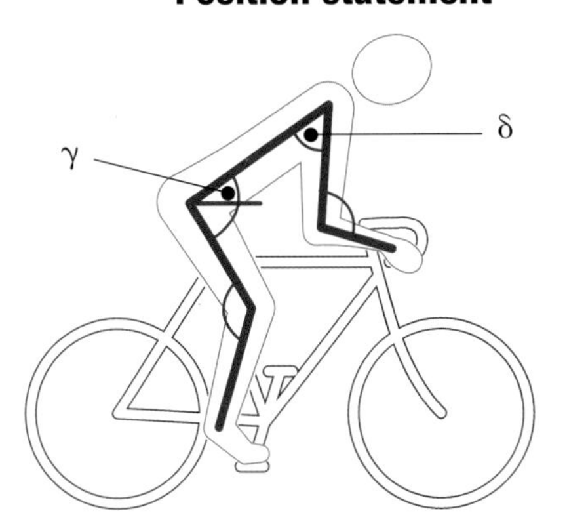

◀ ***Crucial angles*** *Torso and shoulder angles used for defining a cyclist's pose.*

Need to know

Armed with only a watch, a roll-down test allows any cyclist to compare the aerodynamics of different positions on a day when the wind is constant. From a standing start, coast down a straight hill in one position and time the run at a fixed point a few hundred yards down the road. Hold the same pose for two more runs to get an average time. Repeat for a different position and compare the averages. If nothing else changes between the runs, the fastest one is the most aerodynamic.

Which position is the world's fastest?

How can I go with the flow?

The shapes and sizes of the cyclist and their bicycle means that the frontal area of the rider is always bigger than that of any bike. Bikes are relatively narrow and with a small profile, and the rider almost always sticks out at the sides and above it. The easiest way for a cyclist to ride faster without expending any extra energy is to reduce their own frontal area by assuming a more streamlined riding position. This was understood in the late nineteenth century and led to the introduction of drop handlebars.[9] More recently, it has become a vital technique for cyclists racing against the clock in time trials and time-limited world record attempts.

New records were set when drop handlebars were adopted. They allow the rider to retain control of the steering in several different positions. The least effective position is with hands on the top of the bars, but moving the hands forward a little to rest on the brake hoods makes the body crouch a little, reducing frontal area and air resistance. The best shape is achieved with hands right down on the drops. For a conventional bicycle, this remained the most aerodynamic pose until the 1980s, when extensions to the bars allowed riders to stretch their arms farther forward and tuck down even tighter. These aero bars achieved worldwide popularity after Greg Lemond used them on the final stage of the 1989 Tour de France, a time trial finishing on the Champs Elysées, and wiped out a 50-second deficit to snatch overall victory from Laurent Fignon by just 8 seconds in a most dramatic climax.

While aero bars are permitted in some events, cycle sport's governing body, the Union Cycliste Internationale (UCI), has banned other even more aerodynamic positions, such as those used by Graeme Obree. His eponymous pose had the rider crouched with chest resting on the bars and hands tucked under the collar bone. When the UCI legislated against it, he pioneered the "Superman" position, with the rider's arms straight out front and the torso flat. That too was outlawed, but not before Chris Boardman had adopted it to go farther than any cyclist in a single hour.

Need to know

To estimate the total frontal area of a cyclist on an aero time-trial bike:

$$A = 0.218\, H^{0.0293}\, M^{0.425} + 0.650$$

where:

A = the estimated total frontal area (ft^2)

H = the rider's height (ft)

M = the rider's weight (lb)[10]

The coefficient of drag, C_D, is a number that aerodynamicists use to model the complex dependencies of drag on shape, angle in relation to the airflow, and some flow conditions. It's obtained empirically and by modeling and is an indication of the magnitude of an object's air resistance. The lower the value of C_D, the more aerodynamic the object.

The drag force, F_D, on an object moving through dry, 68°F (20°C) air is:

$$F_D = \tfrac{1}{2}\, C_D\, A\, \rho\, V^2$$

where:

F_D = the drag force (lb)

C_D = the coefficient of drag

A = the frontal area of the object (ft^2)

ρ = the density of dry air at 68°F (20°C) and sea-level atmospheric pressure (about 0.076 lb/ft^3 or 1.2 kg/m^3)

V = the speed of the object through the air (ft/s)

The effective frontal area is $C_D \cdot A$, the product of the coefficient of drag and the frontal area.

Aerodynamic positions compared

Sitting up with hands on top of bar

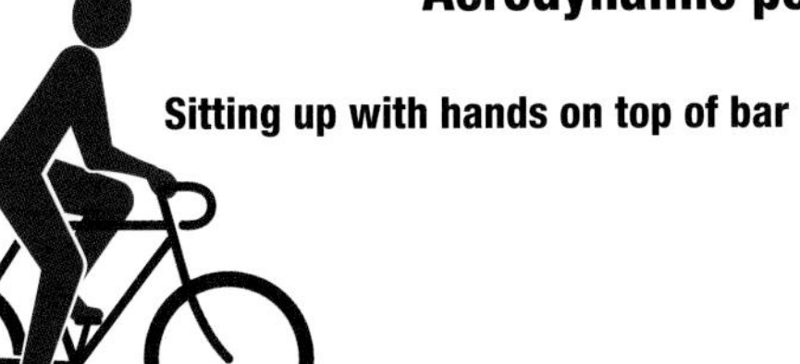

9.254 miles

Bending forward a little with hands on lever hoods

19.872 miles

Tucked down with hands on drops

20.973 miles

Tucked down with hands on extensions

22.095 miles

Obree position

29.808 miles

Superman position

35.030 miles

◀ ***Disappear into your bicycle*** *A rider's position accounts for about 65 to 80 percent of their total aerodynamic drag.[11] Even without an aero helmet, any rider can improve aerodynamics by flattening their arms, torso, and head and tucking in their elbows. This may mean they deliver less power to the pedals, but this is usually offset by the reduction in aerodynamic drag.[12] Recumbent designs were banned from conventional racing early in the twentieth century and, after interim regulations announced in the 1990s, tight limits on all aerodynamic designs were published by the UCI in 2000. More restrictions are currently being enforced. Chris Boardman set the world hour record of 35.030 miles (56.375 km). These are the distances he would have covered if he hadn't adopted the Superman position.*

science in action conquering the mountain

Cycling up hills holds a special attraction for many riders. There's the challenge of defeating the forces of gravity and, once over the summit, the thrill of converting potential energy into a high-speed descent. The two sides of the same hill demand different skills from the climber and the descender, but in both cases science can play a crucial role.

On an ascent, aerodynamics is not so important because of the low speed. Gravity is the major force to overcome; this is directly proportional to the combined mass of the rider and bike. The bike's weight is a smaller proportion of the total for a heavier elite rider (often only 12 percent) than for a lightweight (roughly 17 percent). So the lighter rider expends a bigger part of their energy raising just the bike up the slope. The lightweight, however, has a physiological advantage during the climb, because their relative maximum oxygen consumption (VO_2 max) is greater and outweighs the energy cost of the relatively large bike weight.[13] That's why more compact riders tend to reach the summit first.

On the descent, size is less important than posture, because that other invisible factor, air, can play a crucial role. As speeds can sometimes exceed 50 mph (80 km/h), the rider who tucks into the most aerodynamic crouch will gain time. The course of the route downhill has a bigger influence than it does during the ascent: switchbacks force riders to brake, lose momentum, and reduce any advantage they may have through being more streamlined. More skill and courage are also required to be able to assess the fastest line around a hairpin bend, where the land falls away precipitously at the side of the road. Experience, instinct, and ambition sort the mountain lions from the sheep.

▶ ***Mountain race*** *The mountain stages in the major races such as the Tour de France, Giro d'Italia, and the Vuelta a España can determine who will be the overall winner. Large amounts of time can be lost on the long, slow climbs and can't be recouped during the descents, which are of a much shorter duration and have, on average, a smaller variation in speeds among riders. Physiological strength is the most important factor during the climb, but on the descent psychological strength can help riders to round bends at awesome speeds. Here, Tejay van Garderen of the United States paces Australia's Cadel Evans, winner of the 2011 Tour de France, in stage 16 of the 2012 Tour. Van Garderen held on to the white jersey for leading the young rider classification for the rest of the tour and finished fifth in the general classification; his teammate Evans finished seventh.*

ŠKODA
ŠKODA
BMC
bmc

How does an aerodynamic helmet work? ↪ Should I wear a hat to get ahead?

Need to know
The contribution of the helmet to a rider's total aerodynamic drag is between 2 and 8 percent, depending on its shape.[14] Compared to a standard road helmet, an aero version can cut up to 90 seconds off a 25-mile (40-km) time trial, an improvement of about 3 percent.[15]

When a cyclist adopts their most aerodynamic position, the most important factor affecting their own air resistance, irrespective of the bicycle they are riding, is what they are wearing. Just as the profile of an aircraft's nose is crucial to its efficiency, the design of a cyclist's helmet plays a significant part in the rider's ability to ride most efficiently.

Helmets can help shape the ideal flow of air over and around the top of the rider. The aim is to keep the air moving smoothly as far back along the bike and rider as possible. So the front of the helmet must have rounded contours that guide the air smoothly around the head and encourage regular, laminar flow without turbulent eddies and vortices, which cause drag and slow the rider. The solution is to taper the shell, without any abrupt changes in line, and extend it to fill the gap between the back of the cyclist's head and their shoulder blades. When the air finally leaves the tail of the helmet, it should continue in a laminar flow along the rider's flattened back.

Unfortunately, the advantage of wearing a high-tech, scientifically designed helmet is lost if the rider doesn't hold their head at the optimum angle. A small dip of the head can raise the tail from between the shoulders and upset the airstream at the neck and shoulders. This loss of laminar flow might sometimes be balanced aerodynamically by a reduction in effective frontal area, but it hardly compensates for the investment made in an expensive helmet. Similarly, normal head movements such as turning to look behind can also make the air flow split and become turbulent, lowering efficiency.

Advantages of an aerodynamic helmet at different relative wind angles

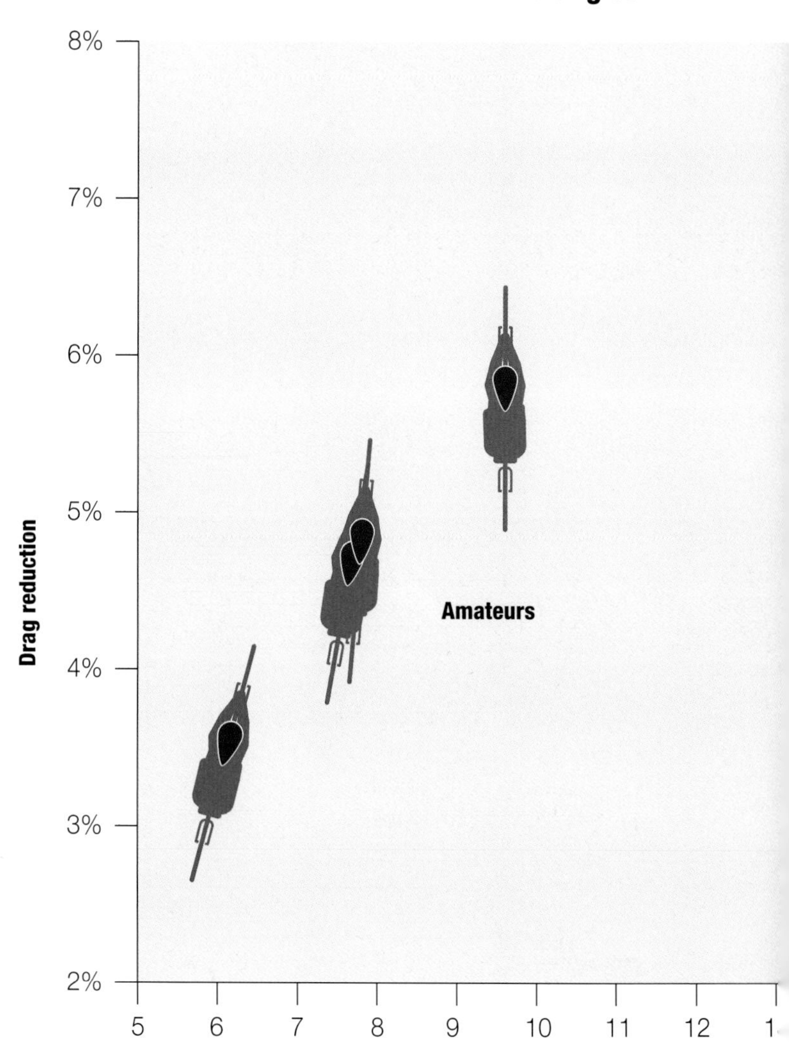

Effect of helmet surface roughness

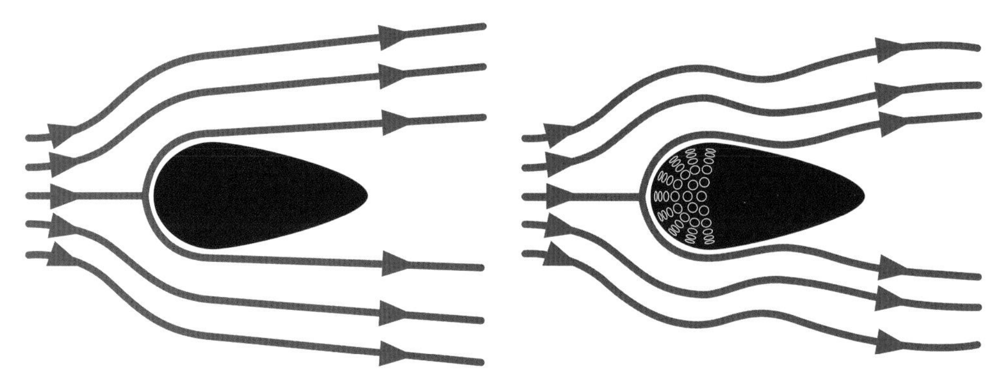

◀ ***Why dimples?*** *Most aero helmets have smooth surfaces to encourage laminar flow, but some are textured. The intentional roughness of the surface acts like the dimples on a golf ball, creating minute surface turbulence. As with a golf ball, it turns out that this turbulent air stays attached to the surface longer, so the separation of the boundary layer from the surface is delayed. Several helmet manufacturers offer dimpled surfaces, although scientific data proving the benefits is scarce.*

Relative headwind velocity 30 mph (13.4 m/s)

Professionals

◀ ***Don't go out without your hat*** *Wind tunnel tests have shown that the overall drag force that works against a cyclist and that requires them to pedal harder can be reduced by wearing an aero helmet. The benefits diminish as the angle between the rider's direction and the wind increases. In the worst test case scenario, with a wind angle of 15 degrees, an aero helmet on an amateur rider cut drag by about 3.6 percent. The best helmet on a pro with no crosswind cut drag by more than 7 percent.*[16] *Consequently, a rider's power savings during a 25-mile (40-km) time trial can be greater than 20 ft·lb/s (30 W).*

15 16 17 18 19 20 21 22 23 24 25

Power saving over 25 miles (ft·lb/s)

Holding the head position

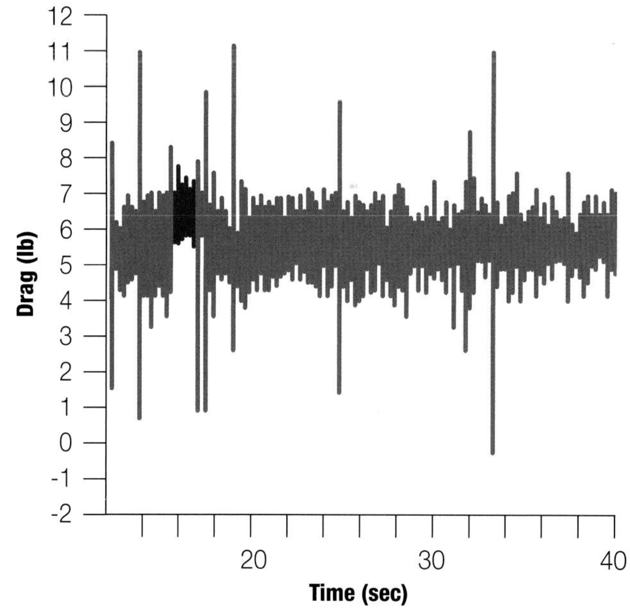

▲ ***Don't look down*** *The advantage of an aero helmet diminishes if the rider drops his or her head so that the helmet's tail rises. The graph shows how drag increased during a wind tunnel experiment—from 4½–7 lb to 5½–8 lb (20–30 to 25–35 N), at the 17 second mark—when the rider moved their head momentarily.*[17]

What difference does an aero frame make?

What's with the funny bike?

Engineers have been improving the aerodynamics of bikes for decades and it is a continuing, iterative process. However, when one component is changed to smooth local airflow, it can upset the aerodynamics of another part. Wind tunnels and computational fluid dynamics (CFD) show what should work best; test rides confirm or trash those results. It is an expensive process, but it has made bikes in general sleeker—and created a niche sector of exotic aero machines.

One component is the bicycle fork blade. This has been flattened and squeezed into an aerofoil profile to reduce its overall frontal area. Blades as narrow as (¼ in) 6 mm can save dozens of seconds on a time trial. Some forks with a single blade have been used to reduce frontal area, with the most extreme designs appearing on track bikes where the surface conditions are ideal (on less than perfect roads they are so fragile they could snap).

Handlebars, another key area for sleek design, come in a variety of aero designs. The most sophisticated are for triathlon, time trial, and track. These have elbow rests that allow the rider to tuck down, with their back almost flat and arms extended. With hands out front and close together, and elbows tucked in, the air is split to run smoothly around the arms and body. Thin, horizontal, airfoil-section bars cause minimum turbulence, and gear or brake cables can be routed inside them, out of the airflow. Frame tubes have aero sections to promote laminar flow, and some seat tubes curve closely around the rear wheel so no turbulence can creep in behind. Aero brakes are designed to sit within the confines of the frame, so they don't add to the frontal area and cause extra drag.

Negative drag

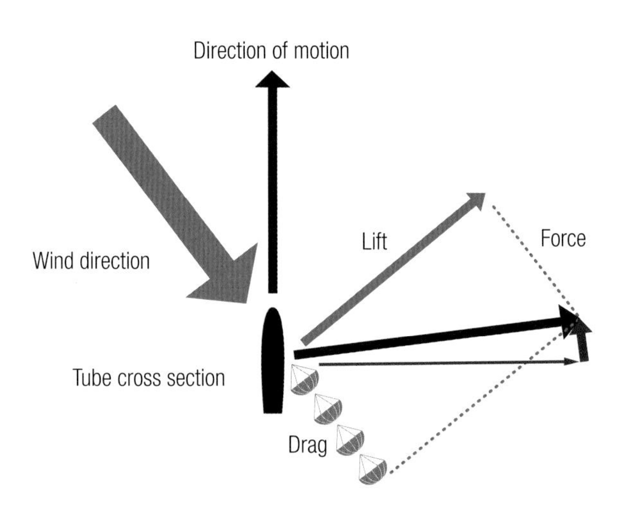

▲ ***Cycling close to the wind*** *In some circumstances the combination of wind speed, direction, and tube design can result in negative drag—otherwise known as thrust. This welcome phenomenon is familiar to sailors who trim their sails to move close to the wind. The wind creates a drag force in the same direction as the wind (blue parachutes), but aerodynamic effects from the wind moving across the shaped surface also create a "lift" force to the side (thin blue arrow). If this sideways force is big enough, the net force—the vector sum of the two effects, shown in black—may actually point a little forward.*

Now, if the shape is constrained to move in just one direction (a sailboat by its hull, a bicycle by its wheels), the only part of that net force that matters is the part that points in the direction of motion. Under the conditions shown here, that part (the thick red arrow) points forward—and gives the bike or boat a forward push.

This trick can be accomplished on bikes that have larger surfaces with tube profiles that extend into the main triangle, although such a design might make it less stable in crosswinds. One bike maker, Trek, claims the profile of its airfoil frame tubes can create this thrust under certain wind conditions and help the rider.

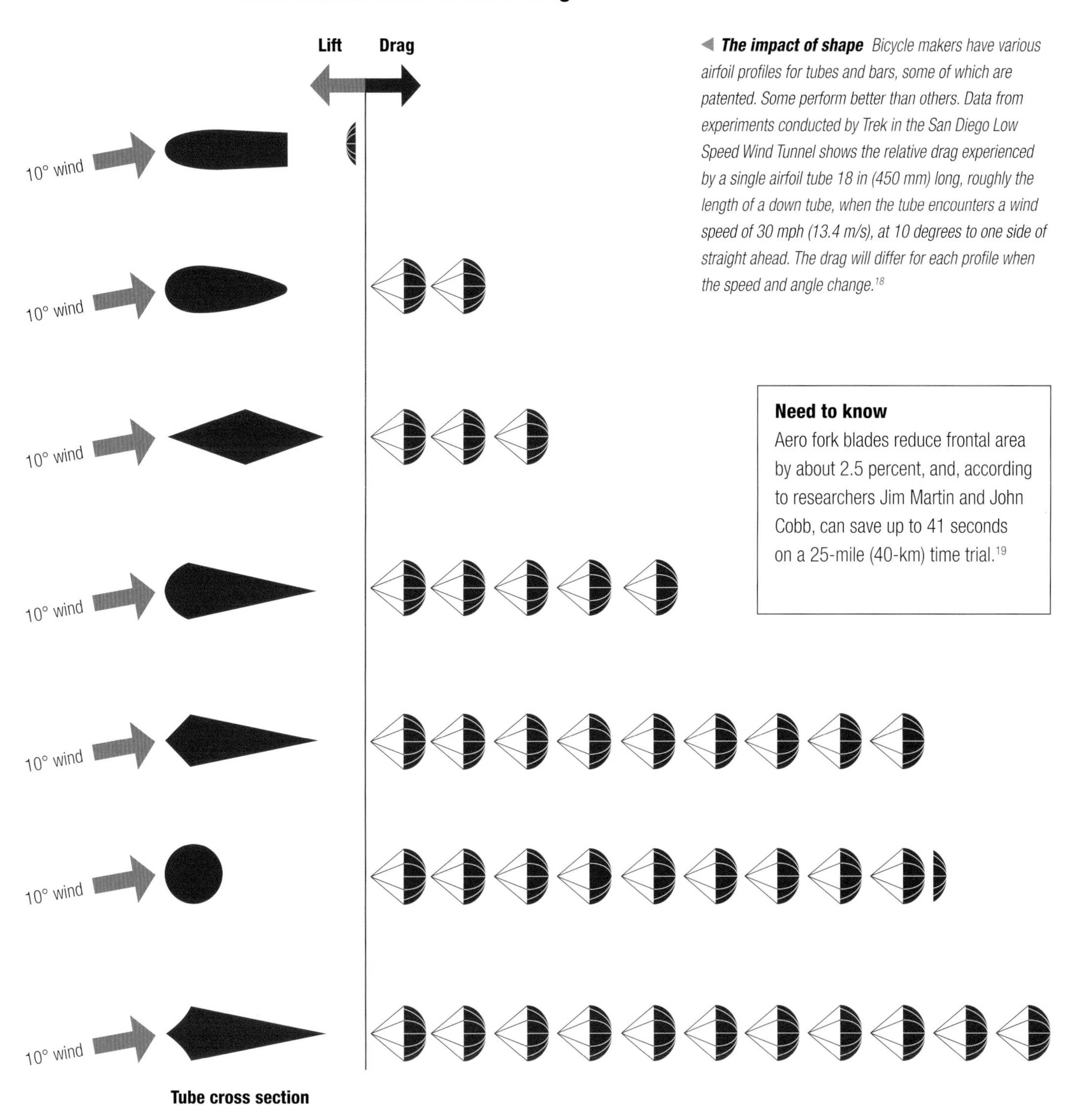

◀ ***The impact of shape*** *Bicycle makers have various airfoil profiles for tubes and bars, some of which are patented. Some perform better than others. Data from experiments conducted by Trek in the San Diego Low Speed Wind Tunnel shows the relative drag experienced by a single airfoil tube 18 in (450 mm) long, roughly the length of a down tube, when the tube encounters a wind speed of 30 mph (13.4 m/s), at 10 degrees to one side of straight ahead. The drag will differ for each profile when the speed and angle change.*[18]

Need to know

Aero fork blades reduce frontal area by about 2.5 percent, and, according to researchers Jim Martin and John Cobb, can save up to 41 seconds on a 25-mile (40-km) time trial.[19]

equipment: track

Velodromes were first built in the late nineteenth century as commercial entertainment venues where spectators could pay to watch professionals race unhindered by the rules and hazards of the open road. By the 1920s, track racing was one of the most popular sports in the United States, and the weather was excluded by enclosing the tracks and support facilities in buildings. New kinds of contests were developed to suit the circuits, including pursuit racing, six-day events, and the Madison, named after Madison Square Garden in New York City, where one of the early velodromes was built. Their popularity waned after World War II, but a revival in cycling has led to new tracks being built in North America, Europe, Asia, and Australasia.

Indoor tracks allow riders to cycle faster. The surface is smooth and predictable, the corners are banked to maximize speed, and the wind cannot hamper the riders' efforts because it is excluded. The faster the velodrome, the more attractive it is to riders and spectators alike, so the desire for speed has driven design and construction for some 130 years.

With two straights and two corners, the key to speed has been to get banking angles perfect. Get these right and cyclists can concentrate on reaching speeds of 55 mph (89 km/h) without having to worry about steering. The banking for 250 m (273 yd) velodromes (the track length required to host World Championship or Olympic races) has settled at around 12 percent along the straights and 43 percent on the corners. The transitions from bend to straight are usually made as wide as

▼ ***In for a spin*** *With a perfect surface, no hills to climb, and no wind to speak of, the benefits of advanced aerodynamic equipment and clothing can be massive for track cyclists.*

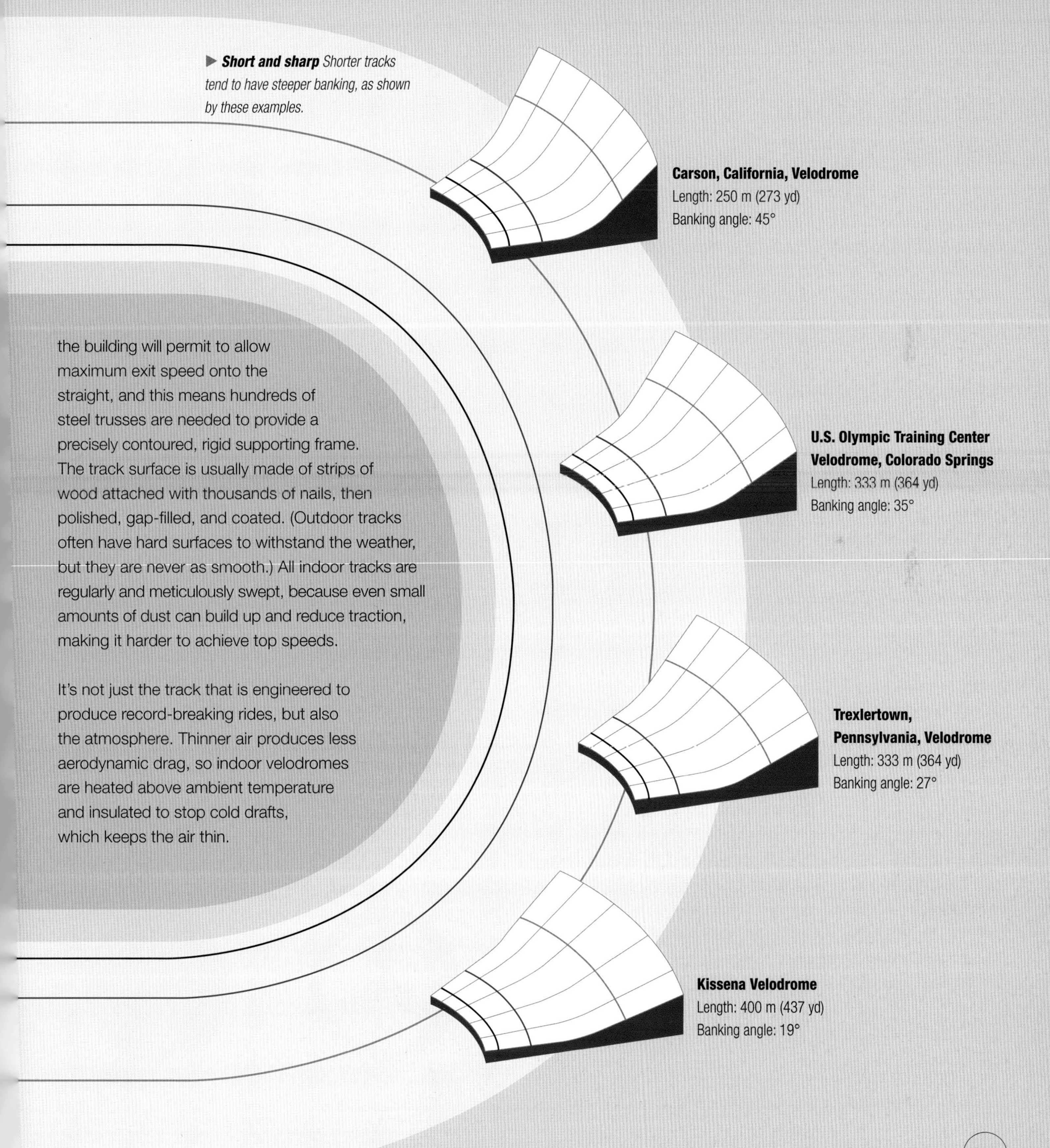

▶ ***Short and sharp*** *Shorter tracks tend to have steeper banking, as shown by these examples.*

the building will permit to allow maximum exit speed onto the straight, and this means hundreds of steel trusses are needed to provide a precisely contoured, rigid supporting frame. The track surface is usually made of strips of wood attached with thousands of nails, then polished, gap-filled, and coated. (Outdoor tracks often have hard surfaces to withstand the weather, but they are never as smooth.) All indoor tracks are regularly and meticulously swept, because even small amounts of dust can build up and reduce traction, making it harder to achieve top speeds.

It's not just the track that is engineered to produce record-breaking rides, but also the atmosphere. Thinner air produces less aerodynamic drag, so indoor velodromes are heated above ambient temperature and insulated to stop cold drafts, which keeps the air thin.

How do wheels affect aerodynamics? → Are my wheels slowing me down?

A wheel with wire spokes whisks the air into a frenzy as it turns. It promotes eddies, jets, and vortices that meddle with the air streaming past the bike, draining the rider's energy. It slows the bike down, but this effect can be reduced with clever design. An aero wheel is the answer.

There are three kinds of aero wheel—tension-spoked, compression-spoked, and disc. Tension wheels can have as few as 12 wire spokes.[20] To reduce turbulence, aerodynamic tension spokes with flattened profiles less than 0.2 in (5 mm) across are used, and rims can be up to 4 in (100 mm) deep for the same reason. Compression-spoked wheels, with three, four, or five spokes, are made of carbon fiber composite. The same material is used for solid disc wheels, which can be flat, convex, or lenticular. Solid front wheels are used only on indoor tracks because the crosswinds outdoors can affect steering significantly.

Wheel aerodynamics is becoming an increasingly sophisticated area of research. Wheels experience two kinds of drag, and they are also susceptible to variations in wind direction. The forward movement through the air causes translational drag, sometimes referred to as linear or body axis drag, and the spinning around the axle causes rotational drag. Surprisingly, the drag on discs and compression-spoked wheels can actually decrease in mild crosswinds as a result of aerodynamic lift.[21] The makers of aero wheels invest a lot in wind tunnel trials to assess the behavior of designs in different wind conditions. As a result, some wheel sets cost as much as top-quality frames. Computational flow dynamics (CFD) is also used, because it can predict flows and forces at hard-to-instrument locations.

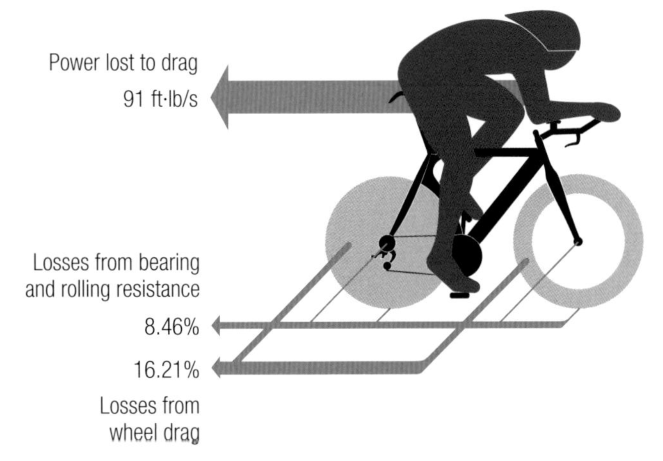

▲ ***Wheels getting you down?*** *Wheels can account for a significant proportion of the power lost, according to comparative wind tunnel measurements of riders wearing aero helmets and using time-trial bikes fitted with aero bars. The importance of wheel aerodynamics decreases as the cyclist accelerates, and the impact of rolling resistance from tires and and bearing inefficiencies in the wheels and bottom bracket diminishes even more dramatically. Yet wheels with less air resistance make it easier to reach those higher speeds. The diagrams show the total power lost to drag, and the percentages of this associated with bearing and rolling resistance and with wheel drag for several riding speeds.*[22]

▶ ***Wheel revolution*** *The standard 36-spoke tension wheel with a slim rim is rarely seen on high-end, time-trial bikes today. Sleek aero models are better at minimizing air disturbance.*

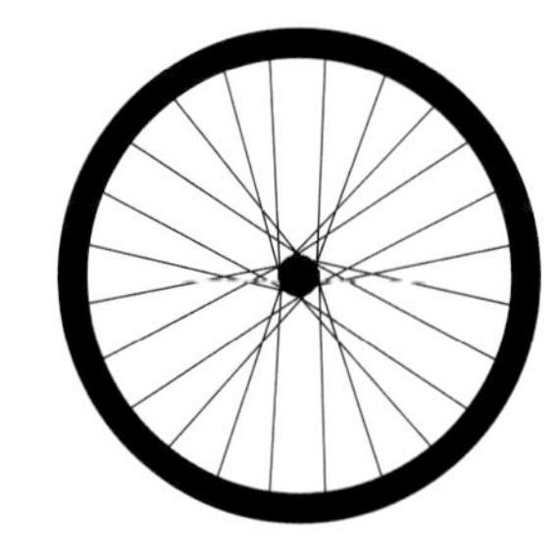

Aero rim and 24 tension spokes

Deep aero rim and three compression spokes

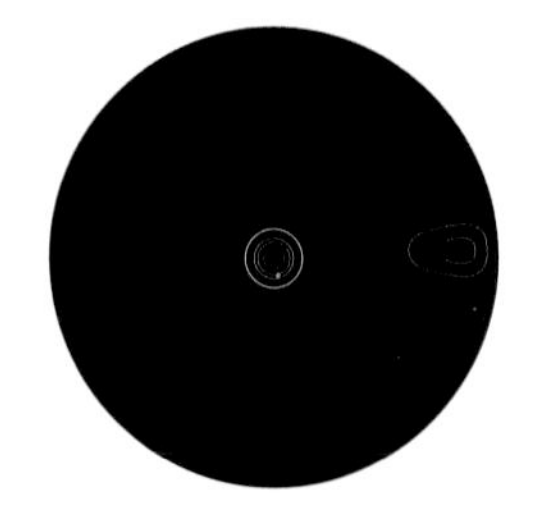

Carbon disc

Contribution of wheels to power loss from drag

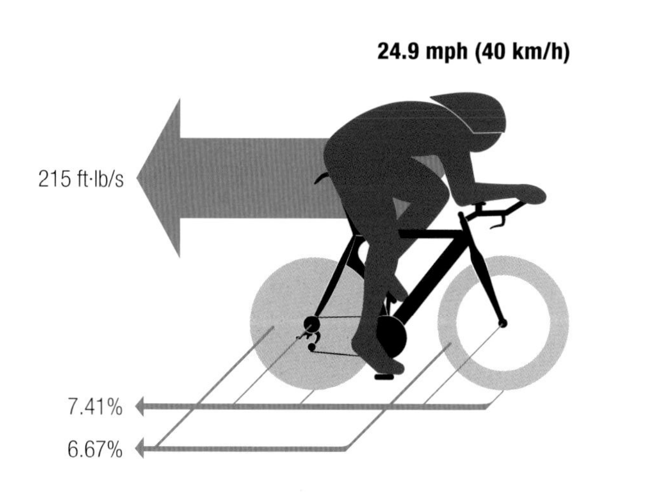

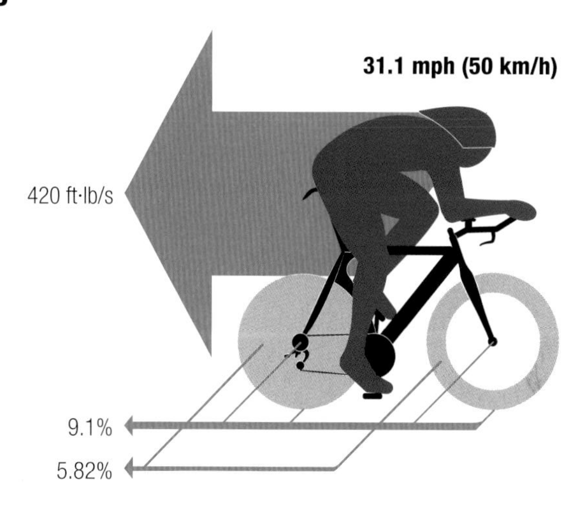

Contributions to drag by wheel components

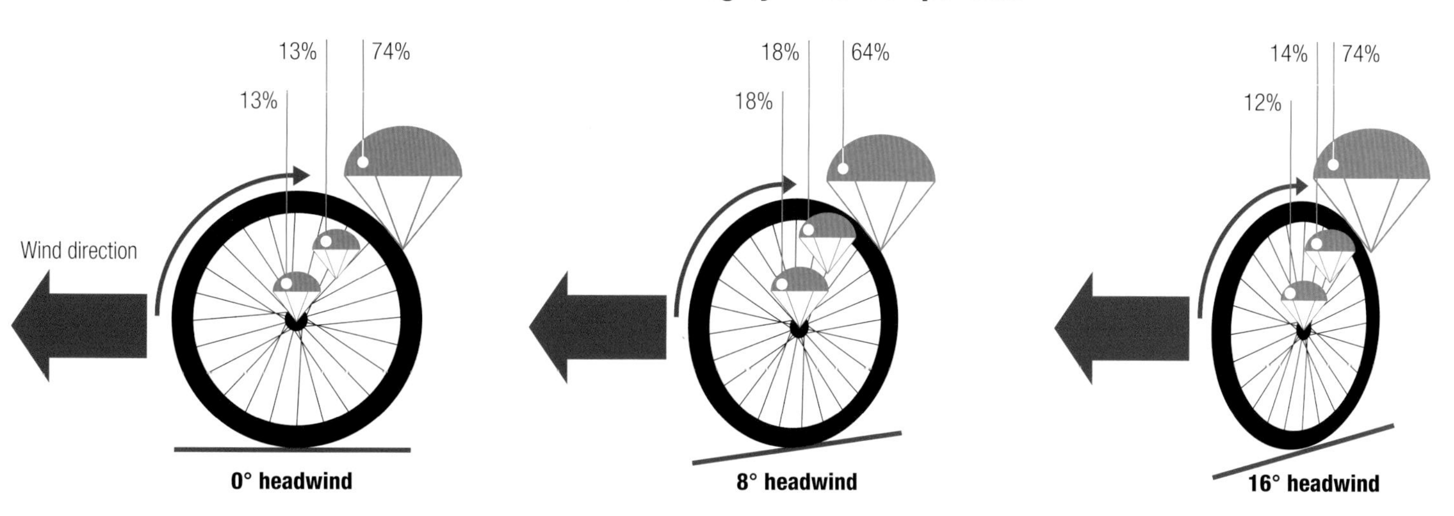

▲ ***Wind in my wheels*** *How much translational drag is caused by the different parts of a wheel as it moves forward through the air? It's not feasible to instrument a spinning wheel to find out, so Intelligent Light used computational fluid dynamics to model a Zip 404 front aero wheel with a Continental tubular tire at different headwind angles. They found that the drag force caused by the tire and rim was about five times greater than that from the hub or the spokes. Interestingly, the total wheel drag dipped when the headwind angle was between 8 and 10 degrees, owing almost entirely to improved rim and tire aerodynamics.*[23]

Need to know

Wind tunnel tests have shown that a deep rim-section, aero tension wheel and a compression wheel can produce 60 percent less drag than a standard 36-spoke wheel in a direct headwind. A disc wheel can cut drag by 70 percent.[24] As the angle, or yaw, of the crosswind increases, drag gets stronger on the tension wheel as the spokes become more exposed to the airflow. This happens to a lesser extent with compression wheels. The drag on a disc wheel can rise suddenly as the wind angle shifts.

What is the advantage of riding in the slipstream of another bike?

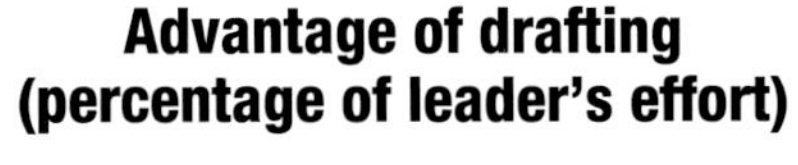

There is one way of harnessing the effects of aerodynamics that doesn't cost anything—riding in the slipstream of another cyclist, also known as drafting. Put simply, it means the rider in front does all the hard work of pushing the air aside while the following rider sits behind. It's a tactic seen in every road race in the form of the peloton—this cuts the workload greatly for the following riders.

Scientists have analyzed the airflow and forces involved when two or more riders draft in the wake of a leader. Early experiments involved cyclists freewheeling in line; more recent work has used wind tunnels, although this tends to reveal only the maximum benefits—the bikes are fixed in a single optimum position, a situation that's unlikely on the track or road. Other researchers have measured instantaneous oxygen consumption of actual drafting riders to see how hard each one is working. Whichever method is used, the answer is as clear to the scientist as to any cyclist—it's less tiring behind. Some research indicates that a third-placed rider benefits even more, and that those in the middle of a large peloton do less than half the work of the person at the front because they are also protected from buffeting by crosswinds.

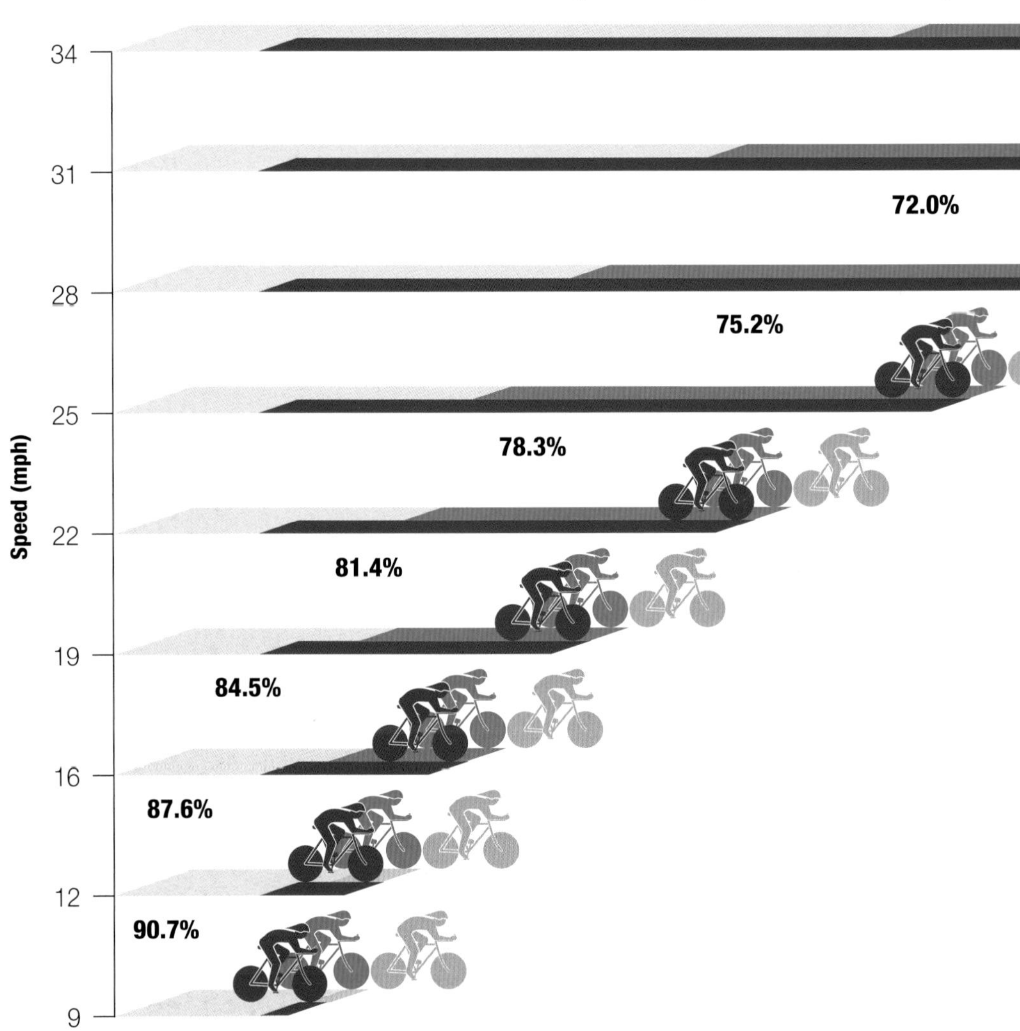

Unusually for research into the science of cycling, the knowledge gained about drafting has had an influence in other subjects. The science and math behind the dynamic organization of pelotons, which arrange themselves organically to reduce the work of most of the members, have been used as a framework to explore how groups of nonhuman creatures, such as flocks of birds, shoals of fish, and huddles of penguins, organize themselves.[25]

65.8%

69.0%

◀ ***The waiting game*** *It's a lot easier to follow than to lead, even for nonelite cyclists. The second rider in a line could be using about 1 percent less energy than the leader for every 1 mph (1.6 km/h) of velocity.[26] So when a pair is pedaling at 25 mph (40 km/h), the unskilled wheelsucker could be saving 25 percent in energy compared to the leader. This leaves the drafting rider with greater reserves available for later in the race.*

▼ ***Follow the leader*** *The US national team of elite cyclists preparing for the 1996 Olympics used two different tracks, a wind tunnel, and power meters to measure how much easier they found it to cycle behind the leader of their pace line. The second-placed rider used only 61 to 66 percent of the power used by the leader, and the rider in third place used even less, 57 to 62 percent. The fourth rider was almost the same as the third.[27]*

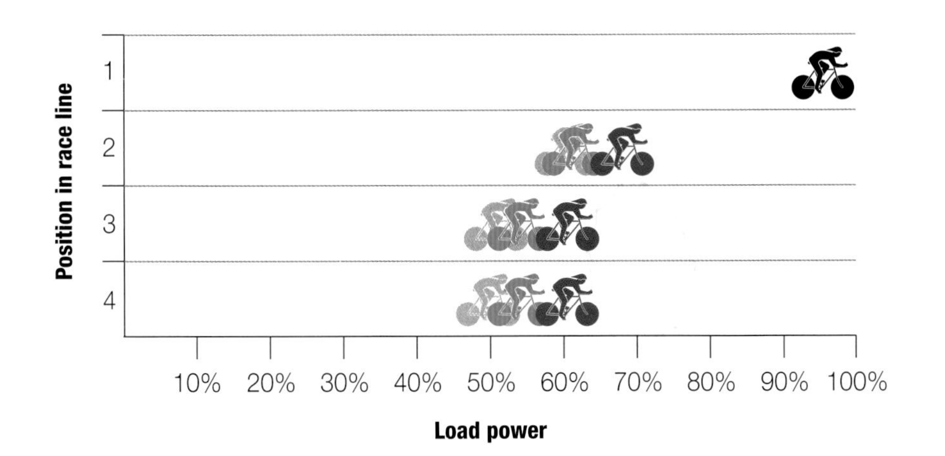

Lead cyclist

Indoor velo

Atlanta velo

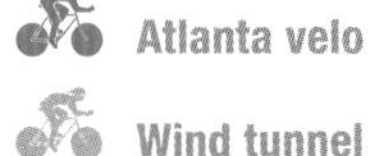
Wind tunnel

CFD model of slipstreaming

▲ ***Right behind you*** *This computational fluid dynamics representation shows how the drafting rider is in a zone where the force of the airflow is far less than that on the leader. There is some turbulence, but its negative effects are more than outweighed by the reduction in air resistance, especially on the bulky areas of the body.*

Need to know

It's best to draft as close as possible, although some benefit can be gained as far behind as 10 ft (3 m).[28] It's not worth drafting at speeds slower than 10 mph (16 km/h).[29] The more aerodynamic the bike in front, the less advantage there is in drafting.

This chapter is about you and how you can become a better cyclist. It explains how the different systems of your body function and how, when treated carefully, they can be improved to work together and give you a better cycling experience. You don't have to go faster or farther but you can raise your ambitions by knowing something about the possibilities revealed by science. It's a happy coincidence that cyclists have become the preferred lab rats of sports science. The bicycle ergometer is a convenient way to run repeatable, controlled experiments indoors, with the participant in a position that allows them to be easily monitored. Every year, scores of volunteers are wired up by researchers and set off on stationary time trials. The knowledge gained has helped deepen our understanding of physiology and endurance, and we are constantly finding out more as fresh results are published. There is a lot more to learn, but this chapter will tell you enough about yourself to make your cycling even more rewarding.

chapter six

the human factor

How does a cyclist's body work? → Am I a machine?

The bicycle both constrains and liberates the body, restricting our choice of postures so our maximum power can flow freely. But unlike the bike, most of our components are hidden and are infinitely more sophisticated. Understanding how our body works helps us as cyclists to ride farther, faster, and for longer without tiring.

The human body contains a host of complex, collaborative, and interdependent systems. Of most interest to cyclists are the nervous, musculoskeletal, circulatory, respiratory, and skin systems. All these systems work together. For example, when a rider wants to speed up, a message from the brain instructs the quadriceps muscles to work harder, which places a greater demand on the heart to supply oxygenated blood fresh from the lungs. As the rider turns the pedals more quickly, the extra work raises their temperature, so the skin produces sweat to keep the body cool. For the rider, all this happens automatically but, just like the bike, there are limits to the performance of each part so an awareness of how they function and what they do increases the cyclist's ability to get full potential from each part.

A bicycle's components can be updated separately for better specific performance, but a key difference between man and machine is that training improves all of the human body; most exercise has a holistic impact. It's obvious that leg muscles develop more than arms through cycling, but there are also invisible benefits for the cardiovascular and respiratory systems. Likewise, sprinting practice not only promotes muscle growth suitable for sudden acceleration but may also quicken reactions. The vast number of physiological and biochemical activities, from the obvious leg movements to submicroscopic nutrient metabolization, can be tuned and trained to make cycling faster, easier, and more rewarding.

Lungs

The lungs transport oxygen from the air into the bloodstream. The total lung capacity of adult males is around 370 in³ (6 liters) and of adult females 290 in³ (4.7 liters).

Breathing rate increases from 12 to 15 times a minute at rest to 20 or more during exercise, while the volume of air expired per minute is increased by up to 40 percent.

Muscles

We can voluntarily contract about 650 skeletal muscles to make the body move. Together these muscles make up almost half the weight of a human being.

With exercise, muscle oxygen consumption can increase up to 70 times the resting value. Trained muscles will have up to 50% more capillaries than untrained muscles, providing them with more oxygen.

Bones

The bones of our skeleton are the reinforcing elements that give humans structure. An adult has 206 bones and the skeleton of a male adult weighs 9 to 13 lb (4 to 6 kg), while that of a female weighs 7 to 9 lb (3 to 4 kg).

Exercise increases bone strength and density.

The human machine

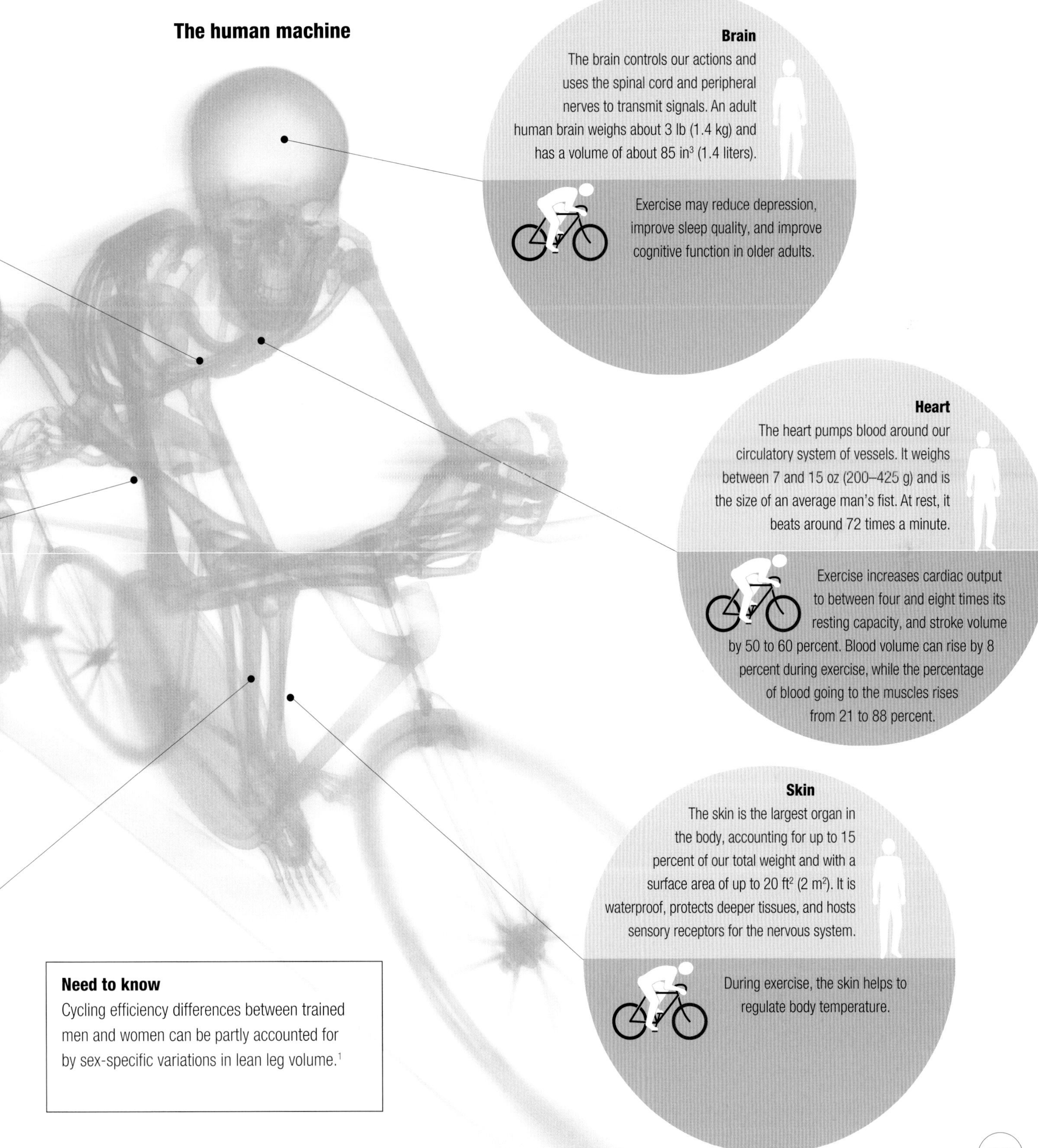

Need to know

Cycling efficiency differences between trained men and women can be partly accounted for by sex-specific variations in lean leg volume.[1]

How does my brain help me ride?

Can I swerve like a pro racer?

Point at which the rider spots a hazard

Picture this. Riding with friends on an open road, the wind in your hair, warm sun on your face, the sweet scent of blossom in the air, and the quiet hum of tires on the blacktop. A rider falls into your path and you instinctively steer around them. How did you react so quickly?

All the body's activities are coordinated by the nervous system, which is made up billions of nerve cells, called neurons, which are interconnected and controlled by the brain. The brain receives and processes information from all parts of the body and sends back instructions. Information travels between neurons in waves of chemical and electrical changes, using potassium and sodium ions to create signals, called nerve impulses. They trigger our reactions, so that we can avoid riding into fallen companions.

There are two components to every physical reaction—the time it takes for the nervous system and brain to identify that there's a hazard, and the time it takes for the muscles to move. It is well known that being fit means that you react faster—and that you react faster still while exercising. By monitoring pedaling riders, scientists discovered that the ability of the nervous system and brain to react does not speed up at all. The instructions still travel at 358 mph (160 m/s) during exercise—it's the muscles that react more quickly if they are exercising.

It's as if muscle tissue is more alert while exercising. In fact, muscle reaction time improves by more than 15 percent while you are exercising.[2] So when the sensory neurons relay to the brain information captured by the eyes showing that there is a hazard on the road, the brain sends an impulse through motor neurons to make the arm and hand muscles turn the handlebars in good time to avoid a collision.

Need to know

The brain contains more than 100 billion neurons, each communicating with thousands of others. Our brain knows what decision we have made as much as 400 milliseconds before we ourselves become aware of having made it.

Humans can only be fully conscious of four or five elements of a scene or event at any one time.[3] So it is important to focus on the priorities—the road and other road users.

Reaction times of fit and unfit riders

▼ ***Fit for purpose*** *The reaction times of fit riders are significantly faster than those of less fit riders. It's a physiological improvement brought about through a regular, good level of exercise. The reaction time differences are measured in tens of milliseconds, but that's long enough to avoid a collision with a sudden obstacle. A rider who hasn't exercised is significantly slower to react and less able to avoid the hazard.*[4]

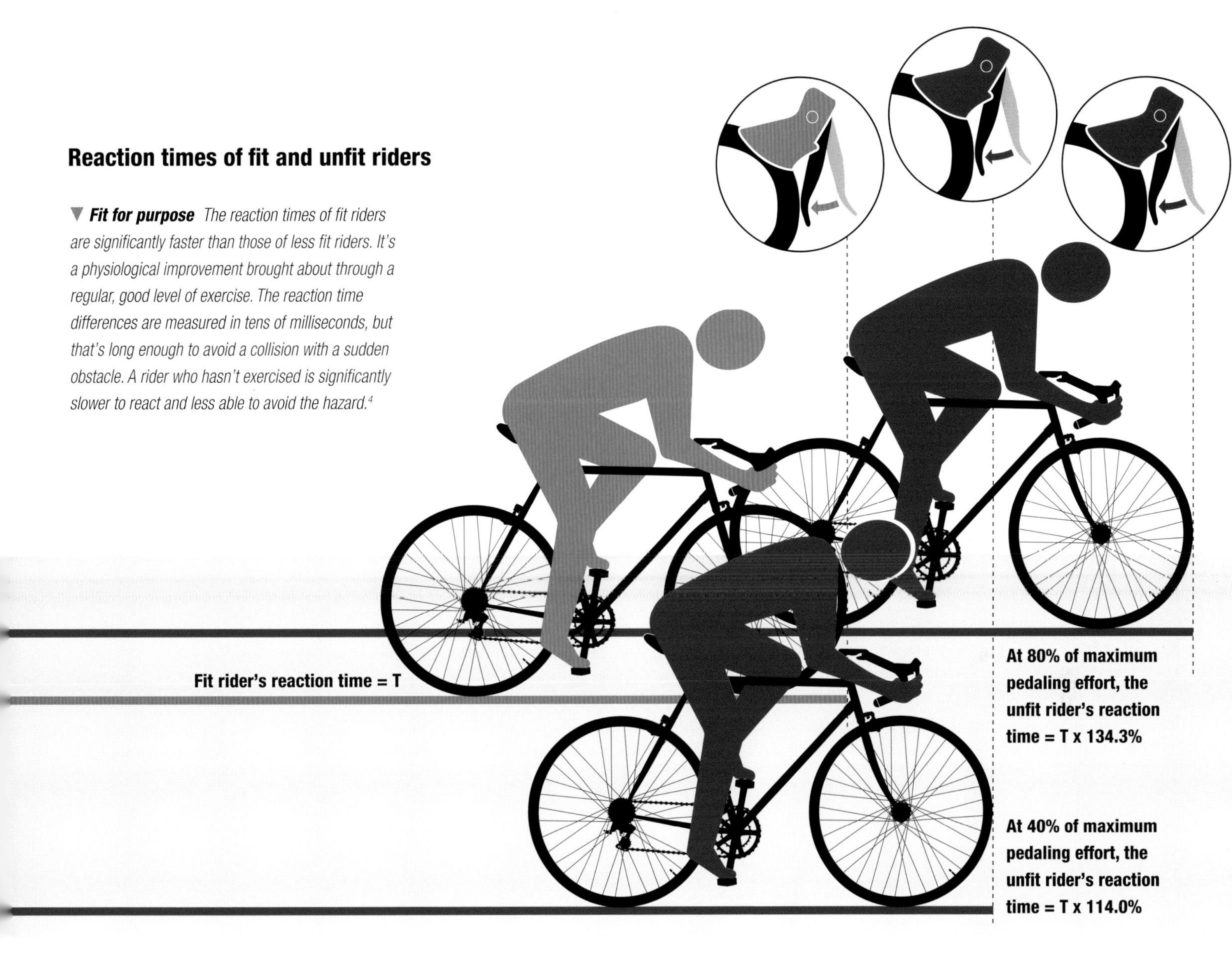

Coming up behind

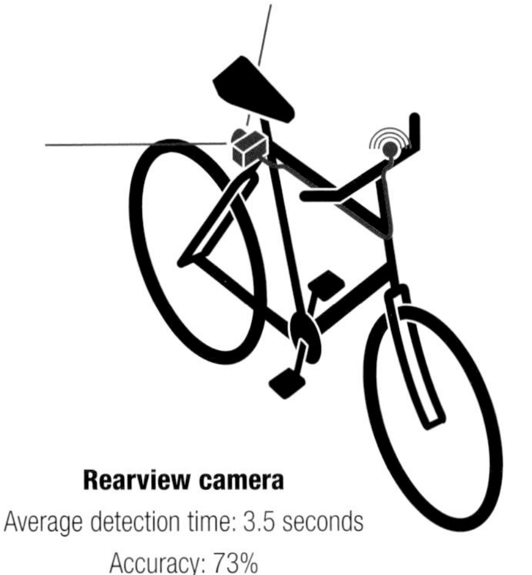

Rearview camera
Average detection time: 3.5 seconds
Accuracy: 73%

◀ ▶ ***Sixth sense*** *Technology could give cyclists more time to react by detecting events that human senses cannot. Cyclists don't have eyes in the back of their head and can't always hear vehicles approaching from behind, but two trials of new technology could change all that. The first was a rearview camera with computing power to flash an alert on the bike's computer whenever a vehicle approached from behind.*[5] *The second trial used a similar system but instead of vision it detected approaching vehicles by their sound.*[6]

Rearview microphone
Average detection time: 1.4 seconds
Accuracy: 81%

How do a cyclist's muscles behave?

Where do I need to bulge most?

Skeletal muscles make our bodies move, mostly by articulating our joints. They react to nerve impulses that turn thought into action. Muscle tissue can exert only a pulling force; it cannot push and so, fortunately, the cyclic aspect of pedaling gives each muscle a chance to recover before the next stroke. At the ends of each muscle, connective tissue sheaths meld together to form tendons, which are anchored to bone. A tendon behaves like a spring to modify the forces, so our muscle and bone are less likely to develop damage if we work them too hard. Nonetheless, vigorous weight training can double or triple muscle size (but disuse can shrink muscle size by 20 percent in two weeks).[7] Used intelligently, muscles can get a mountain biker up a steep hillside or propel a sprinter across the winning line at 35 mph (55 km/h).

The main cycling muscles are in the calf in the lower leg and the quadriceps and hamstrings in the upper leg. These muscles contract in a sequence that lifts the leg at the hip, bends the knee, and flexes the ankle to create the pedaling action. The demands on each muscle change as the rider pedals and also when the rider adjusts his or her position for best comfort, aerodynamics or power. The most significant change happens when the rider stands on the pedals to maximize power, such as when attacking a climb. Then, the gluteus maximus in the buttock has to work harder on the upstroke to lift the leg. Also, the tibialis anterior at the front of the shin greatly increases in effort to raise the foot, which has flexed more at the ankle, and to be ready to transmit the pedaling down force again.

Power drivers

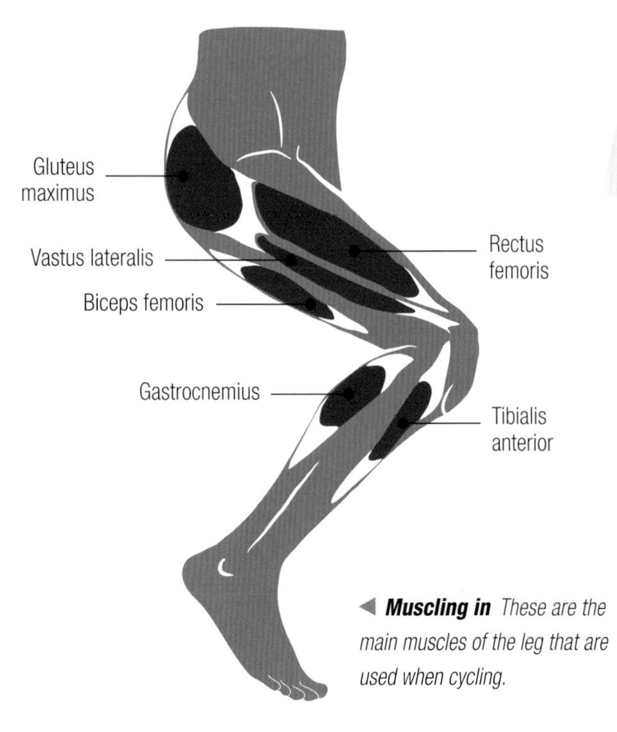

◀ ***Muscling in*** *These are the main muscles of the leg that are used when cycling.*

Need to know

Muscle makes up about 40 percent of body weight and is the most abundant tissue in the human body.

Adults have at least 650 skeletal muscles in their bodies.

▶ ***Rising power*** *Climbing hills demands more work, and sometimes it's necessary to stand on the pedals to produce enough force and maintain speed. This calls for a change in the effort from muscles because of the changes in posture, and the weight of the body has to be carried by the legs instead of the saddle. Different muscles peak in their use at different points in the rotation of the cranks, but the work done by the gluteus maximus (in the buttock) and tibialis anterior (close to the shinbone) is clearly much greater when standing than when sitting.*[8]

Muscle use during a climb (standing and seated)

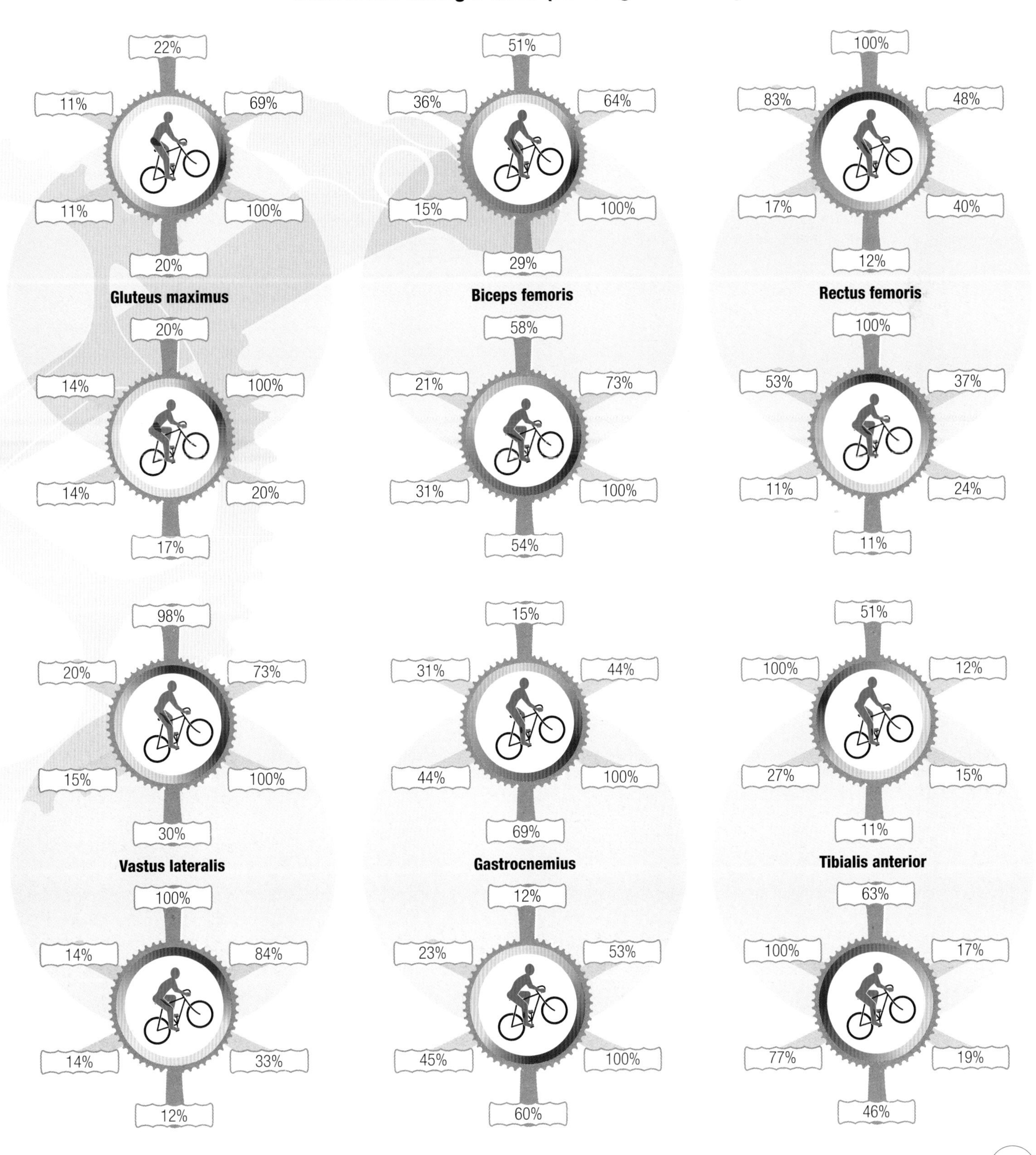

equipment: clothing

"Beware all enterprises that require new clothes," American philosopher Henry Thoreau warned. Within 30 years of his death, the cycle craze had begun and, with it, a market for new garments that were comfortable, protective, and attractive for cyclists. A nineteenth-century sociopolitical movement, the Rational Dress Society, urged Victorian women to swap their corsets and long skirts for the bloomers and divided skirts that made cycling easier. Natural textiles made of wool, cotton, and leather were replaced by synthetics during the second half of the twentieth century, improving the functionality of cycle clothing. Weight was reduced, and insulation and breathability have increased along with durability and comfort.

◀ ***Club class*** *(ca. 1880)*
Despite the frequent risk of going headfirst over the front of a high-wheeler, a cloth cap was the only protection donned by the first club men. It's a garment that has survived largely unchanged since the sport's inception, although it's banned from pro competitions.

◀ ***Divide and unrule*** *(1897)*
Divided skirts and bloomers, a necessity for safe cycling, were also adopted as a mark of women's rebellion against the restrictions of nineteenth-century fashion.

▶ ***Loose fitting*** *(1925)*
The natural fiber clothing worn by Tour de France double winner Ottavio Bottecchia just about kept him warm, but the creases and wrinkles were not aerodynamic.

Fabric and cut distinguish cycle clothing from other apparel, with sports cyclists wearing the most extreme garments. Close cutting maximizes aerodynamics and stops material from flapping into moving parts. Materials that resist dirt, are impermeable to wind, dry quickly, and retain their shape are favored. Mixtures of natural fibers, such as the Brilliantine cotton and worsted wool of the nineteenth century, were the best options. The first vulcanized rubber was used as rain gear.

Synthetics changed everything. In 1959, scientists developed a polyurethane–polyurea copolymer and branded it as Spandex (Lycra in Europe). By the 1970s, it had replaced the traditional alpaca wool shorts and chamois leather inserts. W.L. Gore and other manufacturers produced breathable, waterproof gear, with pores too small for droplets to enter but large enough for vapor to exit. Textile freshness has been enhanced by a better understanding of bacteria and how to combat them with impregnated deodorizing chemicals. Science has also proved that one natural fiber used for cycling garments can do the same thing without chemical intervention: Merino wool has a combination of microbes and morphology that absorbs and holds odors.[9]

Protective hats were made first of pith strips and then of leather, evolving quickly into "hairnets." Today, helmets are mass-produced with hard polymer shells and foam linings that meet certificated impact standards. Buttons may still feature on leisure gear but serious clothing has zippers and Velcro. Race clothing is now skintight and nineteenth-century riders would regard it merely as garish underwear.

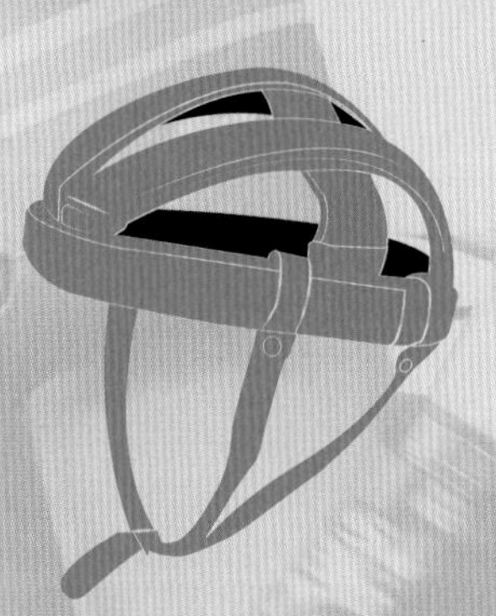

▲ ***Soft option*** *(ca. 1969)*
The padded leather "bunch of bananas" offered riders throughout the twentieth century little more protection than the fruit itself. It weighed little, was well-ventilated, and may have given some a comforting but false sense of security.

▼ ***Hard headed*** *(1978)*
The first purpose-made plastic shell helmet became available, descendents of which are now mandatory in professional road races. Arguments rage about their benefits and disadvantages for everyday cyclists.

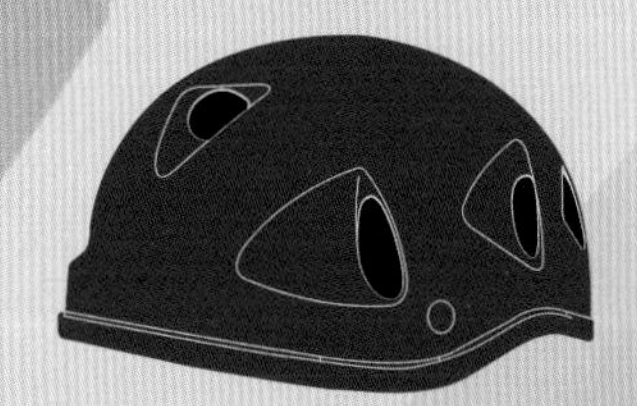

► ***Body armor*** *(2007)*
Downhill mountain bikers race down slopes used by skiers in winter, but at a time when there's no snow to cushion any impacts. Full body armor and visored helmets are needed.

▲ ***Skintight*** *(2012)*
One-piece, body-hugging, and stretchable, the road racer's skinsuit is as aerodynamic as the human body. It's been outlawed in some mountain bike race categories because it offers little protection.

How are muscles fueled? → Am I pumping ions?

Everything needs energy to move, and the muscles of cyclists are no exception. The key is a biochemical called adenosine triphosphate (ATP), a molecule that stores and releases energy and is found in large quantities in most organisms. Supplies of ATP are provided from different sources within the body to power different intensities of exercise. The bonus is that they can each be replenished in a variety of clever ways, although at different rates. It's the rates of recharging that limit how quickly or how far we can cycle.

ATP is a molecule of adenosine, a compound derived from nucleic acid, with a tail of three inorganic phosphate (P) ions attached. When the bond connecting one of the phosphates breaks, energy is released to be used by the muscle. This reaction happens millions of times a second to release enough energy to keep your muscles moving. Each muscle hoards enough ATP for only about five seconds of maximum effort, such as accelerating a mountain bike up and over a van-size rock.[10] After depleting its own stash of ATP, the muscle has to call on a store of phosphocreatine (PCr) molecules to help out. These donate stored phosphate ions to adenosine diphosphate (ADP) to recharge the battery, but only for less than 10 seconds, just enough to allow a sprinter to go flat out for the final 150 yards to the line. So for protracted exertion, such as when you're trying to break away from the peloton, ATP has to be synthesized in other ways. If oxygen is not available (anaerobic exercise), glucose in the blood is broken down to make ATP. This works fine, but only for a few minutes—acidic by-products then make muscles hurt, and hydrogen ions cause fatigue. The steadiest way to produce ATP is during aerobic exercise, when oxygen is available to the cells. It's a slower but longer-lasting process that utilizes proteins, fats, and carbohydrates to produce copious quantities of ATP without unwanted byproducts. This is what's happening in the legs of cycle tourers, and it keeps them riding all day long.

Training and diet help optimize ATP synthesis so that the cardiovascular system and muscle cells can support more intense and longer exercise aerobically before the less efficient anaerobic pathways are needed. Training and diet also promote faster recovery from extreme bursts of activity, such as sprinting, by helping muscles clear inhibiting acids. Oral sugar supplements during a ride can delay the energy trough called "bonk," the condition when all muscle glycogen has been depleted.[11]

▼ ***Powerhouse of the cell*** *When the bond holding a phosphate (P) ion to ATP is broken, energy is released. The resulting adenosine diphosphate (ADP) and P are later recombined by enzymes (protein catalysts) from our food and drink to synthesize more ATP when the muscle needs it.*

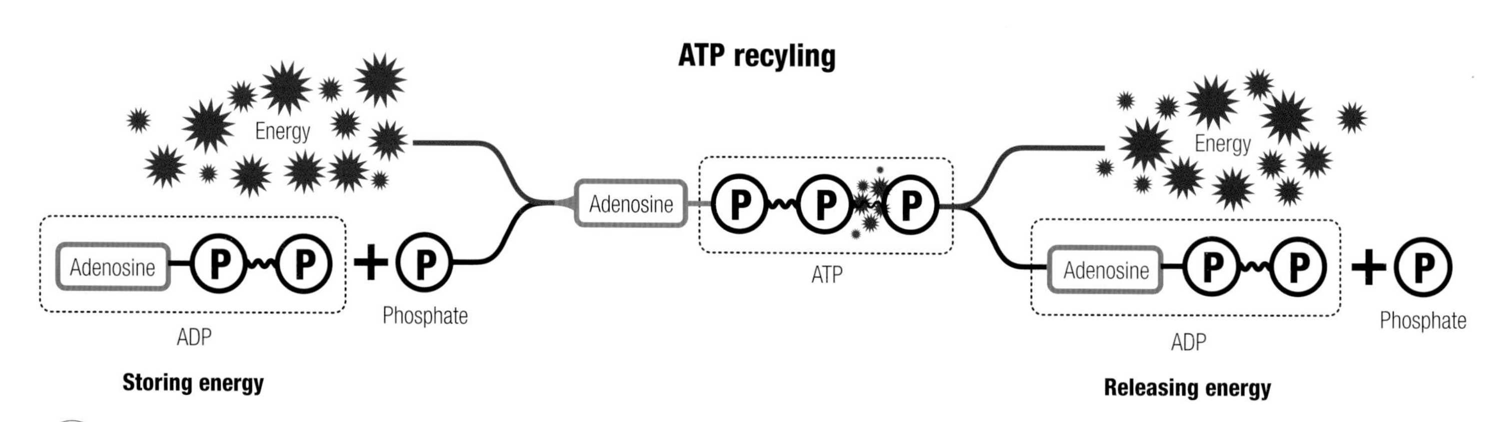

Aerobic or anaerobic?

▼ ***Sustaining physical activity*** *Muscles are fueled by energy released when ATP breaks down, but only a little is available instantly. There are several ways that more can be synthesized over time, and this is what limits the intensity and duration of exercise.*

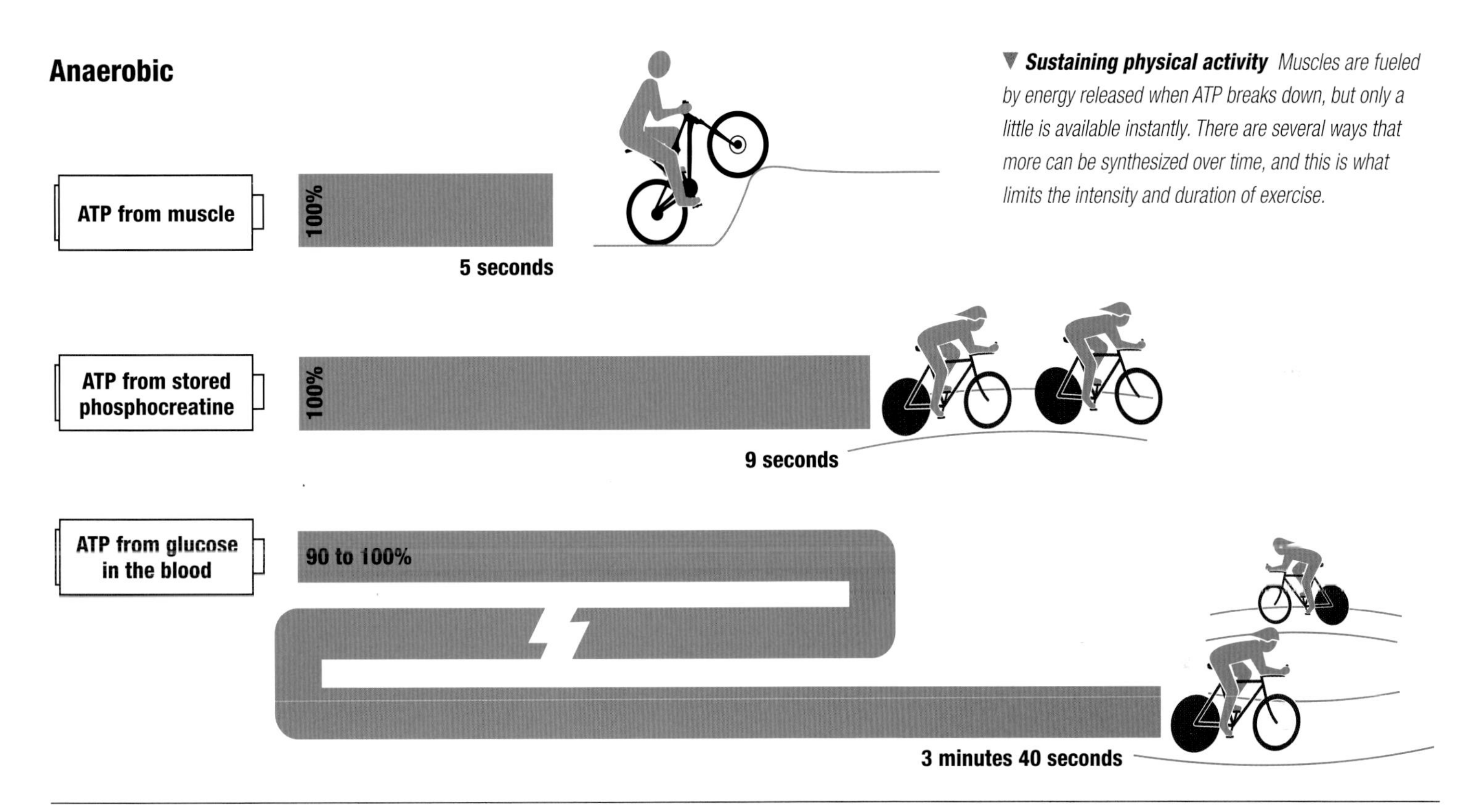

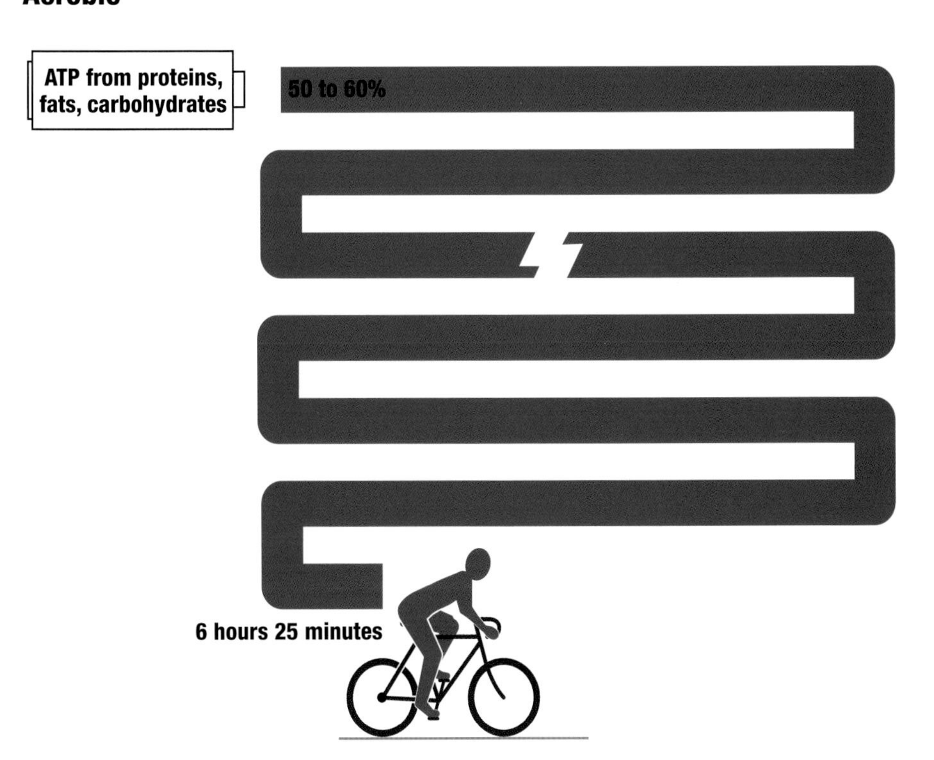

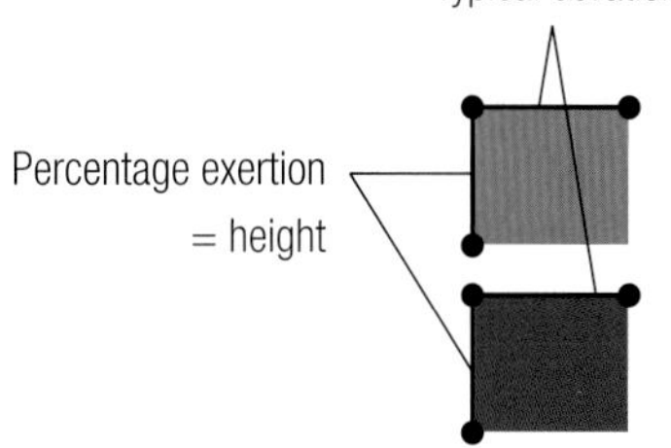

Need to know

The human body, on average, contains 9 oz (250 g) of ATP, equivalent to the energy contained in one AA battery.

Every day, a person recycles ATP equivalent to about half their own body weight—but this can double with physical activity.[12]

What are fast and slow twitch?

IIb or not IIb?

Human muscles contain two different kinds of skeletal muscle fibers—slow twitch and fast twitch. Their names say it all. Slow muscle fibers are best suited to extended periods of activity with low levels of force, such as long-distance riding. Fast twitch fibers are better suited to short periods of vigorous activity such as kicking for the line in a sprint.

Slow (red) and fast (white) fibers behave differently because they have different ways of sourcing and using energy. Slow twitch fibers can quickly access supplies of ATP, the molecule that releases energy, and this makes them able to work for longer without getting tired. They're what keep cyclists going for hours when commuting, leisure riding, or touring. Fast twitch fibers, on the other hand, get most of their ATP from a less efficient process that quickly causes pain and fatigue. That's what limits a cyclist's ability to maintain hill climb attacks and ferocious sprints. The maximum contraction velocity of a single slow twitch fiber is approximately one-tenth that of a fast twitch fiber.[13] Fast twitch fibers are more likely to be injured through strain.[14] Most muscles contain both slow and fast twitch fibers and although the ratio is thought to be genetically determined, it can be changed through training.[15] In fact, there are two kinds of fast twitch fiber—IIa, which operates both aerobically and anaerobically, and IIb, which is active for only a short time, works anaerobically, and tires sooner. Riders who want to increase their count of type IIb fast twitch fibers to help them sprint can start weight training. The process of muscle change is counterintuitive, because weight training can convert type IIb fibers to the more versatile type IIa when sprinters really want a lot of IIb. The trick, however, is to stop training a little while before competition, because then the body seems to reverse the conversions and overshoot, giving the muscle perhaps twice the level of IIb fibers as before training.[16]

▼ ***Switch twitch*** *Training can convert one kind of muscle fiber into another, particularly the two types of fast twitch fiber, IIa and IIb. The capacity to change from fast to slow fibers is not so clear.*[17]

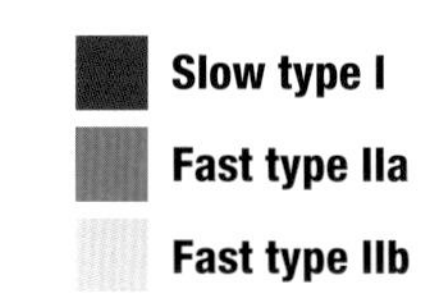

Percentage of muscle fiber types

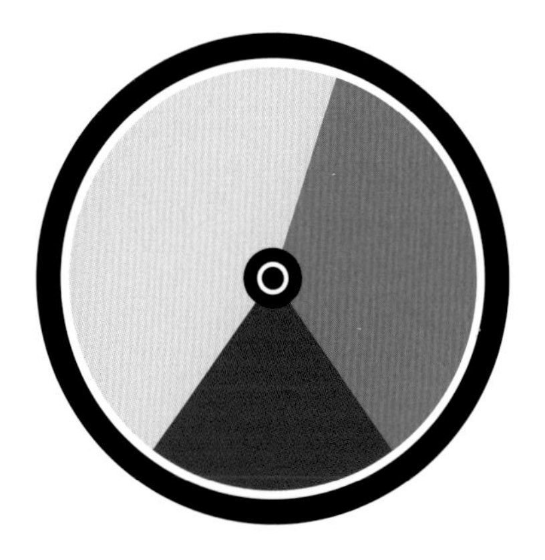

World-class sprinter

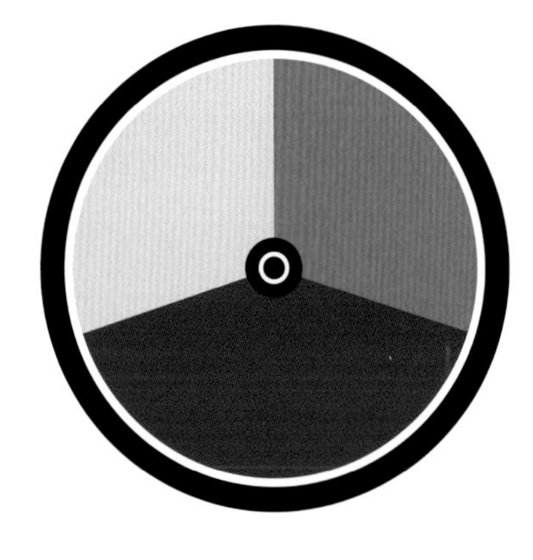

Average couch potato

Average active person

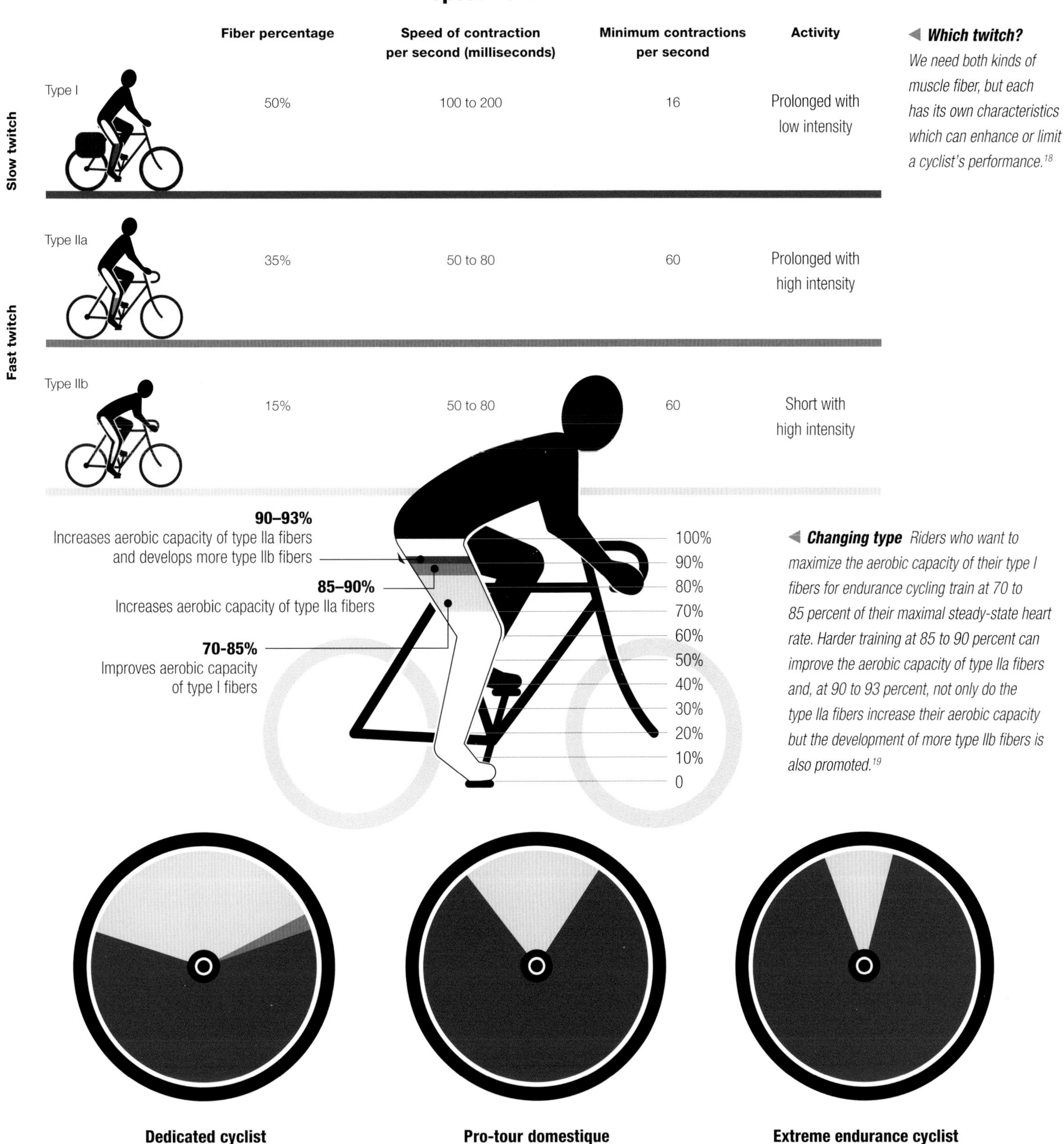

Speed versus distance

	Fiber percentage	Speed of contraction per second (milliseconds)	Minimum contractions per second	Activity
Type I	50%	100 to 200	16	Prolonged with low intensity
Type IIa	35%	50 to 80	60	Prolonged with high intensity
Type IIb	15%	50 to 80	60	Short with high intensity

◀ ***Which twitch?*** *We need both kinds of muscle fiber, but each has its own characteristics which can enhance or limit a cyclist's performance.*[18]

◀ ***Changing type*** *Riders who want to maximize the aerobic capacity of their type I fibers for endurance cycling train at 70 to 85 percent of their maximal steady-state heart rate. Harder training at 85 to 90 percent can improve the aerobic capacity of type IIa fibers and, at 90 to 93 percent, not only do the type IIa fibers increase their aerobic capacity but the development of more type IIb fibers is also promoted.*[19]

science in action

it's all in the head

Winning cyclists must have self-belief, but they should be wary of believing every thought they have. Research shows that the brain lies to the body and prevents it from fulfilling its potential.[20] The alerts to slow down or stop take the form of fatigue and pain, and are created by the brain because it thinks the body might be damaged if the exercise continues. Cyclists who know from practice that they can ignore the warnings are able to ride through the "pain barrier" to finish faster, although they will be utterly depleted. Belief in external feedback can also help a rider improve even when it contradicts their own sensory systems. On a hot ride, cyclists who were tricked into believing that temperatures were only moderate actually rode faster than those told the true figures.[21]

This means that the right psychological preparation for competition can be as important as physical conditioning. The word "Team" now often precedes a squad's name to deliberately reinforce the emotional benefits of being a member of a group. Giving riders comparative feedback during training rides boosts their performances significantly.[22] Choosing exactly the right training music for each rider is a growing field for experts.[23] It's also been shown that when riders familiarize themselves with the route and challenges of a race in advance, they produce a superior performance.[24] There is a host of research into how athletes can associate emotional states with optimal training so that similarly good performances can be triggered through emotions during competition.[25] Those emotions could be anger, anxiety, and fear, or they could be excitement and joy.[26] However, if they've been linked to the best training paces, then switching them on during a race can spark the best ride.

▶ ***Winning psychology*** *The difference between first and second can be psychological, and teams invest increasing amounts of research into the benefits of training feedback, music, and the links between emotional state and optimal training pace. Here, in stage 2 of the 2012 Tirreno-Adriatico race in Italy, Sky team rider Mark Cavendish of Britain wins a sprint finish after a lead-out from teammate Edvald Boasson Hagen. His victory was helped by the physical and mental cocoon provided throughout the stage by his team.*

sky

How does breathing affect cycling? Can I get some air?

Hard cycling demands rapid, deep breathing. That's because the oxygen most cells need to function must be replenished by the respiratory system. This draws in air by contracting and flattening the diaphragm muscle to reduce pressure in the thoracic cavity of the chest. As the intercostal muscles lift and expand the rib cage, the lungs expand so fresh air flows in. Then the diaphragm relaxes, the chest cavity becomes smaller, and the lungs are squeezed to expel the used breath. Breathing is largely subconscious—although it can be affected by emotions and thought processes—so we can go riding without thinking about breathing at all.

But what about the quality of the air we breathe? Rural cyclists certainly breathe cleaner air, but even riding daily through heavy traffic is better for respiration than driving.[27] Through exertion, a cyclist breathes about two to three times as much air as the motorist, but the air is not as polluted as it is inside a vehicle.[28] In some cities, cycle commuters can be exposed to more black carbon soot than pedestrians because they're closer to car exhausts, so it's better not to follow vehicle paths too closely.[29] Luckily, physical exercise strengthens the body's ability to resist the effects of pollution.

Some aspects of breathing cannot be changed. Total lung capacity depends on weight, sex, and age: tall people tend to have larger lungs than shorter people, and females have smaller lungs than males. Lung capacity decreases with age, so that a cyclist at the age of 80 will have only about half the lung capacity of someone of 20. Whatever a rider's lung size, only a fraction of the total capacity is used unless the exercise is strenuous. While it's not possible to make lungs grow bigger, research suggests that specific muscle exercises can improve breathing efficiency, which, in turn, helps cyclists perform better. Experiments in which cyclists have followed training routines to strengthen breathing muscles have made them up to 5 percent faster.

Trained to perform

Before training

Peak work output
250 ft·lb/s (340 W)

Maximum ventilatory capacity
54.0 VO_2

Time-trial time
47 minutes 6 seconds

The benefits of respiratory training

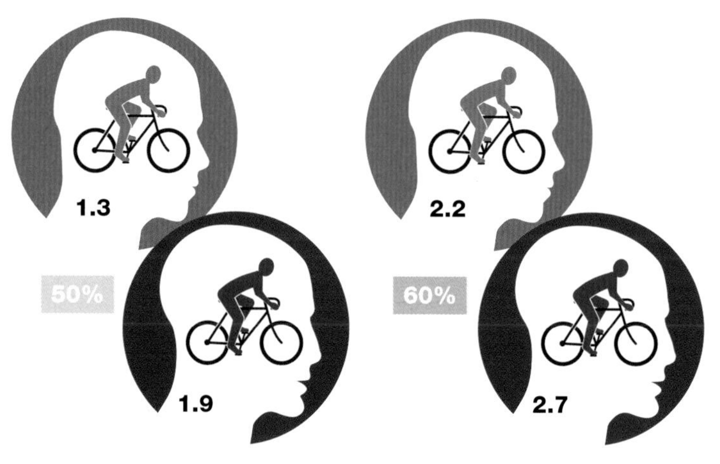

▼ ***Inspiring research*** *The performance of fit young cyclists was measured before and after they were trained to improve their respiratory muscles. They completed a task that simulated a time trial. Not only did their ventilatory capacity increase significantly but they were about 5% faster in the simulated time trial. A control group given no training and a third group given placebo training showed no such gains.*[30]

After training

Peak work ouput
257 ft·lb/s (348 W)

Maximum ventilatory capacity
55.5 VO_2 max

Time-trial time
44 minutes 54 seconds

Need to know

- Breathing rate at rest is about 12 to 15 breaths a minute, increasing to 20 or more during physical exertion.
- The air we breathe in is 21 percent oxygen; the air we breathe out is 16 percent oxygen.
- The adult human has a lung volume of 350 in^3 (5.8 liters), on average.
- The efficiency of breathing is reduced if a cyclist tucks down too tight or wears a poorly fitted backpack.
- Respiration is a carbon dioxide factory. The air we breathe in is just 0.038 percent CO_2, but we exhale air which is 4 percent CO_2.
- One breath at rest takes in 30 in^3 (0.5 liters) of air, of which we retain 4 percent (1.2 in^3 or 0.02 liters) of oxygen per breath.
- At rest we may breathe as few as 12 times a minute, therefore adding 15 in^3 (0.24 liters) of oxygen per minute to our blood.
- At rest we take at least 17,000 breaths a day, shifting about 300 ft^3 (8,500 liters) of air in the process.

▼ ***Airheads*** *Whatever difference respiratory training may make to cycling performance, it can make tough rides seem easier. After being trained to improve their breathing, riders felt that pedaling was easier at all workloads. The scores represent the riders' perceptions of how hard it was to sustain increasing workloads before (red cyclists) and after (blue cyclists) inspiratory muscle training. The higher the number, the harder it was perceived to maintain the workload. The percentages indicate how hard they were actually working in terms of their maximum work rate.*[31]

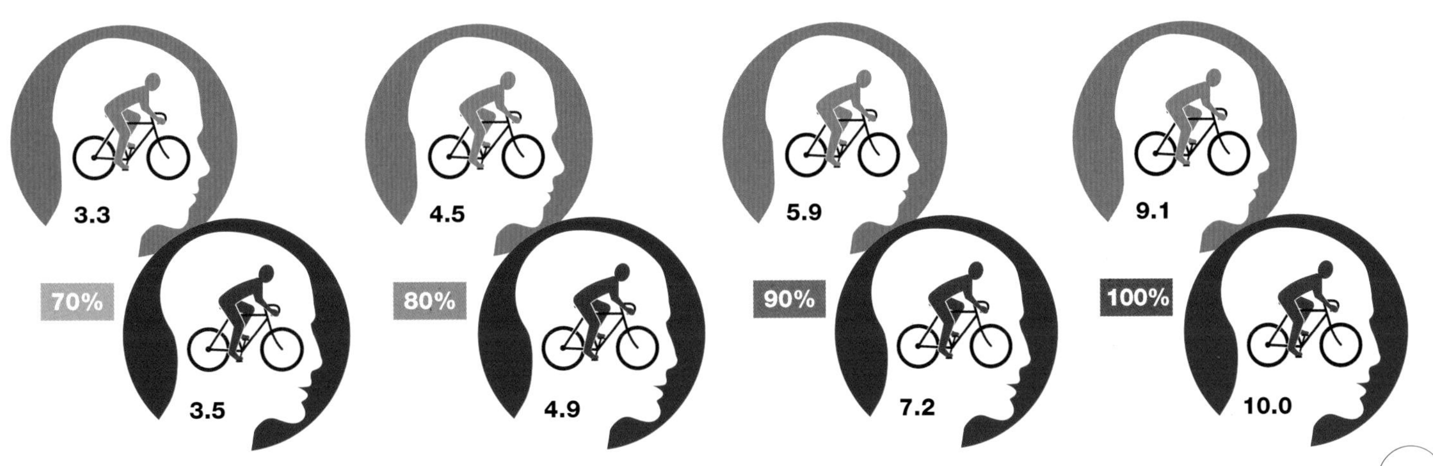

What are the effects of altitude training?

Will I get a nosebleed?

Altitude adds complexity to training and the benefits are uncertain. The performance of some riders improves—but for others there is no gain at all. Reaction to different atmospheric conditions, pressure, and temperature varies between individuals and even for the same rider on separate occasions. The science is still being worked on, so the one definite about altitude training is: it needs careful monitoring.

It's generally the case that rides that start low and pass through high altitudes can be difficult for riders who haven't acclimatized, and not just physically. Research on elite cyclists showed they were significantly more angry, tired, confused, depressed, and tense, and less lively on a two-hour ride in a chamber simulating the atmosphere at 8,200 ft (2,500 m) than during similar rides at sea level. Emotions do play an important role in a rider's ability to fulfill their potential, and although some athletes can harness the extra anger they feel when riding at unusual altitudes, the research suggests that acclimatization before long mountain stages may be of more help overall.[32]

Although the proportion of oxygen in the air breathed in at altitude is the same as at sea level, the lower barometric pressure means not as much goes into the blood. The capacity to exercise diminishes as a rider ascends and mountain sicknesses such as headache, dehydration, or even pulmonary edema can be triggered. That's why it's advisable to reduce training intensity during the first 7–10 days at altitude. Some riders use special tents and chambers to simulate the conditions at altitude by increasing the atmospheric nitrogen content at the expense of oxygen. The attraction of altitude riding for some is that when they return to lower levels they can tolerate low blood oxygen levels better and cycle harder and faster—especially in short-duration contests such as track sprints.[33] Studies suggest high-quality, altitude-inspired performance is sustained at sea level for up to four weeks, although the exact mechanisms and timing are poorly understood as yet.

The effect of altitude on mood

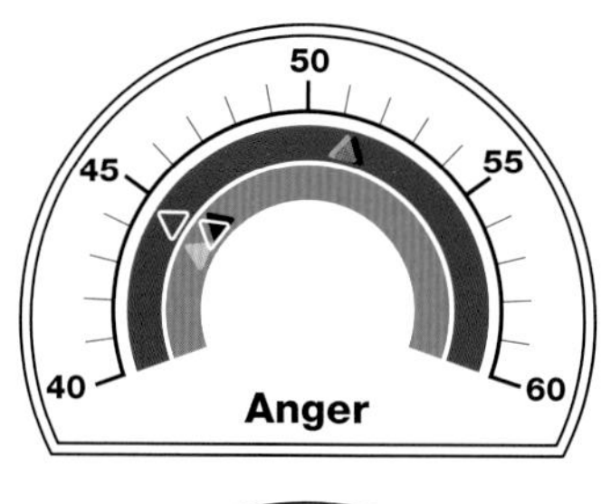

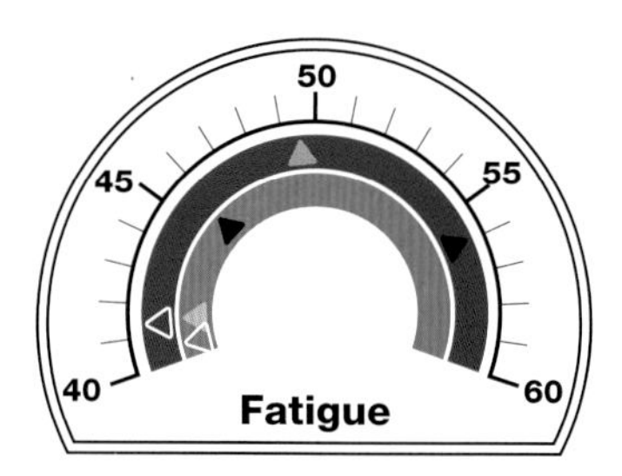

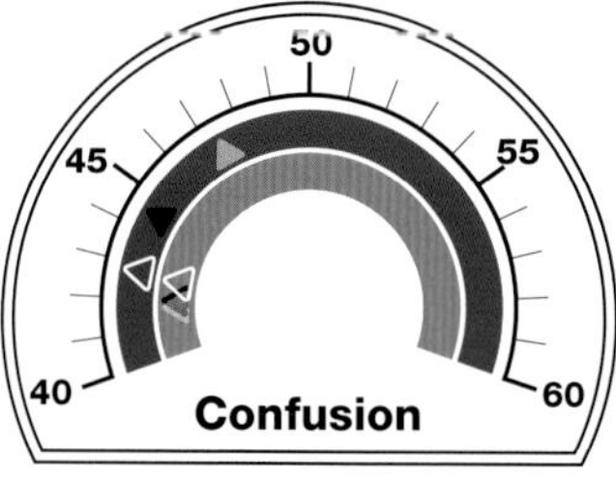

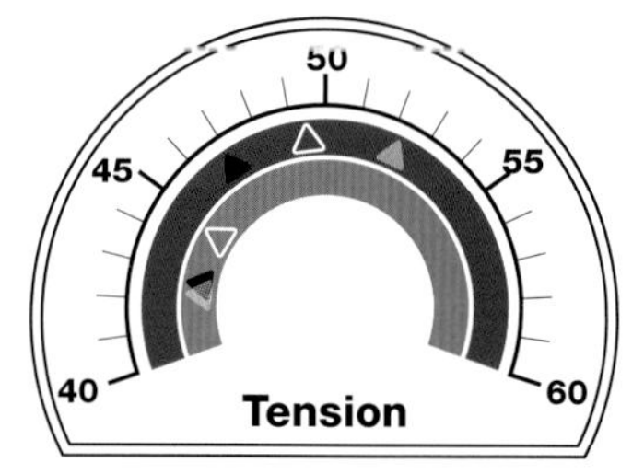

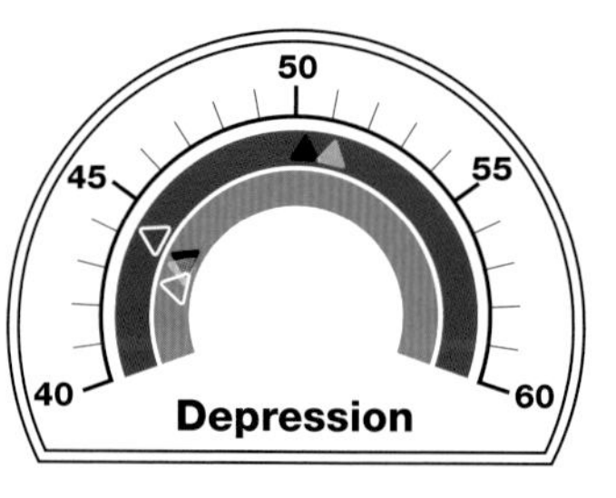

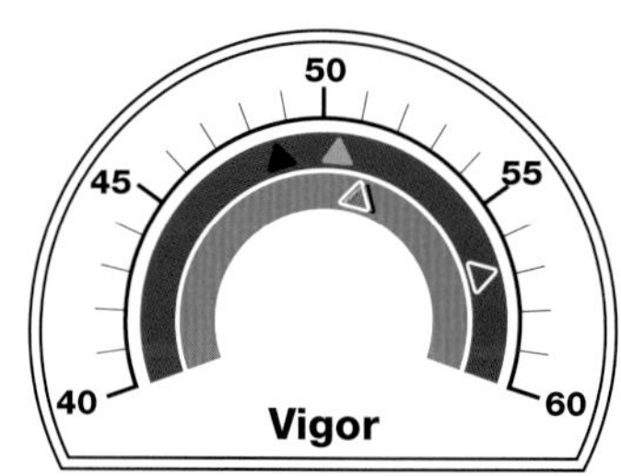

▲ ***Measuring mood*** *The mood states of eight elite male cyclists were assessed before and during two-hour ergometer rides under low-level conditions and at a simulated altitude of 8,200 ft (2,500 m). Mood was tested using the Brunel Mood Scale; the higher the score, the more intense the mood state. At altitude, riders enjoyed a relatively strong sense of vigor before the ride began, but once the experiment was underway, this soon fell below the usual level felt during a low-level ride.*[34]

Pretest
1 hour
2 hours
Normal altitude
8,200 ft (2,500 m)

The highs and lows of training

▼ ***The high life*** *It's not always necessary to move to mountaintops to start altitude training. Hypoxic tents simulate a high-altitude atmosphere, and supplemental oxygen at altitude can help simulate low-level conditions. Artificial techniques such as these are mixed with combinations of high- or low-altitude living and training to produce various approaches, each with its own practical and physiological advantages and disadvantages.*[35]

2 Live high but train as if at low altitude by using supplemental oxygen

Advantages

- Can remain in one training venue.

Disadvantages

- The use of oxygen cylinders means only training on an ergometer or rollers is practical.

1 Live high but train low using natural altitude

Advantages

- Potential for a greater tolerance of hypoxia and endurance performance.
- Living and training in a natural environment.

Disadvantages

- Have to descend for each training session and ascend afterward.

3 Live high, train high

Advantages

- Maximized opportunity for enhancing hypoxia tolerance.
- Remain in one training venue.

Disadvantages

- Potential loss of training quality due to reduced oxygen flux in high-intensity training sessions.

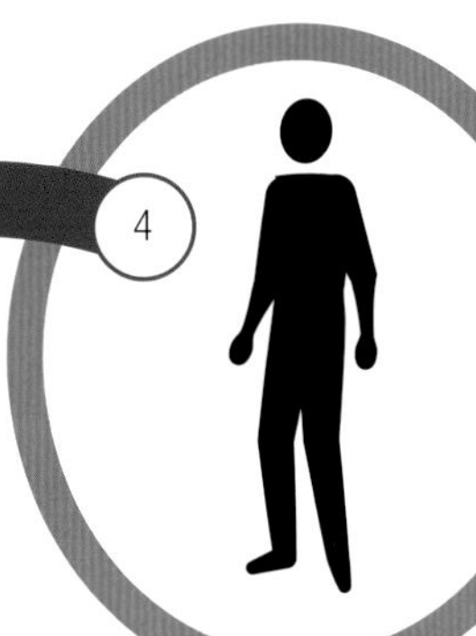

4 Simulate living high by staying in a hypoxic tent but train low outside it

Advantages

- Can remain in home training venue all year round.

Disadvantages

- Difficult to accumulate enough hours in the tent because more than 12 hours per day is recommended.

5 Live low but simulate training high in a hypoxic tent

Advantages

- Potential for enhanced training quality.
- No loss of oxygen flux in high-intensity training sessions at sea level.

Disadvantages

- Training limited to static exercise on ergometer or rollers.
- Duration of exposure to simulated altitude conditions is insufficient for acclimatization.

Is my heart critical for peak performance? → Will cycling break my heart?

Getting oxygen, glucose, and nutrients to tissues and muscles, removing carbon dioxide, and disposing of acid and other waste products is the job of the cardiovascular system. The heart and lungs are part of the cardiovascular system—and both are very important for cyclists.

The heart is a muscle triggered to contract by regular, automatic nerve impulses. These contractions pump oxygenated blood through the heart's four chambers (the right and left ventricle and the right and left atrium) and around your body. The more it is exercised, the more effective a heart is. Cycling, like any other exercise, can't make it exceed its maximum rate, which can vary between individuals from 150 beats per minute (bpm) to 220 bpm. However, regular riding can enlarge the muscle of the left ventricle, the chamber that pumps freshly oxygenated blood into the arteries. This means a larger volume of blood flows around the body with each beat, so a big heart can deliver more oxygen, glucose, and nutrients to tissues that need them. Also, being bigger, it can rest at a lower rate than before exercise. When the resting heart rate is low, the range up to maximum heart rate is greater, so more blood can be circulated, refreshing vital muscles in the legs and other tissue. The resting rate for an average person is 60 to 80 bpm, although it can be much lower—results from the pre-2004 Tour de France medical tests showed that the average rider had a significantly lower resting heart rate.[36]

Low heart rate is a by-product of training, and so is longer life expectancy. In fact, any cycling that is of at least average intensity, when heart rate is raised to at least half its maximum, decreases the chances of dying from heart disease. That pulsing muscle in your chest gets stronger and is better able to respond to the stresses you put on it every extra day you ride.

Cycling speeds and risk of heart disease

Slow
Medium
Fast

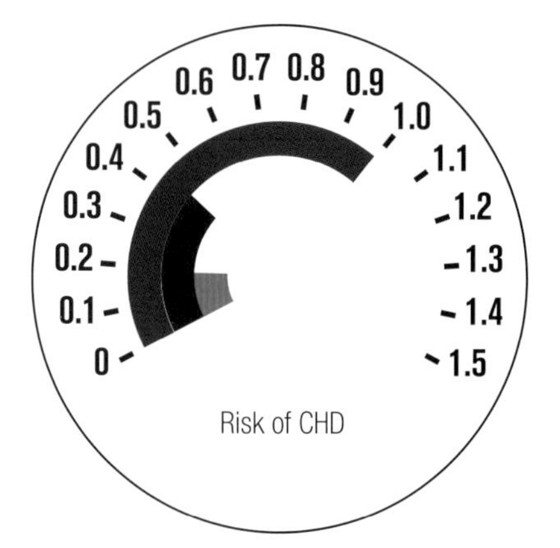

Less than 30 minutes per day

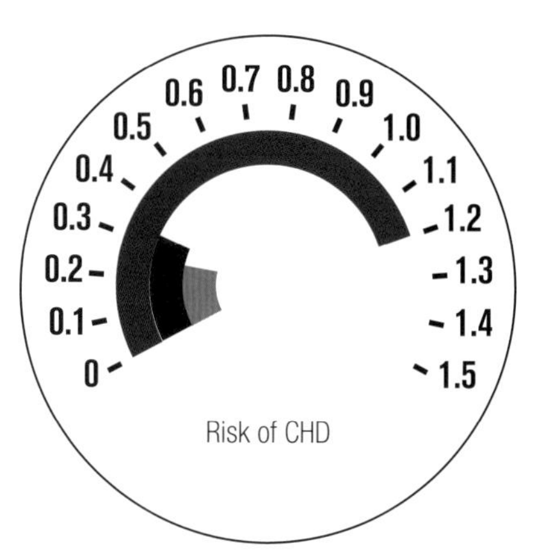

30 to 60 minutes per day

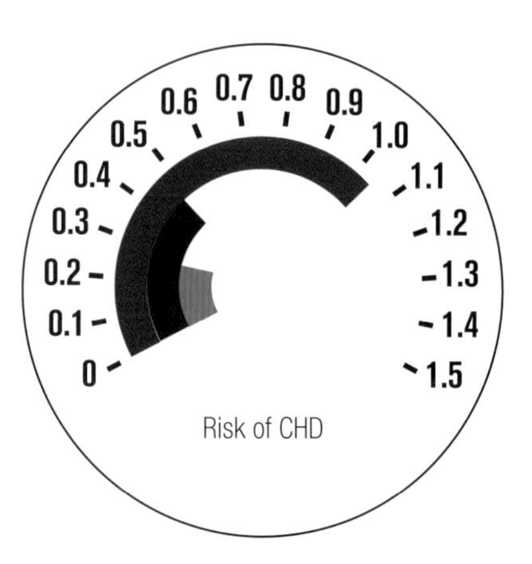

More than 60 minutes per day

▶ ***Lower risks*** *Cycling protects against the risk of coronary heart disease, no matter how long you cycle each day—but cycling faster is better. The data from a large population was calculated using a slow cyclist who rides just a few minutes each day as the baseline. (The data also takes into account other factors such as smoking, drinking, and age, which is why it shows high heart disease death risks for people who ride slowly and for longer—they may be less healthy and less physically able already.) The fastest cyclists over each time period have the best heart health.*[37]

Cycle harder, live longer

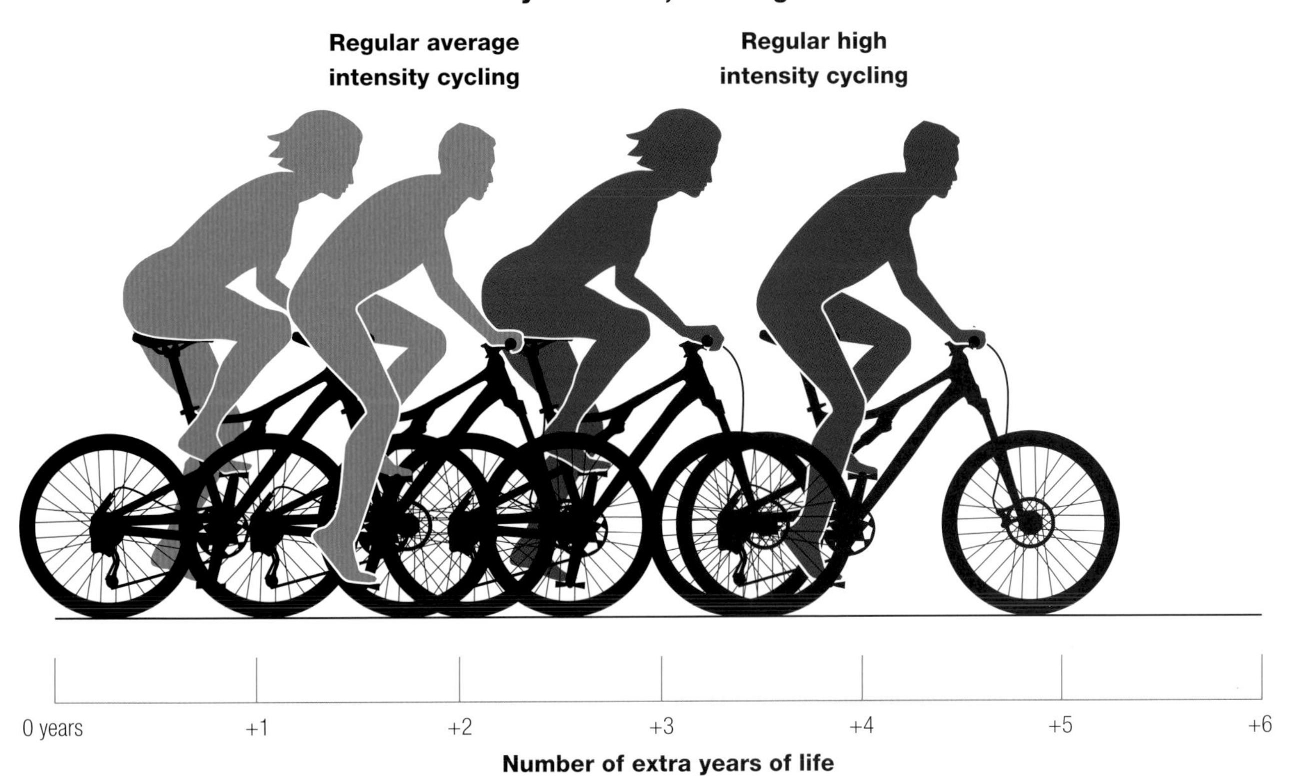

▼ ***Below average*** *The resting rate for an average rider is 60–80 bpm. Before the 2004 Tour de France, the average resting heart rate of the riders was 51 bpm and in the 1990s Miguel Indurain's was an extraordinarily slow 28 bpm.*[38]

▲ ***Heart health*** *Scientists discovered that regular cycling improves heart health, but it is the intensity of exercise which is important, not the duration. They tracked more than 5,000 men and women over 18 years and found that those who cycled harder live significantly longer.*[39]

How low can you go?

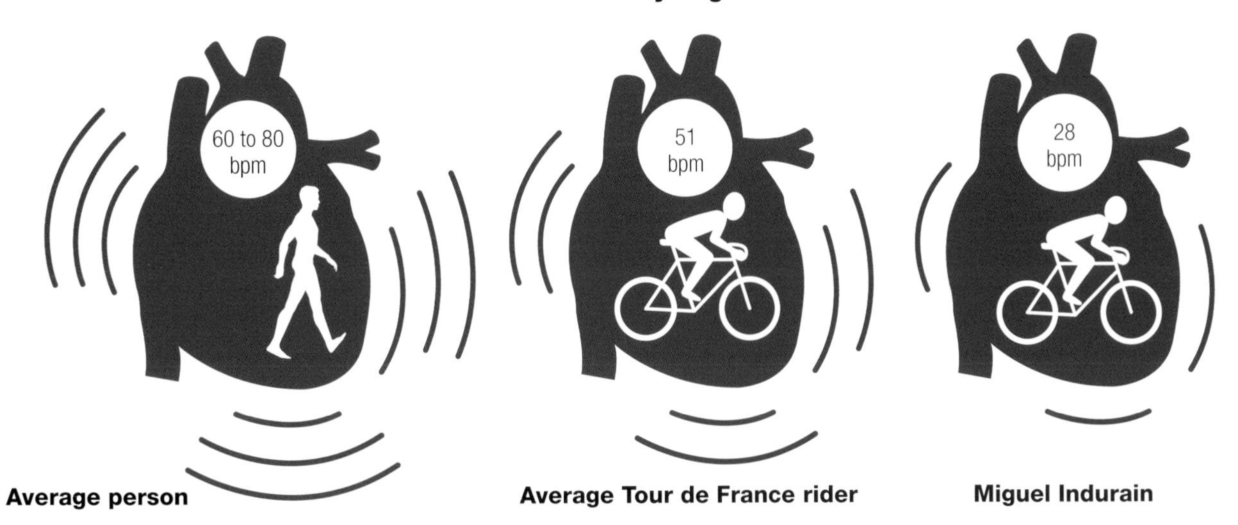

How does "blood boosting" work?

When will the cheating stop?

Scientific knowledge helps cyclists to ride farther, faster, and for longer. It underpins training programs, diet, strategy, and equipment. Unfortunately, it can also be used to help alter the blood of competitors so they can outperform their rivals.

The level of oxygen supplied to muscles can limit the amount of power a cyclist produces. Red blood cells deliver the oxygen to muscles, so ambitious riders have long been keen to raise the levels of these cells. Initially, this was done with a transfusion, known as blood doping. This transfusion could be of someone else's blood or some of their own extracted previously. This practice was largely abandoned in the 1990s, when biochemists synthesized erythropoietin (EPO), a hormone that stimulates bone marrow to produce more red blood cells. Both doping and EPO are now against the rules of competitive cycling, and stringent checks are made to expose cheats. The benefits of EPO are enormous—it can increase oxygen uptake by 8.1 percent, boost peak power output by 13 percent, and extend performance time to exhaustion by a massive 54 percent—all within four weeks of use.[40] These findings probably explain why the speed and power of extraordinary endeavors in major races have never been matched since EPO testing began in 2002.

After EPO detection was introduced, blood doping returned to favor, but new checks have been introduced to counter the trend. Of course, cyclists vulnerable to their own ambition or the greed of their team may be tempted to try any of the hundred or so new EPO-like performance-enhancing drugs that experts say are currently undetectable.[41] Instead of competing against each other, these riders would be using their bodies as labs for experiments. Meanwhile, the cycling regulators are engaged in a different race altogether, to work out how to spot the cheats before they claim unfair glory.

Need to know

The body contains about 5 quarts (4.7 liters) of blood, with about 28,000 billion red blood cells that each live for 100–120 days.

Every blood cell contains about 250 molecules of hemoglobin, each of which can carry four oxygen molecules.

The heart pumps about 5 in^3 (80 ml) of blood at each beat.

> **"The cycling regulators are engaged in a different race altogether, to work out how to spot the cheats before they claim unfair glory.**

The results of changes to EPO testing

▼ ***Cheating hearts*** *Samples from Tour de France riders have revealed extraordinary changes in the levels of chemicals that indicate EPO use and blood doping. The level of immature red blood cells, called reticulocytes, is the giveaway. In 2001, almost one in seven samples showed extreme levels of reticulocytes and, of these, the vast majority were abnormally high—a symptom of EPO use. Urine testing for EPO began in time for the 2003 Tour, and still almost one in eight samples showed extreme reticulocyte levels. Yet, curiously, the majority were now abnormally low—an indication of blood doping. When the biological passport was introduced in 2008 to monitor a wide range of biochemicals over long periods, the proportion showing extreme levels of reticulocytes plummeted to less than 1 in 25 samples and this hit rate fell farther for the next two years, when nearly all samples were abnormally low. By 2011, only 2 percent of samples were abnormal.[42] The optimistic interpretation is that EPO use has been largely abandoned, while only a tiny handful of riders are still desperate enough to resort to blood doping.[43]*

Key

1% of the samples with levels of immature red blood cells within the normal range

1% of the samples with extremely high levels of immature red blood cells

1% of the samples with extremely low levels of immature red blood cells

2001

2003

2008

Does skin really affect riding efficiency?

Should I shave my eyebrows as well?

The increasingly popular World Naked Bike Ride is a form of political activism, but it also highlights how significant skin is. Skin defends us from physical, chemical, and microbial attacks, and protects vulnerable tissues and organs. It regulates our body temperature and uses sunshine to make vitamin D. It may be less fashionable than Spandex, but it constantly renews itself and lasts a lifetime.

A network of nerve cells in your skin gathers valuable data about airflow, temperature, and pressure changes while you are riding. If the data reveals a headwind, this can prompt you to adopt a more streamlined position. If your skin senses a tailwind, you might sit up. Pressure receptors in the contact points of the feet, hands, and buttocks alert riders to the best path for a smoother ride. Off-road riders learn from the pain receptors in the skin not to ride too close to bushes or tree branches.

Skin also plays an important role in temperature regulation. There is evidence that well-trained cyclists subconsciously alter the intensity of their exercise according to skin temperature.[44] If the ambient temperature drops (perhaps if the sun goes behind a cloud), skin can insulate the body. The dermis-based arrector pili muscles cause body hairs to stand up and trap a layer of insulating warm air. Conversely, on hot rides your skin acts as a radiator, emitting excess heat. When the ambient temperature is high, eccrine glands all over the body can release salty water (sweat), which evaporates and promotes heat loss. It's more comfortable to wear shorts and leave your legs uncovered so the heat can escape. Shaving your legs won't promote heat loss, but it does make it easier to treat wounds and get massages that make your skin feel better.

Muscles and skin temperature

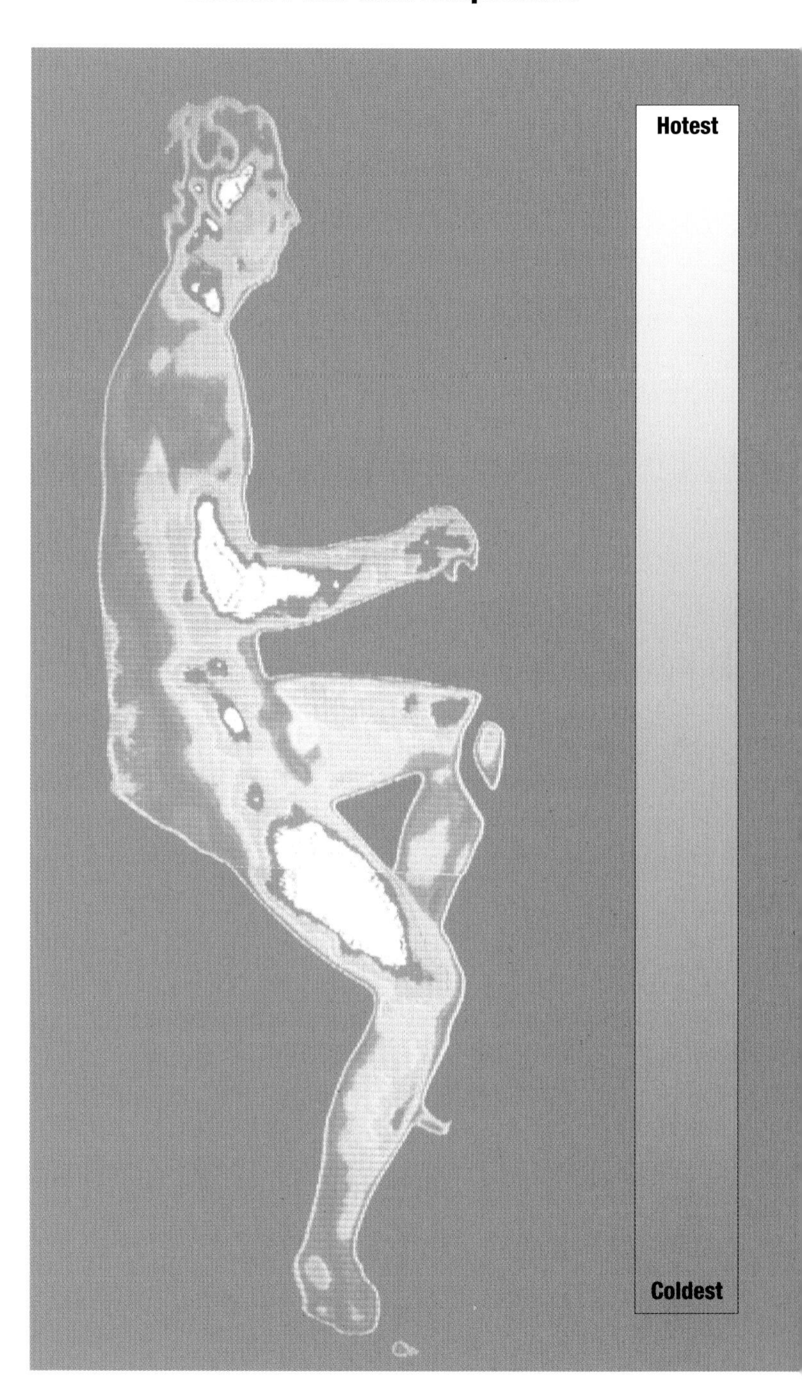

Skin temperature and power output

Rider starts hot

Rider starts cold

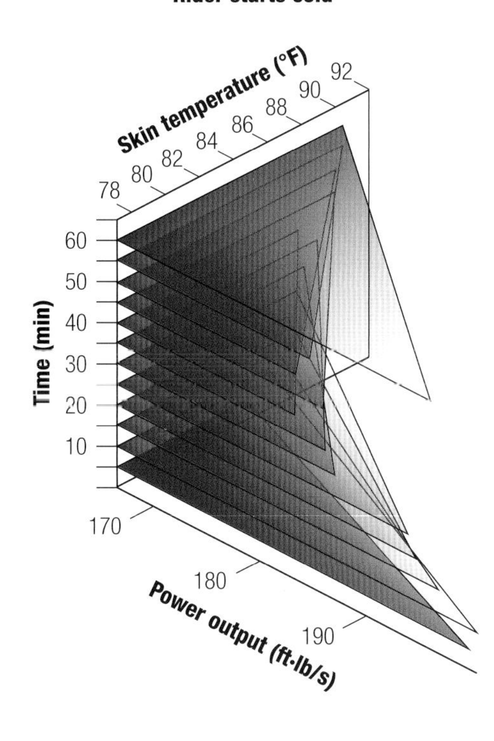

◀ ***Surface protection*** *These thermograph images show the variation in skin temperature of a man on an exercise bike; the muscles warm up through work, and this heat is emitted via the skin. The coldest areas are darker, while the hotter areas are in white.*

▲ ***Sensitive skin*** *Skin temperature can influence how hard a cyclist works, at least for the first 15 minutes of a ride. A sophisticated experiment that used water jackets to change a rider's skin temperature without altering the ambient temperature showed that a rider who starts with hot skin generates significantly less power than a rider who starts cold. The initial discrepancy quickly disappears, however, and the power outputs of the two riders gradually converge and meet at 50 minutes, even though the trends in their skin temperature are in opposite directions.*[45]

equipment: the human body

The bicycle is designed to fit the body of Homo sapiens, a complex organism that has evolved over millions of years. Even if our earliest ancestors had invented the wheel and fabricated metals, they would have had difficulty riding a modern bike because they themselves were built differently. They had skeletons, muscles, and brains sized and optimized for another, more primitive, world. Chimpanzees are among our closest relatives, but a chimp can't rotate its thumb to squeeze it against their fingertips—so unscrewing tire valve caps and tightening spoke nipples would be a problem. Even the earliest hominids may not have had this ability.

The dividing line between hominids and great apes is our larger brain size and the ability to stand upright. The earliest evidence of a species that could walk without using its knuckles is Orrorin tugenensis, which lived six million years ago. The size of a chimp, Orrorin tugenensis could possibly have mastered a child's tricycle, although one strengthened to withstand its

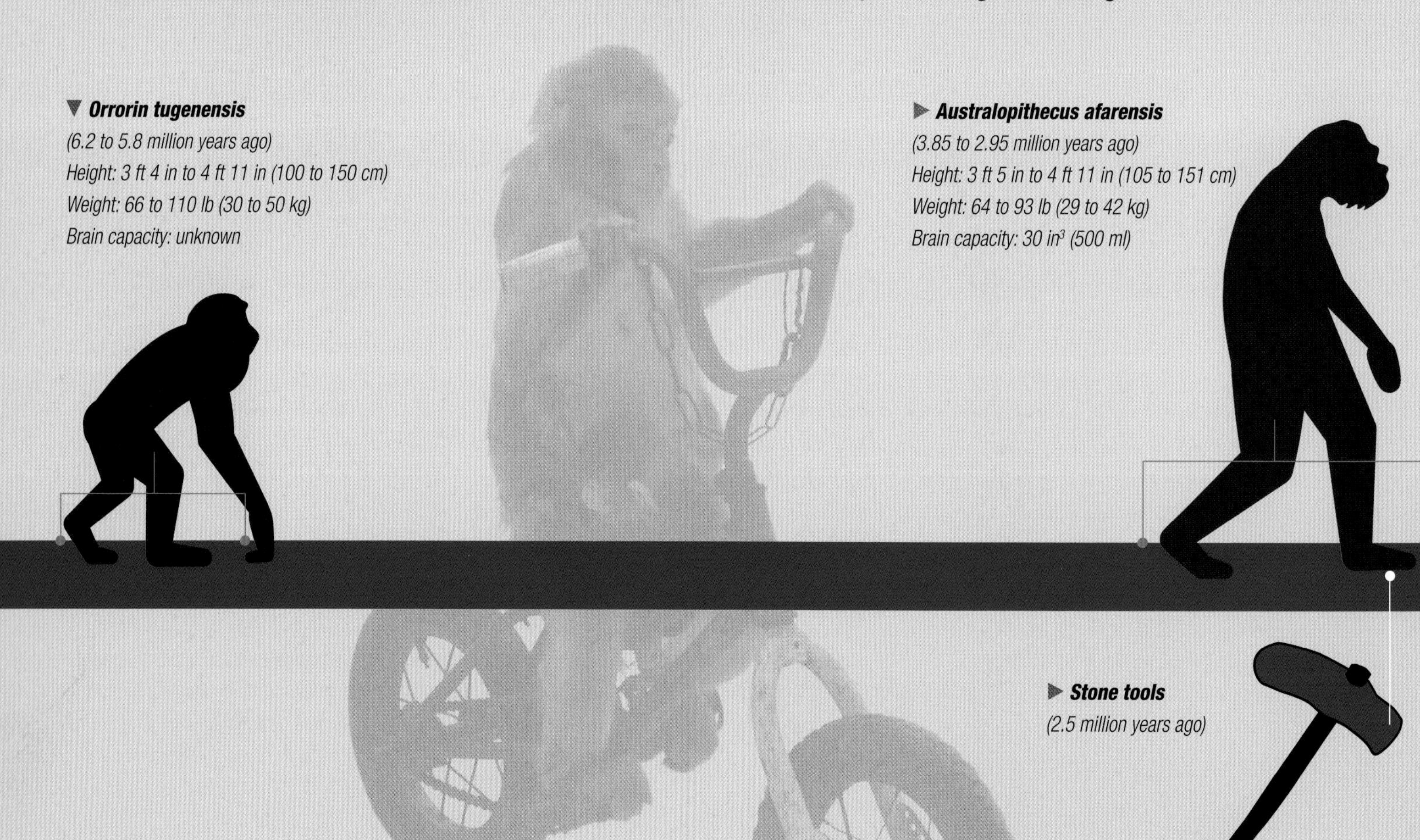

▼ ***Orrorin tugenensis***
(6.2 to 5.8 million years ago)
Height: 3 ft 4 in to 4 ft 11 in (100 to 150 cm)
Weight: 66 to 110 lb (30 to 50 kg)
Brain capacity: unknown

► ***Australopithecus afarensis***
(3.85 to 2.95 million years ago)
Height: 3 ft 5 in to 4 ft 11 in (105 to 151 cm)
Weight: 64 to 93 lb (29 to 42 kg)
Brain capacity: 30 in³ (500 ml)

► ***Stone tools***
(2.5 million years ago)

65–110 lb (30–50 kg) weight. Three million years ago, Australopithecus afarensis emerged in Eastern Africa. This species survived for almost a million years with a brain one-third the size of ours, with an average male height of 4 ft 11 in (151 cm) and females at 3 ft 5 in (105 cm). With a relatively long torso and limbs all the same length, the frame of their bike would need to be long and low. A recumbent might have fitted.

The oldest early humans with proportions similar to ours, and potentially able to cycle, were Homo erectus. Their legs were elongated and their arms shorter compared to the length of the torso. A little lighter than average modern humans, they ranged from 4 ft 9 in to 6 ft 1 in (145–185 cm) tall, and could not only walk but also probably run. They became extinct some 143,000 years ago. The short and stocky Neanderthal (Homo neanderthalensis) of Europe, who may have still been around 30,000 years ago, might slog through a modern cyclo-cross race, but would lose out to its longer-lasting neighbor Homo sapiens, who first evolved 200,000 years ago and eventually took to cycling 150 years ago.

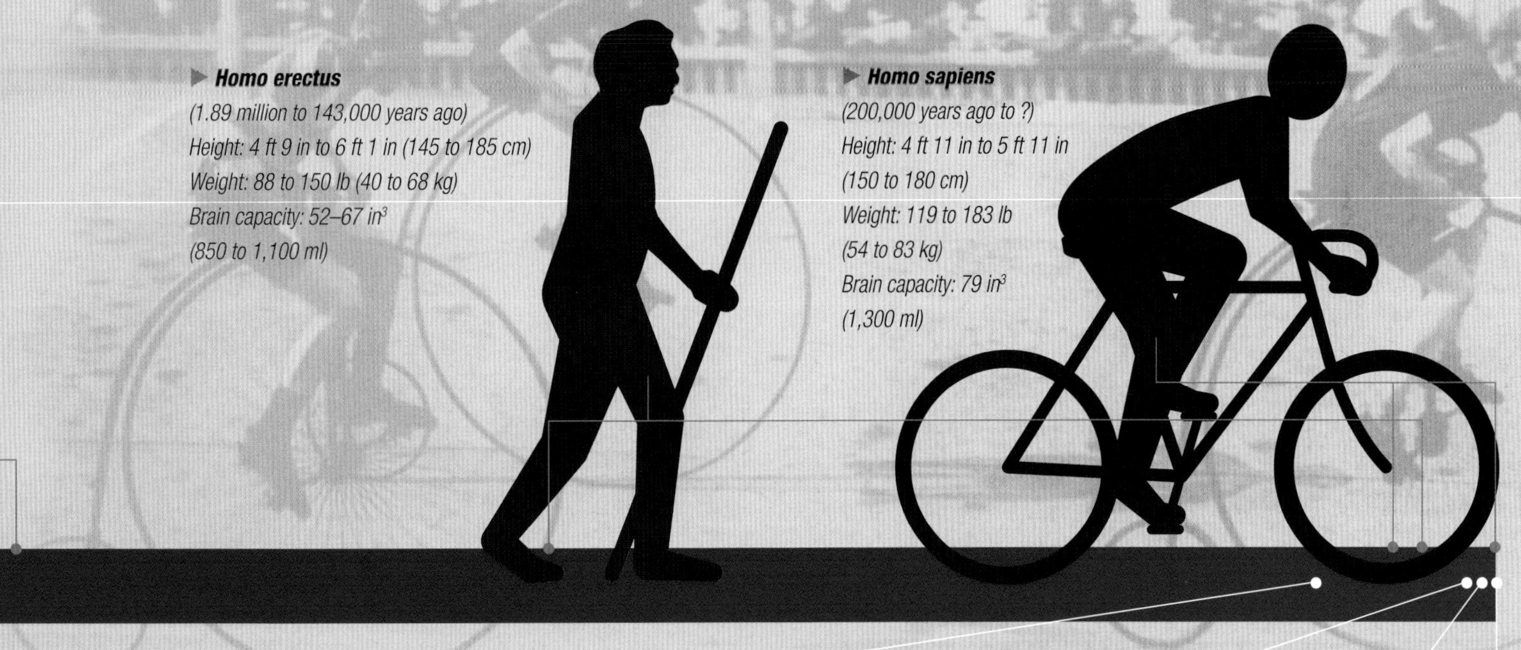

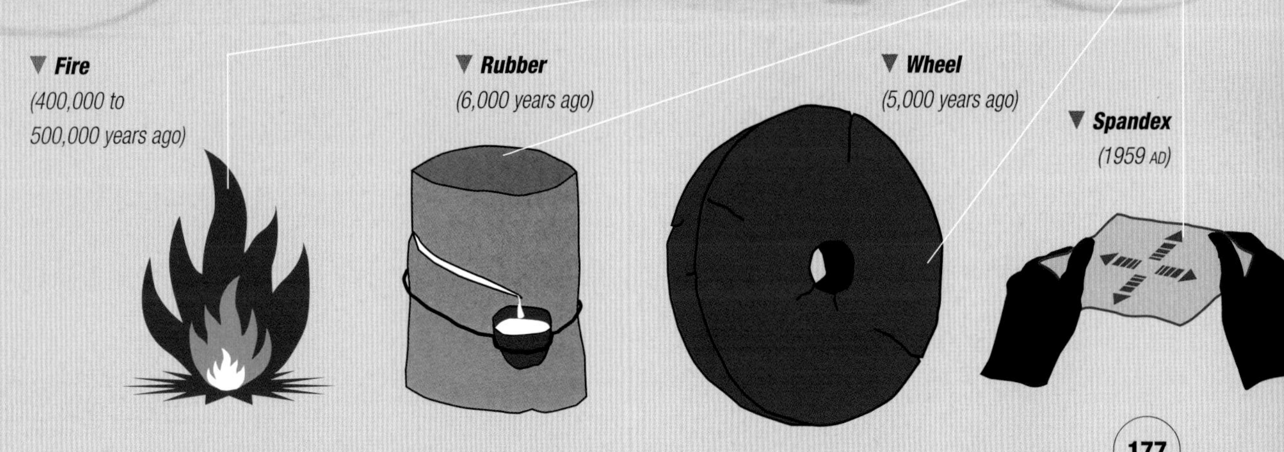

What factors most affect riding speed?

What will make me go faster?

The information in this book will help you as a cyclist to make better decisions about what, where, when, and how you ride. While most cyclists want only to arrive at their destination safely, with a modest amount of effort and at a reasonable cost, others strive for strength and speed. In both cases, a knowledge of the science underpinning riding a bike makes it easier to strike the optimum balance.

There is a peloton of options for riders who are determined to use their energy to cycle faster—and this book could be twice as long yet still not cover all the potential methods. National and professional cycling teams have full-time staff dedicated to identifying tiny details to make riders go quicker, but, for the average rider, too much information can distract from what they like best—riding their bikes. So we'll end with a summary of the most accessible alternatives and the likely gains they could bring over a ride of 25 miles (40 km). Don't take the recommendations as immutable and infallible. By now, you'll have gathered that every cyclist, every route, and every ride is different every day—so, while cycling obeys the laws of nature, there are too many variables to be 100 percent certain the chosen input will produce the desired result. After all, when was the last time you accurately predicted a puncture?

Bad luck plays a part in cycling as it does in all aspects of life, but this research shows overwhelmingly what a champion from another sport, golfer Arnold Palmer, said: "The more I practice, the luckier I get." Training conditions the body to give the best performance possible, and allows you to enjoy the ideal way to travel—cycling.

▶ ***Go slow*** *It's too easy to go slower than you want. You don't need to have read this book to know that it's harder to pedal a heavier bike and an overweight body than lighter versions of both. Interestingly, however, the time losses incurred from extra weight pale in comparison to those from an unaerodynamic shape.*

▶ ***Train hard or invest?*** *How much faster could you ride if you had access to the best training and the top equipment? Taking the example of a 25-year-old male weighing 154 lb (70 kg) riding a 25-mile (40-km) time trial course on a 22 lb (10 kg) bike with aero wheels and an aero helmet, it's clear that the novice without money is better off training hard. If they have cash to spend, then investing in aerodynamics will reap rewards, as it does for the elite cyclist. The novice cyclist is assumed to be traveling at 20.45 mph (32.91 km/h) for a total time of 1 hour 12 minutes 56 seconds, while the elite cyclist will cover the 25 miles in 52 minutes 2 seconds at an average speed of 28.66 mph (46.12 km/h).*[46]

Potential time losses for a 25-mile time trial

Novice cyclist

Elite cyclist

-0:25

Gaining weight from 154 to 161 lb (70 to 73 kg) and increasing drag area

-6:49

Changing from elbows on time trial handlebars to hands on brake hoods

Changing from elbows on time trial handlebars to hands on drops

Changing from aero wheels to standard wheels

Potential time savings for a 25-mile time trial

Changing from a round steel-tube frame to an aero frame

Changing from a round steel-tube frame to an aero frame and adopting an optimized body position

Changing from no training to training at maximum effect

Ingestion of caffeine combined with carbohydrate-electrolyte solutions

Novice cyclist

Changing from no training to training at average effect

Ingestion of carbohydrate-electrolyte solutions

Elite cyclist

Changing from no training to training at minimum effect

Altitude training

Changing from a 22 lb (10 kg) bike to a 15 lb (7 kg) bike

Losing weight from 154 to 148 lb (70 kg to 67 kg) and reducing drag area

Notes

CHAPTER 1

1. Bicycle efficiency from James B. Spicer, Christopher J. K. Richardson, Michael J. Ehrlich, Johanna R. Bernstein, Masahiko Fukuda, and Masao Terada, "Effects of Frictional Loss on Bicycle Chain Drive Efficiency," *Transactions at the ASME* 123 (December 2001); walking efficiency from "Walk Without Waste," http://www.abc.net.au/science/news/health/HealthRepublish_232296.htm accessed 13 July 2012.

2. Nick Carey, "Establishing Pedestrian Walking Speeds," (Portland State University, 2005).

3. Firas Massaad, Thierry M. Lejeune, and Christine Detrembleur, "The Up and Down Bobbing of Human Walking: a Compromise Between Muscle Work and Efficiency," *Journal of Physiology* 582 (2007).

4. David Gordon Wilson, *Bicycling Science*, 3rd ed. (Cambridge, MA: MIT Press, 2004).

5. Rainer Pivit, http://lustaufzukunft.de/pivit/aero/formulas.html accessed on 10 July 2012.

6. Margaret A. McDowell, Cheryl D. Fryar, Cynthia L. Ogden, and Katherine M. Flegal, "Anthropometric Reference Data for Children and Adults: United States, 2003–2006," *National Health Statistics Reports* 10 (US Department of Health and Human Services, 22 October 2008).

7. Umbra Fisk, "Umbra on the Impacts of Biking," http://grist.org/article/ten-speed-demon/ accessed 13 July 2012.

8. Shreya Dave, "Life Cycle Assessment of Transportation Options for Commuters" (MIT, February 2010).

9. Mike Berners-Lee, "What's the Carbon Footprint of . . . Cycling a Mile?" http://www.guardian.co.uk/environment/2010/jun/08/carbon-footprint-cycling accessed 13 July 2012.

10. Mirjan E. Bouwman, "An Environmental Assessment of the Bicycle and Other Transport Systems" (University of Groningen, October 2000).

11. Dave, "Life Cycle Assessment."

12. Ralph S. Paffenbarger Jr., Robert Hyde, Alvin L. Wing, and Chung-Cheng Hsieh, "Physical Activity, All-Cause Mortality, and Longevity of College Alumni," *New England Journal of Medicine* 314 (6 March 1986).

13. L. Andersen, P. Schnohr, M. Schrol, and H. Hein, "All-Cause Mortality Associated with Physical Activity During Leisure Time, Work, Sports, and Cycling to Work," *Archives of Internal Medicine* 160 (2000).

14. W. Tuxworth, A. M. Nevill, C. White, and C. Jenkins, "Health, Fitness, Physical Activity and Morbidity of Middle Aged Male Factory Workers," *British Journal of Industrial Medicine* 43 (1986).

15. Penny Gordon-Larsen, Janne Boone-Heinonen, Steve Sidney, Barbara Sternfeld, David R. Jacobs Jr., and Cora E. Lewis, "Active Commuting and Cardiovascular Disease Risk: the CARDIA Study," *Archives of Internal Medicine* 169, no. 13 (13 July 2009).

16. Jeroen Johan de Hartog, Hanna Boogaard, Hans Nijland, and Gerard Hoek, "Do the Health Benefits of Cycling Outweigh the Risks?" *Environmental Health Perspectives* 118 no. 8 (August 2010).

17. "Final Local Implementation Plan, Chapter 5 LIP Proposals: Cycling" (Haringey Council, October 2006).

18. Johan de Hartog et al., "Health Benefits of Cycling."

19. Ibid.

20. Ibid.

21. Ibid.

22. Tuxworth et al., "Health of Factory Workers."

23. "Implementation Plan: Cycling" (Haringey Council).

24. J. S. Mindell, S. J. Watkins, and J. M. Cohen, eds., Health on the Move 2: Policies for Health-Promoting Transport (Transport & Health Study Group, 2011).

25. World Health Organisation, "Distribution of Road Traffic Deaths by Road User," http://apps.who.int/ghodata?vid=51312 accessed 24 June 2012; Kurt Van Hout, "Literature Search Bicycle Use and Influencing Factors in Europe" (Instituut voor Mobiliteit, July 2008); European Commission, http://epp.eurostat.ec.europa.eu/tgm/table.do?tab=table&language=en&pcode=tps00001&tableSelection=1&footnotes=yes&labeling=labels&plugin=1 accessed 24 June 2012; "Annual Statistical Report 2011" (European Road Safety Authority, March 2011); David R. Bassett Jr., John Pucher, Ralph Buehler, Dixie L. Thompson, and Scott E. Crouter, "Walking, Cycling, and Obesity Rates in Europe, North America, and Australia," *Journal of Physical Activity and Health* 5 (2008).

26. http://www.milkpresents.com accessed 13 July 2012.

27. Ross Tucker, "The Limit to Cycling Performance: Can Physiology Flag Doping?" http://www.sportsscientists.com/2010/07/cycling-performance-what-is-possible.html accessed 13 July 2012.

28. Doug Malewicki (http://www.canosoarus.com).

29. C. Kyle, "Mechanical Factors Affecting the Speed of a Cycle," in *Science of Cycling* edited by Edmund R. Burke, (Champaign: Human Kinetics, 1986).

30. J. G. Seifert, D. W. Bacharach, and E. R. Burke, "The Physiological Effects of Cycling on Tandem and Single Bicycles," *British Journal of Sports Medicine* 37 (2003).

31. http://www.kreuzotter.de/english/espeed.htm accessed 26 October 2011.

32. Seifert et al., "Physiological Effects."

CHAPTER 2

1. Reynolds Technology Ltd.

2. A. M. Lovatt, H. R. Shercliff, and P. J. Withers, "Material Selection and Processing," http://www-materials.eng.cam.ac.uk/mpsite/physics/introduction/default.html accessed on 22 December 2011 (data courtesy of Granta Design Ltd.).

3. Data primarily from "Tube Material and Processes" (Reynolds Technology Ltd., 2009); http://www.ami.ac.uk/courses/topics/0123_mpm/index.html accessed 23 May 2012.

4. Triathletes information from http://www.pinarello.it/else/2011_technical_data.pdf accessed 23 May 2012; commuter information from Robert Penn, *It's All About the Bike* (London: Particular Books, 2010).

5. Penn, *It's All About the Bike*, 77.

6. http://www.ebicycles.com/article/bicycle-frame-size-charts.html accessed 23 May 2012.

7. "Die Steifigkeiten der getesteten Rennräder im Detail," http://www.roadbike.de/test/bikes/die-leichtesten-renner-der-welt-im-test-zwoelf-modelle-unter-6-8-kilo.371802.9.htm?skip=5 accessed on 23 May 2012.

8. Jason Woznik, "Crank test #4," www.fairwheelbikes.com/forum/viewtopic.php?f=77&t=7916 accessed on 25 May 2012.

9. John F. Cowell, John B. Cronin, and Michael R. McGuigan, "Time Motion Analysis of Supercross BMX Racing," *Journal of Sports Science and Medicine* 10 (2011).

10. Hunter Allen and Andrew R. Coggan, "Fatigability and BMX Performance at the Olympic Level," http://www.trainingandracingwithapowermeter.com/2010/05/fatigability-and-bmx-performance-at.html accessed on 13 April 2012.

11. James Pritlove, Michael Reid, J. Michael, Alex Lee, and D. G. E. Robertson, "Comparative Study Between Suspension and Nonsuspension Front Forks on Mountain Bikes," *Proceedings of the Third North American Conference on Biomechanics* (1998).

12. J. Titlestad, T. Fairlie-Clarke, M. Davie, A. Whittaker and S. Grant, "Experimental Evaluation of Mountain Bike Suspension Systems," *Acta Polytechnica* 43 no. 5 (2003).

13. Ibid, figure 5.

14. J. Seifert, M. Luetkemeier, M. Spencer, D. Miller, and E. Burke, "The Effect of Mountain Bike Suspension Systems on the Energy Expenditure, Physical Exertion, and Time Trial Performance During Mountain Bicycling," *International Journal of Sports Medicine* 18 (1997).

15. Pritlove et al., "Suspension and Nonsuspension Front Forks."

16. Joel Fajans, "Steering in Bicycles and Motorcycles," *American Journal of Physics* 68 (7 July 2000).

CHAPTER 3

1. "CERN Experiments Observe Particle Consistent with Long-Sought Higgs Boson," http://press.web.cern.ch/press/PressReleases/Releases2012/PR17.12E.html accessed 16 July 2012.

2. "Pedal Power Probe Shows Bicycles Waste Little Energy," http://www.jhu.edu/news_info/news/home99/aug99/bike.html (John Hopkins University Office of News and Information, August 1999), accessed on 3 July 2012.

3. http://www.engineeringinspiration.co.uk/brakecalcs.html accessed on 17 July 2012.

4. James B. Spicer, Christopher J. K. Richardson, Michael J. Ehrlich, Johanna R. Bernstein, Masahiko Fukuda, and Masao Terada, "Effects of Frictional Loss on Bicycle Chain Drive Efficiency," *Transactions at the ASME* 123 (December 2001).

5. David R. Bassett Jr., Chester R. Kyle, Louis Passfield, Jeffrey P. Broker, and Edmund R. Burke, "Comparing Cycling World Hour Records, 1967–1996: Modeling with Empirical Data," *Medicine and Science in Sports and Exercise* 31 (1999).

6. Tim Maloney, "Ernesto Colnago 50th Anniversary Interview," http://fry.cyclingnews.com/sponsors/italia/2004/colnago/?id=colnago5 accessed 27 June 2012.

7. Dave Perry, "Bike Cult," http://www.bikecult.com/bikecultbook/sports_recordsHour.html; "The Hour Record," http://www.wolfgang-menn.de/hourrec.htm; and "The Hour Record 2010," http://www.velominati.com/tradition/the-hour-record-2010; all accessed 17 June 2012.

8. Mehul P. Patel, T. Terry Ng, Srikanth Vasudevan, Thomas C. Corke, and Chuan He, "Plasma Actuators for Hingeless Aerodynamic Control of an Unmanned Air Vehicle," *Journal of Aircraft* 44 (July/August 2007).

CHAPTER 4

1. Jong-Peel Joo, Soo-Jung Moon, Yoon-Ho Ahn, Seung-Gon Oh, Jin-Bong Kim, and Sung-Jae Lee, "Prediction of Cycle Shoe Performance in Relation to Outsole Materials Based on Biomechanical Testing and Finite Element Analysis," http://www.asbweb.org/conferences/2005/pdf/0310.pdf accessed 27 July 2012.

2. E. M. Hennig and D. J. Sanderson, "In-Shoe Pressure Distributions for Cycling with Two Types of Footwear at Different Mechanical Loads," *Journal of Applied Biomechanics* 11 (1995).

3. Nathan E. Jarboe and Peter M. Quesada, "The Effects of Cycling Shoe Outsole Material on Plantar Stress," http://www.tulane.edu/~sbc2003/pdfdocs/0405.PDF accessed 27 July 2012.

4. Henning and Sanderson, "In-Shoe Pressure Distributions."

5. R. R. Davis and M. L. Hull, "Measurement of Pedal Loading in Bicycling: II. Analysis and Results," *Journal of Biomechanics* 14 (1981).

6. G. Mornieux, B. Stapeleldt, A. Gollhofer, and A. Belli, "Effects of Pedal Type and Pull-Up Action During Cycling," *International Journal of Sports Medicine* 29 (2008).

7. T. Korff, L. M. Romer, I. Mayhew, and J. C. Martin, "Effect of Pedaling Technique on Mechanical Effectiveness and Efficiency in Cyclists," *Medicine and Science in Sports and Exercise* 39 (June 2007).

8. Henning and Sanderson, "In-Shoe Pressure Distributions."

9. Joo et al., "Shoe Performance in Relation to Outsole Materials."

10. "Bicycle Efficiency and Power—or, Why Bikes Have Gears," http://users.frii.com/katana/biketext.html accessed 11 June 2012.

11. Penn, *It's All About the Bike*, 93.

12. Dinh Quang Truong, Kyoung Kwan Ahn, Le Duy Khoa, and Do Hoang Thinh, "Development of a Smart Bicycle Based on a Hydrostatic Automatic Transmission," *Journal of Advanced Mechanical Design, Systems, and Manufacturing* 6 (2012).

13. Chris Juden, "Internal Gears Q&A," http://www.ctc.org.uk/DesktopDefault.aspx?TabID=3816 accessed 11 June 2012.

14. Kraig Willett, "Wheel Performance," http://biketechreview.com/reviews/wheels/63-wheel-performance accessed 16 March 2012.

15. Tom Anhalt, "Why Wheel Aerodynamics Can Outweigh Wheel Weight and Inertia," http://www.slowtwitch.com/Tech/Why_Wheel_Aerodynamics_Can_Outweigh_Wheel_Weight_and_Inertia_2106.html accessed 16 March 2012.

16. John S. Allen, "Check Spoke Tension by Ear," http://bikexprt.com/bicycle/tension.htm accessed 11 June 2012.

17. Ian Smith, "Bicycle Wheel Analysis," http://www.astounding.org.uk/ian/wheel/index.html accessed on 13 March 2012; graphic based on an FEA analysis by Hannes Zietsman (www.pyrolights.co.za).

18. Personal communication from Hans Erhard Lessing, 6 October 2011.

19. "Friction and Coefficients of Friction," http://www.engineeringtoolbox.com/friction-coefficients-d_778.html accessed 12 June 2012.

20. Roman Beck (Beck Forensics), http://www.beckforensics.com/BrakingGraphs.pdf accessed 7 June 2012.

21. Based on an average of various reviews at http://www.bikeradar.com accessed 7 July 2012.

22. Wilson, *Bicycling Science*.

23. Eight million rotations are predicted in ibid.

CHAPTER 5

1. Wilson, *Bicycling Science*.

2. Mark Jermy and Lindsey Underwood, "Forces and Speed," http://www.sciencelearn.org.nz/Science-Stories/Cycling-Aerodynamics/Forces-and-speed accessed 29 March 2012.

3. S. R. Bussalari and E. R. Nadel, "The Physiological Limits of Long-Duration Human Power Production," *Human Power* 7 (1989).

4. C. R. Kyle and M. D. Weaver, "Aerodynamics of Human-Powered Vehicles," *Proceedings of the Institution of Mechanical Engineers, Part A: Journal of Power and Energy* 218 (2004).

5. L. Underwood, J. Schumacher, J. Burette-Pommay, and M. Jermy, "Aerodynamic Drag and Biomechanical Power of a Track Cyclist as a Function of Shoulder and Torso Angles," *Sports Engineering* 14 (2011).

6. Lindsey Underwood and Mark Jermy, "Optimal Hand Position for Individual Pursuit Athletes," *Procedia Engineering* 2 (2010).

7. Ibid.

8. Underwood et al., "Aerodynamic Drag and Biomechanical Power."

9. Kyle and Weaver, "Aerodynamics of Human-Powered Vehicles."

10. Pierre Debraux, Frederic Grappe, Aneliya V. Manolova, and William Bertucci, "Aerodynamic Drag in Cycling: Methods of Assessment," *Sports Biomechanics* 10 (September 2011).

11. Firoz Alam, Aleksander Subic, and Aliakbar Akbarzadeh, "Aerodynamics of Bicycle Helmets," *Engineering of Sport* 7 (2008).

12. Randall L. Jensen, Saravanan Balasubramani, Graham Brennan, Keith C. Burley, Daniel R. Kaukola, James A. LaChapelle, and Amir Shafat, Power output, Muscle Activity, and Frontal Area of a Cyclist in Different Cycling Positions," (XXV ISBS Symposium, 2007).

13. Erik W. Faria, Daryl L. Parker, and Irvin E. Faria, "The Science of Cycling: Factors Affecting Performance Part 2," *Sports Medicine* 35 (2005).

14. Alam et al., "Aerodynamics of Bicycle Helmets."

15. http://www.rudyprojectuk.com/products/helmets/wingspan/black-white/unisize.html accessed 19 June 2012.

16. Stephanie Sidelko, "Benchmark of Aerodynamic Cycling Helmets Using a Refined Wind Tunnel Test Protocol for Helmet Drag Research" (MIT, 2007).

17. Lindsey Underwood and Mark Jermy, "Wind Tunnel Testing of Cyclists," http://www.sciencelearn.org.nz/Science-Stories/Cycling-Aerodynamics/Wind-tunnel-testing-of-cyclists accessed 2 April 2012.

18. Paul Harder, Doug Cusack, Carl Matson, and Mike Lavery, "Airfoil Development for the Trek Speed Concept Triathlon Bicycle" (Trek, 2010).

19. Jim Martin and John Cobb, "Bicycle Frame, Wheels, and Tires," in *High-Performance Cycling*, ed. Asker Jeukendrup (Champaign: Human Kinetics, 2002).

20. C. R. Kyle, "New Aero Wheel Tests," *Cycling Science* 3 (1991), cited in R. A. Lukes, S. B. Chin, and S. J. Haake, "The Understanding and Development of Cycling Aerodynamics," *Sports Engineering* 8 (2005).

21. J. P. Broker, "Cycling Power: Road and Mountain," in *High-Tech Cycling* 2nd ed., ed. E. R. Burke (Champaign: Human Kinetics, 2003).

22. J. K. Moore, "Aerodynamics of High Performance Bicycle Wheels" (University of Canterbury, New Zealand, 2008).

23. "Studying Bicycle Wheel Aerodynamics with Computational Fluid Dynamics" (Intelligent Light, 2010).

24. G. S. Tew and A. T. Sayers, "Aerodynamics of Yawed Racing Cycle Wheels," *Journal of Wind Engineering and Industrial Aerodynamics* 82 (1999), cited in Lukes et al., "Development of Cycling Aerodynamics."

25. Hugh Trenchard, "Energy Savings in Bicycle Pelotons, a General Evolutionary Mechanism and a Framework for Group Formation in Eusocial Evolution," presented at the Eighth International Conference on Complex Systems, 2011.

26. Personal communication from Hugh Trenchard.

27. J. P. Broker, "Cycling Power: Road and Mountain," in Burke, *High-Tech Cycling*, 147–174.

28. T. Olds, "The Mathematics of Breaking Away and Chasing in Cycling," *European Journal of Physiology* 77 (1998).

29. P. Swain, "Cycling Uphill and Downhill," *Sportscience* 2 (1998), cited in Trenchard, "Energy Savings in Bicycle Pelotons."

CHAPTER 6

1. James Hopker, Simon Jobson, Helen Carter, and Louis Passfield, "Cycling Efficiency in Trained Male and Female Competitive Cyclists," *Journal of Sports Science and Medicine* 9 (2010).

2. Karen Davranche, Boris Burle, Michel Audiffren, and Thierry Hasbroucq, "Physical Exercise Facilitates Motor Processes in Simple Reaction Time Performance: An Electromyographic Analysis," *Neuroscience Letters* 396 (2006).

3. Benjamin Libet, "Unconsciousness: Cerebral Initiative and the Role of Conscious Will in Voluntary Action," *Behavioural and Brain Science* 8 (1985); T. Nørretanders, *The User illusion: Cutting Consciousness Down to Size* (New York: Viking, 1998).

4. J. Brisswalter, R. Arcelin, M. Audiffren, and D. Delignieres, "Influence of Physical Exercise on Simple Reaction Time: Effect of Physical Fitness," *Perceptual and Motor Skills* 85 (1997).

5. Stephen Smaldone, Chetan Tonde, Vancheswaran K. Ananthanarayanan, Ahmed Elgammal, and Liviu Iftode, "The Cyber-Physical Bike: A Step Towards Safer Green Transportation," presented at HotMobile 2011 (2011).

6. Vancheswaran Koduvayur Ananthanarayanan, "Audio Based Detection of Rear Approaching Vehicles on a Bicycle" (Rutgers, 2012).

7. Jesper L. Anderson, Peter Schjerling, and Bengt Saltin, "Muscle, Genes and Athletic Performance," *Scientific American*" (September 2000).

8. Li Li and Graham E. Caldwell, "Muscle Coordination in Cycling: Effect of Surface Incline and Posture," *Journal of Applied Physiology* 85 (1998).

9. R. H. McQueen, R. M. Laing, C. M. Delahunty, H. J. L. Brooks, and B. E. Niven, "Retention of Axillary Odour on Apparel Fabrics," *Journal of the Textile Institute* 99 (2008).

10. Wilson, *Bicycling Science*.

11. http://www.cptips.com/bscphys.htm accessed 12 July 2012.

12. Susanna Törnroth-Horsefield and Richard Neutze, "Opening and Closing the Metabolite Gate," *Proceedings of the National Academy of Sciences* 105 (2008).

13. Anderson et al., "Muscle, Genes and Athletic Performance."

14. T. J. Noonan and W. E. Garrett Jr., "Muscle Strain Injury: Diagnosis and Treatment," *Journal of the American Academy of Orthopaedic Surgeons* 7 (July 1999).

15. http://exrx.net/ExInfo/Muscle.html accessed 12 July 2012.

16. Anderson et al., "Muscle, Genes and Athletic Performance."

17. Ibid.

18. http://exrx.net/ExInfo/Muscle.html accessed 12 July 2012; http://michaeldmann.net/mann14.html accessed 12 July 2012; and Dario Frederick, "Making the Most of Muscle," *Velo News* 33 (December 20, 2004).

19. http://exrx.net/ExInfo/Muscle.html, accessed 12 July 2012.

20. Timothy D. Noakes, "The Central Governor Model in 2012," *British Journal of Sports Medicine* 46 (2012).

21. P. C. Castle, N. Maxwell, A. Allchorn, A. R. Mauger, and D. K. White, "Deception of Ambient and Body Core Temperature Improves Self Paced Cycling in Hot, Humid Conditions," *European Journal of Applied Physiology* 112 (January 2012).

22. A. R. Mauger, A. M. Jones, and C. A. Williams, The Effect of Non-Contingent and Accurate Performance Feedback on Pacing and Time Trial Performance in 4-km Track Cycling," *British Journal of Sports Medicine* 45 (2011).

23. *The Sport and Exercise Scientist* 28 (British Association of Sport and Exercise Sciences, Summer 2011).

24. J. Swart, R. P. Lamberts, M. I. Lambert, E. V. Lambert, R. W. Woolrich, S. Johnston, and T. D. Noakes, "Exercising with Reserve: Exercise Regulation by Perceived Exertion in Relation to Duration of Exercise and Knowledge of Endpoint," *British Journal of Sports Medicine* 43 (2009).

25. B. Baron, F. Moullan, F. Deruelle, and T. D. Noakes, "The Role of Emotions on Pacing Strategies and Performance in Middle and Long Duration Sport Events," *British Journal of Sports Medicine* 45 (2011).

26. *The Sport and Exercise Scientist* 29 (British Association of Sport and Exercise Sciences, Autumn 2011).

27. J.H. van Wijnen, A.P. Verhoeff, H. W. Jans, and M. van Bruggen, "The Exposure of Cyclists, Car Drivers and Pedestrians to Traffic-Related Air-Pollutants," *International Archives of Occupational and Environmental Health* 67 (1995).

28. Jette Ranka, Jens Folkeb, and Per Homann Jespersena, "Differences in Cyclists and Car Drivers Exposure to Air Pollution from Traffic in the City of Copenhagen," *Science of The Total Environment* 279 (2001).

29. C. Nwokoro, N. Mushtaq, C. Harrison, M. Ibrahim, I. Dickson, O. Hussain, Z. Manzoor, C. Ewin, I. Dundas, and J. Grigg, "Inhaled Black Carbon in the Lower Airways of London Cyclists," presented at the European Respiratory Society's Annual Congress, Amsterdam 2011 (2011).

30. Paige Holm, Angela Sattler, and Ralph F. Fregosi, "Endurance Training of Respiratory Muscles Improves Cycling Performance in Fit Young Cyclists," *BMC Physiology* 4 (2004).

31. Lee M. Romer, Alison K. McConnell, and David A. Jones, "Effects of Inspiratory Muscle Training on Time Trial Performance in Trained Cyclists," *Journal of Sports Sciences* 20 (2002).

32. Andrew M. Lane, Gregory P. Whyte, Rob Shave, Sam Barney, Matthew Stevens, and Matthew Wilson, "Mood Disturbance During Cycling Performance at Extreme Conditions," *Journal of Sports Science and Medicine* 4 (2005).

33. *The Sport and Exercise Scientist* 30 (British Association of Sport and Exercise Sciences, Winter 2011).

34. Lane et al., "Mood Disturbance During Cycling Performance."

35. *The Sport and Exercise Scientist* 30 (British Association of Sport and Exercise Sciences, Winter 2011).

36. http://www.guardian.co.uk/sport/2004/jul/26/tourdefrance2004.tourdefrance1 accessed 12 July 2012.

37. Peter Schnohr, Jacob Marott, Jan Jensen, and Gorm Jensen "Intensity Versus Duration of Cycling, Impact on All-Cause and Coronary Heart Disease Mortality: The Copenhagen City Heart Study," *European Journal of Preventive Cardiology* 19 (February 2012).

38. http://www.heart.org/HEARTORG/Conditions/HighBloodPressure/PreventionTreatmentofHighBloodPressure/Physical-Activity-and-Blood-Pressure UCM_301882_Article.jsp#.TtUN_2BzjgQ accessed 12 July 2012; William Fotheringham, "Fotheringham's Tour de France Trivia," http://www.guardian.co.uk/sport/2004/jul/26/tourdefrance2004.tourdefrance1 accessed 12 July 2012; and Keith Moore, "How is Bradley Wiggins Different from the Average Man?" http://www.bbc.co.uk/news/health-18959642 accessed 31 July 2012.

39. Schnohr et al., "Intensity Versus Duration of Cycling."

40. C. Juel, J. J. Thomsen, R. L. Rentsch and C. Lundby, "Effects of Prolonged Recombinant Human Erythropoietin Administration on Muscle Membrane Transport Systems and Metabolic Marker Enzymes," *European Journal of Applied Physiology* 102 (2007).

41. John Mehaffey, "Doping—Undetectable New Blood Boosters available Says Expert," http://www.reuters.com/article/2012/03/15/doping-epo-idUSL4E8EF70520120315, accessed 31 July 2012.

42. Eric Niiler, "Is the Tour de France Turning Around?" http://news.discovery.com/adventure/tour-de-france-drugs-120629.html accessed 12 July 2012.

43. Mario Zorzoli and Francesca Rossi, "Implementation of the Biological Passport: The Experience of the International Cycling Union," *Drug Testing and Analysis* 2 (November–December 2010), cited in Ross Tucker And Jonathan Duga, "The Biological Passport—Giving Some Context to the Performances in the Tour," http://www.sportsscientists.com/2011/07/tour-de-france-biological-passport.html accessed 30 May 2012.

44. Zachary J. Schlader, Shona E. Simmons, Stephen R. Stannard, and Toby Mündel, "Skin Temperature as a Thermal Controller of Exercise Intensity," *European Journal of Applied Physiology* 111 (2011); Koen Levels, Jos J. de Koning, Carl Foster, and Hein A. M. Daanen, "The Effect of Skin Temperature on Performance During a 7.5-km Cycling Time Trial," *European Journal of Applied Physiology* (2012).

45. Schlader et al., "Skin Temperature as a Thermal Controller."

46. Asker Jeukendrup and Luke Moseley, "Effective Training" in Jeukendrup, *High Performance Cycling*.

Glossary

aerobic Occurring in the presence of oxygen or requiring oxygen to live. In aerobic respiration, which is the process used by the cells of most organisms, the production of energy from glucose metabolism requires the presence of oxygen. Aerobic exercise is exercise that increases the need for oxygen.

air resistance (drag) A resistive force associated with movement through air, and caused by a combination of air friction, turbulence, and pressure effects; it acts in the opposite direction to the motion of an object through the air.

anaerobic In reference to an organism or process, this means it is not dependent on the presence of oxygen. Anaerobic bacteria can produce energy from food in the absence of oxygen. In anaerobic exercise, oxygen is used up more quickly than the body is able to replenish it inside a working muscle.

angular (or rotational) acceleration The time rate of change of the angular velocity of a rotating body; the rate at which rotation is increasing or decreasing.

angular (rotational) velocity A measure of rotational speed; the rate of change of an object's angular position with time.

bonk The sudden loss of energy when a cyclist's body runs out of fuel.

bottom bracket The component that runs through the frame, at the bottom bracket shell, allowing the pedal axle to rotate freely within the frame. It comprises a water-tight bearing unit and a spindle onto which the crankset is located.

brakes There are two main kinds of brake: rim and disc. Rim brakes are the traditional method, using the rim as the braking surface and clamping the pads on each side via a cable pull system. Disc brakes use a braking surface that is not integral to the wheel—a metal disc attached to the hub. The greater power control in wet and muddy conditions makes disc brakes ideal for mountain biking.

breaking force The greatest stress, particularly tension, that a material can withstand before it ruptures.

cadence The rate at which a cyclist turns the pedals, measured in rotations per unit of time, for example, rotations per minute (rpm).

cardiovascular system The organs and tissues, such as the heart and blood vessels, involved in circulating blood and lymph through the body. This system delivers nutrients and other essential material to cells, and removes waste products.

cassette/freewheel The set of cogs on the rear wheel that the chain moves between with the use of a derailleur, allowing the rider to access a range of different gear ratios. Cassettes and freewheels mount slightly differently to the hub, but serve the same purpose. They both contain a ratcheting mechanism that allows the wheel to continue turning even if the rider ceases to pedal and the cassette or freewheel ceases to rotate.

centripetal force A force that acts on a body to keep it moving in a circular path; it is directed toward the center around which the body is moving. For a satellite, the centripetal force is gravity; for a turning bicycle, it is friction between the tires and the ground.

chainrings Part of the crankset, these are toothed rings attached on one side of the bottom bracket and rotated, via the cranks, by the pedals to catch the chain and set the drive train in motion. There are generally between one and three chainrings of different diameters.

chain stays Frame tubes that run parallel to the chain between the bottom bracket shell and the rear drop-outs (notches in the frame into which the rear axle locates) on each side of the rear wheel.

coefficient of friction The ratio of the force required to move two sliding surfaces (such as a tire and the road) over each other to the force holding them together. The higher the coefficient, the better the grip.

composite Solid material that results when two or more substances are combined (physically, not chemically) to create a new material whose properties are superior in a specific application to those of the original substances. The term specifically refers to a structural matrix (such as plastic) within which a fibrous material (such as carbon fibers) is embedded.

compression or compressive force A force applied to an object in such a way as to squash or compress it.

cranks/crank arms Levers that run between the pedals and the bottom bracket and pedal axle, connected to the chainrings.

crankset The name for the combination of the crank, spider, and up to three chainrings on one side of the bottom bracket.

criterium A high speed, multilap, short-course race typically staged in an urban area.

deformation A change in the dimensions of an object resulting from pressure or stress.

derailleur A mechanism by which a bicycle chain is moved, either from one sprocket to another on the rear cassette or freewheel or between the chainrings at the front. Most front and back derailleurs are operated by cables that run to levers on the handlebars. The rear derailleur also incorporates a method of taking up the chain slack caused by the movement between cogs. There are variations, but the majority involve a parallelogram mechanism with a cage holding a sprung pulley system of a jockey wheel and a chain guide, through which the chain travels in an S pattern.

disc wheel A solid wheel or sometimes a disc-shaped fairing that covers a spoked wheel.

domestique A road cyclist who works to help the team and team leader win instead of racing for a personal victory.

down tube Part of the front triangle of a traditional frame, running from the head tube down to the bottom bracket.

drafting A technique by which the rider behind sits in the slipstream of the rider in front, thus saving energy. If a rider does this without "taking a turn at the front," they will be accused of wheelsucking.

drivetrain The description of the entire mechanism that transmits power from the rider to the drive wheel. The most common form of drivetrain uses a metal linked chain running between cogs, the front of which is attached to cranks and pedals and the rear of which is attached to the wheel. There are, however, other systems in use, including belt drives and shaft drives. Most systems also include some form of gearing.

drop-out A notch in the frame or fork blades to hold the wheel axle.

efficiency The ratio of the useful work done by a machine, engine, or other device to the energy supplied to it, often expressed as a percentage.

elastic limit or yield point The greatest stress that can be applied to a material without causing permanent deformation or failure.

failure When a structure fails to withstand the stresses imposed on it. Failure manifests itself in permanent deformation or breaking.

force A quantitative description of the interaction between two physical systems, such as an object and its environment. A force can be directly correlated with changes in an object's motion, specifically its acceleration.

fork The two blades on each side of the front wheel; they hold it in place and can be turned to enable steering.

freewheel *See* cassette/freewheel.

friction A force experienced by objects or substances in contact with one another, and that resists their relative motion.

full suspension A type of mountain bike with suspension in the fork and a shock absorber with links or flex built into the frame to dampen shocks at the rear.

gravity The natural force of attraction exerted by any two objects on one another, for example, between the Earth and objects at or near its surface, tending to draw those objects toward the center of the Earth. An object's weight is the gravitational force on it.

hardtail A type of mountain bike with suspension at the front but not at the rear.

headset The bearing assembly at the top and bottom of the head tube, through which the steerer tube of the front fork runs and rotates.

head tube The part of the frame that contains the headset and through which the steerer tube runs, moving freely, connecting the handlebars to the front fork.

hub The center of each wheel, through which the axle runs and from which the spokes radiate to suspend the rim.

hybrid A type of bicycle that is neither a traditional road bike nor a mountain bike, but incorporates aspects of both; it is often used for commuting or leisure riding.

hypoxia A deficiency in the amount of oxygen reaching the cells and tissues of the body.

inertia The tendency of a body to resist any change in its state of motion (speed or direction), i.e. to remain at rest or in straight-line motion unless acted on by an outside force.

jockey wheel A rotating cog on the rear derailleur arm.

kinetic energy The energy an object possesses because of its motion, equal to $\frac{1}{2}mv^2$, where m is its mass and v is its velocity.

Madison A type of track race; it is named after Madison Square Garden in New York City, where it originated. Only one member of the team races at any one time, and the next rider must be touched before taking over, which often takes the form of a hand-sling.

matter Any substance that has mass and occupies space. All physical objects are composed of matter, in the form of atoms, which are, in turn, composed of protons, neutrons, and electrons.

maximum oxygen uptake (VO_2 max) The greatest volume of oxygen that can be absorbed from the lungs by the blood (also called aerobic capacity), typically expressed as the amount of oxygen taken up per kilogram of body weight per minute of exercise. It is generally considered the best indicator of cardiorespiratory endurance and aerobic fitness.

mechanical advantage A unitless number which expresses the ability of a machine such as a lever or pulley system to multiply force; it is the ratio of the force produced by a machine to the input force applied to it.

metabolism The chemical processes occurring within a living cell or organism that are necessary for the maintenance of life.

moment of inertia A quantity that describes how difficult it is to change the angular motion of an object about a particular axis, i.e. its tendency to resist angular acceleration. For a simple particle, it is the product of its mass and the square of its distance from the axis.

pedal One of the contact points between the rider and the bicycle. A pedal can be a simple, flat platform or can be clipless, where a cleat on the shoe of the rider is required to clip into the pedal. The term "clipless" comes from the fact that until the invention of these pedals, an arrangement of clips and straps was used to maintain constant contact—and, thus, better energy transfer—between pedal and shoe.

polymer A chemical compound made of smaller, identical molecules (monomers) linked together. Some polymers such as cellulose are naturally occurring, while others such as nylon are artificial.

potential energy The energy an object has because of its position. It is called "potential" because it has the potential to be converted into other forms of energy such as kinetic energy. A bicycle at the top of a hill is said to have gravitational potential energy.

power The rate at which work is done or energy is transferred; commonly measured in foot-pounds per second (ft·lb/s) or in watts (W), it is the amount of work done per unit of time.

respiration The act or process of inhaling and exhaling. This also refers to the oxidative process occurring within living cells by which the chemical energy of organic molecules is released in a series of metabolic steps involving the consumption of oxygen and the liberation of carbon dioxide and water.

rim The metal or carbon fiber hoop into which the tire and inner tube are located, connected to the hub of the wheel by the spokes.

rolling resistance In the case of a bicycle tire, the resistance to motion arising from the deformation of the tire and the surface on which it rolls, which results in the loss of some energy to heat. It is expressed as a resisting force.

seat stays Part of the rear triangle of the frame, these run from near the top of the seat tube down to the chain stays and rear drop-outs. Often these tubes run parallel to each other, with a bridge connecting them above the back wheel.

seat tube Part of the frame into which the seat post drops; it runs down to the bottom bracket. The front derailleur is generally attached to the seat tube, and frequently small mounts are brazed onto the seat tube for this purpose.

spider Part of the crankset connecting the crank arm to the chainrings.

spokes Radiating from the hub to the rim of the wheel, modern spokes work by tension, holding the rim in place in relation to the hub. An even tension on all the spokes is required to keep the rim from warping.

spoke nipple This holds a spoke to the rim. It resembles a hollow screw with the thread on the inside, so the head sits in the rim and the spoke screws into the nipple. It can be tightened or loosened to adjust the tension of the spoke and true the wheel.

stem The component that attaches the handlebars to the steerer tube, which is the top section of the front fork that comes up through the frame at the head tube.

stiffness The physical property of being inflexible and hard to bend. The stiffness of a body is a measure of its resistance to changes in shape. *See also* Young's modulus.

strain A measure of the relative or fractional distortion of an object when a force is applied to it.

stress A measure of the distribution of an applied load over a material, expressed as force per unit of area over which the force is applied.

tensile strength *See* ultimate tensile strength.

tension A force applied to an object that tends to stretch or elongate it.

thermal energy The internal energy of an object arising from the kinetic energy of its atoms and/or molecules. When thermal energy flows from an object at a high temperature to an object at a lower temperature, we say that heat is transferred.

top tube Also known as the cross bar, this is the part of the frame that runs from the head tube to the rear of the seat tube.

torque A "twisting force," which causes a change in the rotational motion of a body about an axis.

track cycling This takes place on closed banked tracks or in velodromes where riders compete on fixed wheel, single-speed cycles.

ultimate tensile strength The maximum tensile stress (stretching force per unit of cross-sectional area) that a material can withstand before it narrows and breaks.

viscosity A fluid's resistance to flow.

wheelbase The distance between the centers of the front and rear wheels.

work The product of a force and the distance over which it acts. Work done on an object leads to a change in its energy.

Young's modulus A measure of "bendiness," defined as the ratio of the stress (force per unit of area) applied to a material to its resulting strain (relative distortion). The stiffer a material, the higher its modulus.

Table of measurements

Distance

1 in = 25.4 mm
1 mm = 0.039 in
1 ft = 0.305 m
1 m = 3.281 ft
1 mile = 1.609 km
1 km = 0.621 mile

Speed

1 ft/s = 0.305 m/s = 0.682 mph = 1.097 km/h
1 m/s = 3.281 ft/s = 2.237 mph = 3.6 km/h
1 mph = 1.609 km/h = 1.467 ft/s = 0.447 m/s
1 km/h = 0.621 mph = 0.911 ft/s = 0.278 m/s

Acceleration

1 ft/s^2 = 0.305 m/s^2
1 m/s^2 = 3.281 ft/s^2

Force

1 lb = 4.448 N
1 N = 0.225 lb

Mass

1 slug = 14.594 kg
1 kg = 0.0685 slug

Converting mass to weight on Earth

1 slug = 32.174 lb = 143.117 N = 14.594 kg
1 lb = 0.0311 slug = 4.448 N = 0.454 kg
1 kg = 9.807 N = 0.0685 slug = 2.205 lb
1 N = 0.102 kg = 0.225 lb = 0.00699 slug

Energy

1 ft·lb = 1.356 J
1 J = 0.738 ft·lb

Power

1 ft·lb/s = 0.00182 hp = 1.356 W
1 hp = 550 ft·lb/s = 745.7 W
1 W = 0.738 ft·lb/s = 0.00134 hp

Area

1 ft^2 = 0.0929 m^2
1 m^2 = 10.764 ft^2

Pressure

1 psi = 6894.757 Pa
1 Pa = 0.000145 psi

Volume

1 in^3 = 0.0164 liter
1 liter = 61.0237 in^3
1 ft^3 = 0.0283 m^3
1 m^3 = 35.315 ft^3

Density

1 lb/ft^3 = 16.018 kg/m^3
1 kg/m^3 = 0.0624 lb/ft^3

Viscosity

1 lb·s/ft^2 = 47.88 Pa·s
1 Pa·s = 0.0209 lb·s/ft^2

Temperature

°F = (°C × 9/5) + 32
°C = (°F – 32) x 5/9
K = °C + 273.15
°C = K – 273.15

Dietary energy

1 calorie = 4.187 kJ
1 kJ = 0.239 calories

Abbreviations

in	inch
ft	foot
mm	millimeter (0.001 m)
cm	centimeter (0.01 m)
m	meter
km	kilometer (1,000 m)
s	second
h	hour
oz	ounce
lb	pound
N	newton
kg	kilogram
g	gram
hp	horsepower
W	watt
J	joule
psi	pound square inch (also lb/in^2)
Pa	pascal
kPa	kilopascal (1,000 Pa)
MPa	megapascal (1,000,000 Pa)
°F	farenhite
°C	centigrade
K	kelvin

Index

Page numbers followed by a "g" refer to glossary entries.

Acknowledgments

The author knows he couldn't have written this book without the wisdom of Sally Cranfield and support from Ruby and Vivi Glaskin. Jeremy Torr has been helpful beyond reason. Rory Hitchens got it started and Tony Peake of Peake Asscoiates got it sorted. Tom Kitch at Ivy Press edited patiently and sensitively. UK public libraries were as helpful as ever, particularly those in Acomb, Aldershot, Bewdley, Brighton, Farnborough, Kidderminster, Maidstone, Stourport, and Ulverston. Please support your local library. Thanks go also to the following, all of whom are blameless for the published text: Keith Noronha and Reynolds Technology Ltd, Paul Lew, Director of Technology and Innovation at Reynolds Cycling and CEO of Unmanned Systems Design, Stuart Evans, Mark Jermy, Lindsey Underwood, Hugh Trenchard, Tony Durham, Helen Viner at TRL (Transport Research Laboratory), Geoff Booker at Trykit Conversions, John Cobb, Bob Cranfield and Dr Hans-Erhard Lessing. In addition, Ivy Press would like to thank the following for their help with creating the graphics: Iva Adlerova, Scott Drawer, Riin Gill, Andrew Griffiths, Ewald Hennig, Brandon Kelly, Rob Lewis, Doug Malewicki, Rainer Pivit, Dave Salazar, David Sanderson, Hugh Shercliff, Ian Smith, Jason Woznick, and Hannes Zietsman.

The Ivy Press would like to thank the following for permission to reproduce copyright material: Getty Images: Bryn Lennon (7 and 69), Lionel Bonaventure/Stringer/AFP (11), Carl De Souza/AFP (39 and 144), Joel Saget/AFP (125), Keystone-France/Hulton Archive (132), Getty Images Sport (158), Cancan Chu (176), and Archive Holdings Inc./Archive Photos (177); Fotolia: Oleg Zabielin (22), Hamurishi (30), Apfelweile (48), Pavelsh (76), and Torsten Lorenz (112); Science Photo Library: Carol and Mike Werner, Visuals Unlimited (26), Gustoimages (152), Pasieka (166), and Dr Ray Clark & Mervyn Goff (174); Calfee Design (82); Corbis: Tim De Waele (97) and Benati/Epa (151); Benedikt Saxler/Shutterstock (104); A2 Wind Tunnel www.a2wt.com (129, top); Saddleback saddleback.co.uk (129, middle); and UK Sport Research & Innovation (149).

Every effort has been made to trace copyright holders and to obtain their permission for the use of copyright material. The publisher apologizes for any errors or omissions in the lists above and will gratefully incorporate any corrections in future reprints if notified.